Green Energy and Technology

For further volumes:
http://www.springer.com/series/8059

Nicu Bizon · Hossein Shayeghi
Naser Mahdavi Tabatabaei
Editors

Analysis, Control and Optimal Operations in Hybrid Power Systems

Advanced Techniques and Applications for Linear and Nonlinear Systems

Springer

Editors
Nicu Bizon
Faculty of Electronics, Communication and Computers
University of Pitesti
Pitesti
Romania

Naser Mahdavi Tabatabaei
Electrical Engineering Department
Seraj Higher Education Institute
Tabriz
Iran

Hossein Shayeghi
Department of Technical Engineering
University of Mohaghegh
Ardabil
Iran

ISSN 1865-3529 ISSN 1865-3537 (electronic)
ISBN 978-1-4471-5537-9 ISBN 978-1-4471-5538-6 (eBook)
DOI 10.1007/978-1-4471-5538-6
Springer London Heidelberg New York Dordrecht

Library of Congress Control Number: 2013953200

Printed on acid-free paper

Springer is part of Springer Science+Business Media (www.springer.com)

Dedicated to
all our teachers and colleagues
who enabled us to write this book,
and our family and friends
for supporting us all along

Foreword

Electrical power systems have increasingly become more complex in design and operation, thus increasing their dynamic performances and safety, which are also required. The efficient and reliable designing of hybrid systems in control technology and industrial information are of interest to research institutions and also in industrial sectors.

Renewable energy grids supply the energy demand directly and prevent the requirement for long usual distribution infrastructures. A combination of different and perfect energy generation systems, which consist of renewable energies and fossil fuel generators together with an energy storage system and power conditioning system, is known as hybrid power systems.

Hybrid power systems are generally planned for electrical power generation as great and wide independent electric grids which are operated in distant areas. Hybrid systems include a number of power generation apparatuses that include wind turbines, solar systems (hot water and pumps), photovoltaic, micro-hydro, power backup systems, LED lighting, remote switching, and fossil fuel generators which extend from small to very large systems and grids.

Hybrid power systems supply power to electrical grids whose fuel costs are very high. The cost of fuel can be decreased by using renewable resources of clean energies which promote the quality of energy consumption and living style.

In using alternative energies, more than one source of alternative energy systems and generating power are often considered as hybrid systems. A hybrid power system with alternative energies can help make the most of electricity generation investment. Hybrid dynamic systems contribute in the applications in the fields of process industry, automotive, avionics, communication networks, energy systems, transportation networks, embedded systems, biology, and robotics.

Hybrid systems provide the best features of combined energy resource as well as electrical power quality in a wide power range of demands. They are advanced as integrated small electric distribution systems as mini-grids-based power systems which can be provided to series and parallel basic configurations.

The above aspects are illustrated in this book by the editors and authors, in the following topics: conventional generators; renewable energy resources; local and remote generations; power electronic converters; Barkhausen's equality; state space; conservativity; dissipativity; Tellegen's principle; power system dynamic stability; low frequency oscillations; single-machine infinite-bus; unified power flow controller; optimal tuning; particle swarm optimizing; genetic algorithm; flexible and well-balanced mechanism; damped electromechanical mode; transmission network expansion planning; fuel cell; hybrid power sources; ripple mitigation; spread power spectrum; nonlinear control.

The book chapters and materials are efficient in theoretical and application issues and are highly recommended for studying and considering in the educational and research fields.

Baku, Azerbaijan, April 30, 2013 Arif M. Hashimov

Preface

Hybrid power systems use both conventional generation systems such as thermal and nuclear power systems and also renewable resources such as wind turbine, photovoltaic cell, fuel cell, etc., to supply demand. Conventional systems have serious environmental problems such as emissions and greenhouse gas effects, high cost of fuel and low efficiency, and reliability that can be improved by combination with the renewable energy resources. Usually, hybrid power systems combine two or more energy sources accompanied with energy storage devices to deliver power continually to the DC load or to the AC load via the inverter system.

However, designing and operation of hybrid power systems are complicated as the control systems comprise several subsystems from the stability issue. Hybrid power systems also include problems of loop control and feedback configurations that are solved by conventional and modern control methods.

Hybrid power systems as nonlinear electrical networks should be modeled and analyzed in various operational situations considering the related topological structures. The models use mathematical formulas and physical principles to describe the appropriate system behavior. The optional modeling approaches also denote the parametric and non-parametric identification procedures. In this case the physical structure of the hybrid power system would be known to have adequate results from the physical correctness point of view and concern the observability, controllability, and state minimality of the system representations.

As power demand grows rapidly and transmission and generation fields expanded, the availability of hybrid power systems and related resources are affected by constraints of more loading and stability limitations. Therefore, the stability and dynamic performance of the power system should be improved by Power System Stabilizers (PSSs). The adverse effect of PSSs in voltage profile and server disturbances is damped by using Flexible AC Transmission System (FACTS) controllers, such as Static VAR Compensators (SVCs), Thyristor Control Series Compensators (TCSCs), Static Synchronous Compensators (STATCOMs), and Unified Power Flow Controller (UPFC) as the supplementary signal for main control loops. Voltage Source Converter (VSC)-based High Voltage DC (HVDC) transmission systems are also used in hybrid power systems and provide various forms of modulation for damping power system oscillation and improving dynamic performance.

The UPFC acts as a shunt compensating and a phase shifting device simultaneously, which has the ability to control the hybrid power flow transmission, improve the transient stability, mitigate system oscillation, and provide voltage support. The global optimization techniques and heuristic population-based search procedures like Genetic Algorithms (GA) have been applied to the UPFC-based damping controller parameter optimization. To overcome the highly correlated and large number of optimized parameters, the classic and quantum behaved particle swarm optimization (PSO) techniques are used for optimal tuning of controller parameter to improve optimization synthesis and the speed of algorithm convergence.

Hybrid power systems should be evaluated carefully to reduce the investment cost of transmission system expansion and solve the far location between the generating plants and the load centers. Transmission Network Expansion Planning (TNEP) is an important part of power system planning to minimize the network construction and operational cost while satisfying the requirement of delivering electric power safely and reliably to load centers. The TNEP could be classified as static to determine the information about new required transmission lines, and a dynamic one to verify the optimal expansion design over the whole time restrictions planning period. For solution of TNEP problem, the classic method based on mathematical principles with linearization, and non-classic methods based on random search algorithms are applied.

Hybrid power systems include more energy sources and the hybrid Energy Storage Devices (ESD) as well as Power Dynamic Compensators (PDC) as the embedded power unit to deliver or store energy and enhance the performance of the entire system. Some hybrid power systems use Fuel Cells (FC) topologies with ultracapacitor, battery, and hybrid types in some dynamic load applications. Therefore, the fuel cells-based hybrid power systems are important with regard to designating, modeling, analyzing, and operating including active controlled topologies to obtain both performances in energy conversion and in ripple mitigation.

Advanced Techniques and Applications on Stability, Control, and Optimal Operation of Hybrid Power Systems is a book aimed to highlight the above concepts and challenges by a systematic approach and, therefore, to provide understanding on the hybrid power system structures and topologies, in flow of generation and transmission, designing, modeling, analyzing, operation, and the ways of controlling different systems by using different research methods.

Some of the specialists are joined as authors of the book chapters to provide their potentially innovative solutions and research related to the field of hybrid power systems, in order to be useful in developing new ways in their technologies, design, and operational strategies. Efficient theoretical researches, case analysis, and practical implementation processes are put together in this book that aims to act research and design guides to help the postgraduates and researchers in electrical power engineering and energy systems. In particular, the specific purposes of the subjects also include the knowledge base and applications used by undergraduates with regard to indicating research fields and innovative solutions as the challenges and opportunities for solving problems.

The book presents significant results obtained by leading professionals from the industry, research, and academic fields, and can be useful to various groups in the specific areas analyzed in this book. All works contributed to this book are new, previously unpublished material, or extended versions of published papers in the proceedings of international conferences and transactions on international journals.

"Overview of Hybrid Power System" introduces an overview of hybrid power systems combining renewable energy resources and conventional generators. The chapter describes the components of conventional generators and renewable energy resources, power electronic converter, and elements of control systems. The chapter includes some important objects related to main power plants, generators, renewable energy resources, supervisory control of hybrid power systems and their modeling, control and management as well as their relevant operation strategies. Besides, the chapter introduces storage devices as a part of hybrid power systems and any possible combination of conventional and renewable resources.

The general condition of problems of loops control is given in "Revisiting and Generalizing Barkhausen's Equality". It shows that the fixed point equations over a space of functions are essential for loops control. The chapter discusses that the particular conditions as Barkhausen for the loops especially with at least two subsystems are derived as solutions in the functions space. The entire power of the loop concept and fixed point conditions is described by the differential loops equations and deal with the inverse function and fixed point condition.

"State Energy-Based Approach as a Tool for Design and Simulation of Linear and Nonlinear Systems" presents the detection of the physical interaction correctness in the strictly causal system representations as energy exchanging. The chapter also investigates the proposed solution approach related to the generalization of Tellegen's theorem. The analyzing and synthesizing of linear and nonlinear causal systems is mainly contributed in the chapter. The state observability and controllability properties are proven as the complete analysis of the system behavior. The chapter presents the state space energy and port-Hamiltonian approaches for modeling the physical systems using their energetic behavior and their internal interconnection structure.

The linearized Phillips-Heffron model of a parallel AC/DC power system for studying power system stability is described in "Power Systems Stability Analysis Based on Classical Techniques in Work". The chapter also presents the modeling of a Back-to-Back Voltage Source Converter (BtB VSC) HVDC to damp low frequency oscillations in a weakly connected system using Relative Gain Array (RGA), Singular Value Decomposition (SVD), and Damping Function (DF) as the supplementary controllers. The chapter shows that the SVD analysis evaluates the electromechanical controllability to enhance the dynamic stability of HVDC system via SVC supplementary controller. A UPFC-based linearized power system is also analyzed for increasing the dynamic stability and damping the load angle oscillations by adding an extra signal to its control loops. On-line adaptive controllers are also simulated to decrease rotor speed oscillations and identify power system parameters.

In "Optimal Design of UPFC Based Damping Controller Using PSO and QPSO", damping controller formulation is discussed based on the UPFC and using Classic and Quantum behaved Particle Swarm Optimization technique (CPSO, QPSO). The UPFC is advantageous as a FACTS device related to transient stability improvement, oscillation system mitigation and voltage support providing is also presented in the chapter considering the different objective functions in damping controller designing. The objective functions are related to the damping factor eigenvalues and time domain basis. The chapter shows the performance of controllers using eigenvalue analysis and nonlinear time domain simulation applied to the single machine infinite bus and multimachine power systems.

A review of the researches in the field of TNEP is described in "Application of PSO and GA for Transmission Network Expansion Planning" considering the effects of GA and PSO algorithms. The chapter introduces the TNEP as a large-scale and nonlinear combinatorial optimization problem to determine the number and location of new additional lines to transmission network. As the chapter description, mathematical principles based classic method like linear programming and Bender decomposition, and non-classic or heuristic methods like GAs can be applied for solution of complex problems. The ability to deal with non-convex, nonlinear, integer-mixed optimization problems like the STNEP problem has been demonstrated by global optimization techniques like GA as a random search method.

The control of Hybrid Power Sources (HPS) with some applications is analyzed in "Applications in Control of the Hybrid Power Systems". The chapter proposes an efficient fuel cell/battery HPS topology for high power applications which uses an inverter system directly powered from the appropriate Polymer Electrolyte Membrane Fuel Cell (PEMFC) stack. The chapter discusses about the FC HPS supply inverter systems and PEMFC current ripple in operation of the inverter system and the Low Frequency (LF) ripple mitigation based on the active non-linear control in the fuel cell current ripple. The anti-ripple current generation and control in buck Controlled Current Source related to the output of the HPS is also reviewed. The chapter shows that good ripple performances related to the LF harmonics mitigation are also obtained with the hysteretic current-mode control as the better performances for nonlinear control. The control performances are also shown by simulations and the designed control law is validated using a Fuzzy Logic Controller (FLC). At the end of the chapter a nonlinear control for FC HPS based on bi-buck topology is proposed and analyzed in comparison with the FLC approach. The nonlinear voltage control to stabilize the HPS output voltage with a low voltage ripple is also analyzed, designed, and simulated.

The editors recommend this book as suitable for dedicated and general audiences that include the power system professionals, as well as researchers and developers of energy sectors and the electrical power engineering community. It is expected that readers will be graduates of energy and power engineering degree programs having a basic mathematical background.

As a whole, the book covers both theoretical background and application examples in hybrid power systems, and special and professional fields of electrical energy systems altogether. Because the subject approached in this book is vast in itself, it has been slightly difficult to balance the theoretical and applicative aspects in each of the chapters, so efforts of editors have been made to well cover the essential topics of the book. Specific in-depth further studies are pointed to the dedicated intensive resources of the book subjects for interested readers. Meanwhile, the application and study cases are possibly selected with as much real implications.

Finally, the editors hope that this book will be helpful to undergraduate and graduate students, researchers, and engineers, trying to solve energy and power engineering problems using modern technical and intelligent systems based on theoretical aspects and application case studies.

Pitesti, Romania, May 2013 — Nicu Bizon
Ardabil, Iran — Hossein Shayeghi
Tabriz, Iran — Naser Mahdavi Tabatabaei

Contents

Overview of Hybrid Power System

Amir Ahmarinezhad, Mohammad Abroshan, Naser Mahdavi Tabatabaei and Nicu Bizon

Abstract Hybrid power system (HPS) is an electricity generation system combining renewable energy resources and conventional generators. Several components will be utilized in a hybrid power system namely as: conventional generators (diesel generator, synchronous generator, DC generator for field excitation of synchronous generator), renewable energy resources (such as wind generator, photovoltaic system, etc.), power electronic converter, components' controller, and also a supervisory control system to control the whole power system. In this chapter, an overview of hybrid power system and a brief description of important possible parts of a hybrid power system is introduce.

A. Ahmarinezhad (✉) · M. Abroshan
Department of Electrical Engineering, Islamic Azad University, Shahre Rey Branch, Tehran, Iran
e-mail: ahmarinezhad@gmail.com; ahmarinejad@iausr.ac.ir

M. Abroshan
e-mail: abroshan@aut.ac.ir

N. M. Tabatabaei
Electrical Engineering Department, Seraj Higher Education Institute, Tabriz, Iran
e-mail: n.m.tabatabaei@gmail.com

N. Bizon
Faculty of Electronics, Communication and Computers, University of Pitesti, Pitesti, Romania
e-mail: nicu.bizon@upit.ro; nicubizon@yahoo.com

N. Bizon
Doctoral School, University 'Politehnica' of Bucharest, Bucharest, Romania

N. Bizon et al. (eds.), *Analysis, Control and Optimal Operations in Hybrid Power Systems*, Green Energy and Technology, DOI: 10.1007/978-1-4471-5538-6_1,

1 Introduction

Conventional generation systems such as thermal and nuclear power system are not a proper solution to electrification of remote areas due to economical and technical issues. Beside environmental problems such as emissions and greenhouse gas effects which are caused by conventional generation systems, the high cost of fuel, and construction of these power plants as well as efficiency and reliability are also serious problems of conventional generation systems. Hybrid power systems are an alternative solution to supply remote loads.

Hybrid power systems utilize both conventional and renewable resources to supply demand. As regards this demand is usually at remote point; hence, the local renewable resources should be utilized. Diesel generators as a conventional generation system, wind turbine, photovoltaic cell, fuel cell, and its relevant components, and other renewable-based generation system may be used in a hybrid power system.

Moreover, the integration of renewable energy resources into conventional system can bring reliability and efficiency to supply demand and the cost of components is expected to decline. However, further improvement in design and operation of hybrid power system is still needed.

On the contrary, the most important problem of hybrid power systems is the complicated control system. Although each component has their own control system, a controller should supervise the actions of each individual controller. The action of supervising controller is called supervisory control and this system is called supervisory control system. This system receives the operation status of each component and sends a signal to set the reference point of each component.

In this chapter, at first, the main power plants in large capacity scale are briefly introduced and described. After that generators as the main component of any generation system are introduced in which both AC and DC generators are described briefly. Power transformers are also included in this section. The chapter continues with renewable energy resources and their relevant power plants followed by supervisory control system of a hybrid power system and its management issues.

2 Main Power Plants

Electric power is generally generated using conventional generators based on nonrenewable energy sources such as coal, oil, natural gas, and Uranium. The capacity of these generators is very high, since they are responsible for producing bulk amounts of power necessary to supply the entire loads. However, supply of fuels for these generation technologies is limited and creates the need for looking at alternative renewable fuel sources.

The main component to generate electric power is generator. When a generator is coupled with a prime mover, the electricity is generated. However, the type of power plant is determined by the type of prime move.

2.1 Steam Power Plants

Steam is an important medium to produce mechanical energy. The most important merits of steam is that it can be raised from water which is available in abundance, it does not react much with the materials of power plant equipments, and it is stable at the temperature required in the plant. Steam is used to drive steam engines, steam turbines, etc. A steam power plant must have a furnace to burn the fuel, steam generator, or boiler containing water, main power unit such as an engine or turbine to use the heat energy of steam and perform work, and piping system to convey steam and water.

In addition, the plant requires various auxiliaries and accessories depending on the availability of water, fuel and the service for which the plant is intended.

A steam power plant using steam as working fluid works basically on Rankine cycle. Steam is generated in a boiler, expanded in the prime mover, and condensed in the condenser and fed into the boiler again. The different types of systems and components used in steam power plant are high pressure boiler, prime mover, condensers and cooling towers, coal handling system, ash and dust handling system, draft system, fed water purification plant, pumping system, air preheater, economizer, super-heater, and feed heaters.

The heat produced by burning of coal converts water stored in boiler into steam at suitable pressure and temperature. The generated steam is passed through the super-heater. The superheated steam then flows through the turbine; and consequently the pressure of steam is reduced due to work which is done in the turbine. Steam which leaves the turbine passes through the condenser whose steam pressure depends on flow rate and temperature of cooling water and effectiveness of air removal equipment. Water circulated through the condenser may be taken from the various sources such as river, lake, or sea. If sufficient water is not available, the hot water coming out of the condenser may be cooled in cooling towers and circulated again through the system. Air taken from the atmosphere is first passed through the air preheater, where it is heated by flue gases. The hot air then passes through the furnace. The flue gases passes over boiler and super-heater pipes, after that they flow through the filter and then economizer, and finally they are exhausted to the atmosphere through the chimney.

2.2 Gas Turbine Power Plant

A simple cycle of a gas turbine power plant consists of a compressor, a combustion chamber, and a turbine. The gas turbine obtains its power by utilizing the energy of burnt gases and air whose pressures are high in the range of 4–10 bars. This high pressure produced by compressor which can be centrifugal or axial. Moreover, the compressor is driven by turbine.

To get a higher temperature of the working fluid, a combustion chamber is required where combustion of air and fuel takes place. Beside temperature and pressure, the volume of working fluid is another important parameter. Increasing the volume of the working fluid at constant pressure, or alternatively increasing the pressure at constant volume can increase the power developed by the turbine.

Regarding the fact that the compressor is coupled with the turbine shaft, it consumes some of the power produced by the turbine; hence, the compressor lowers the efficiency. Therefore, the network is the difference between the turbine work and work consumed by the compressor.

The gas turbine power plants which are used in electric power industry are classified into two groups: open cycle gas turbine and closed cycle gas turbine.

2.2.1 Open Cycle Gas Turbine

As is shown by Fig. 1 a simple open cycle gas turbine consists of a compressor, combustion chamber, and a turbine. The compressor takes in ambient air and raises its pressure. Heat is added to the air in combustion chamber by burning the fuel and raises its temperature. The compressor takes in ambient air and raises its pressure. Heat is added to the air in combustion chamber by burning the fuel and

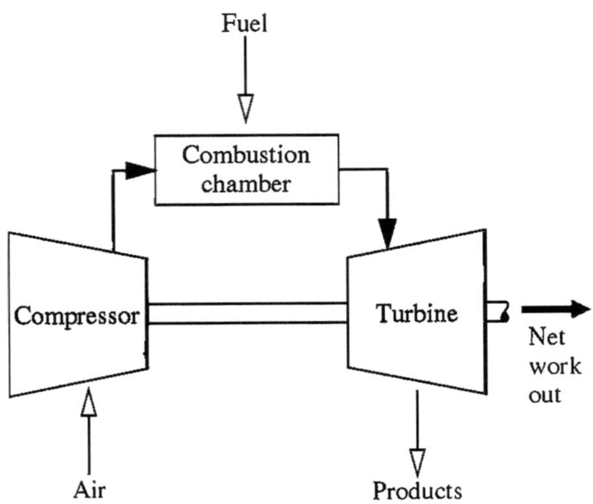

Fig. 1 The schematic diagram of an open cycle gas turbine power plant [2]

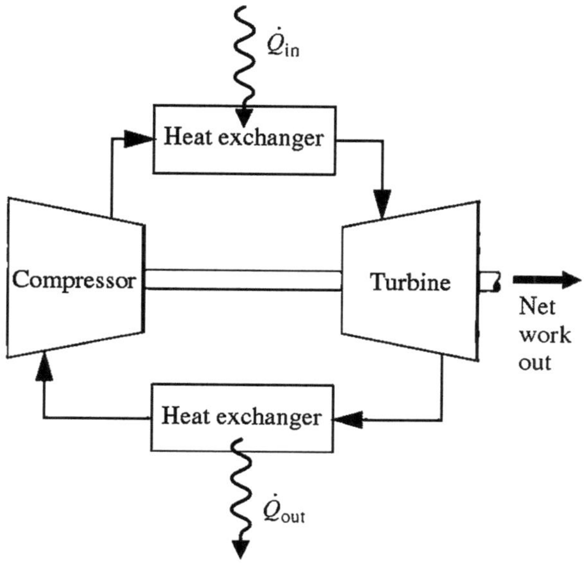

Fig. 2 The schematic diagram of a closed-loop gas turbine power plant [2]

raises its temperature. The hot and high pressure gas then flows to turbine, after that the exhausted steam returns back to atmosphere [1].

2.2.2 Closed Cycle Gas Turbine

As is illustrated in Fig. 2, in closed cycle gas turbine plant, the working fluid (air or any other suitable gas) which comes out from compressor is heated in a heater by an external source at constant pressure. Then, the high temperature and high pressure air results from the external heater is passed through the gas turbine. After that, the gas coming out from the turbine is cooled to its original temperature in the cooler using external cooling source before passing to the compressor. Therefore, the working fluid is continuously used in the system without its change of phase and the required heat is given to the working fluid in the heat exchanger [1].

2.3 Combined Cycle Power Plants

As is described previously, significant amount of thermal energy is wasted at steam output of turbine both in open cycle and closed cycle technologies. If this energy can be utilized, the efficiency of power plans will be enhanced. Generally, the thermal energy of exhausted steam from gas turbine is used to preheat the working fluid of a steam turbine in a combined cycle power plant. There may be various combinations of the combined cycles depending upon the place or country requirements. Even nuclear power plant may be used in the combined cycles.

In a combined cycle power plant, the exhaust of gas turbine which has high oxygen content is used as the inlet gas to the steam generator where the combustion of additional fuel takes place. This combination allows nearer equality between the power outputs of the two units than is obtained with the simple recuperative heat exchanger. For a given total power output, the energy input is reduced (i.e., saving in fuel) and the installed cost of gas turbine per unit of power output is about one-fourth of that of steam turbine. In other words, the combination cycles reveal higher efficiency. The greater disadvantages include the complexity of the plant, different fuel requirements, and possible loss of flexibility and reliability [1].

2.4 Nuclear Power Plants

Nuclear energy is an important source for electricity generation in many countries. In 2003, 19 countries depended on nuclear power for at least 20 % of their electricity generation. France is the country which has the highest portion (78.1 %) of the electricity generated by nuclear power. Though the percentage of nuclear electric power over the total national electricity generation is not very high (about 20 %), the U.S. is still the world's largest producer of electric power using nuclear fuels. Higher fossil fuel prices and the entry into force of the Kyoto Protocol could result in more electricity generated by nuclear power. In the emerging economic regions, such as China and India, nuclear power capacity is expected to grow. However, nuclear power trends can be difficult or even be reversed due to a variety of reasons. The safety of a nuclear power plant is still the biggest concern. And how to deposit nuclear wastes can always be a discussion topic for environmentalists. Moreover, the nuclear fuel (Uranium) is not renewable [2].

The main advantages and disadvantages of nuclear power plants are:

- Space requirement of a nuclear power plant is less as compared to other conventional power plants of equal size.
- A nuclear power plant consumes very small quantity of fuel. Thus fuel transportation cost is less and large fuel storage facilities are not needed Further, the nuclear power plants will conserve the fossil fuels (coal, oil, gas, etc.) for other energy need.
- There is increased reliability of operation.
- Nuclear power plants are not affected by adverse weather conditions.
- Nuclear power plants are well suited to meet large power demands. They give better performance at higher load factors (80–90 %).
- Materials expenditure on metal structures, piping, storage mechanisms are much lower for a nuclear power plant than a coal burning power plant.
- It does not require large quantity of water.

- Initial cost of nuclear power plant is higher as compared to hydro or steam power plant.
- Nuclear power plants are not well suited for varying load conditions.
- Radioactive wastes if not disposed carefully may have bad effect on the health of workers and other population.
- Maintenance cost of the plant is high.
- It requires trained personnel to handle nuclear power plants.

2.5 Hydroelectric Power Plant

Today, hydropower is still the largest renewable source for electricity generation in the world. In 2002, more than 18 % of the world electricity was supplied by renewable sources, most of which comes from hydropower. The world hydroelectric capacity is expected to grow slightly due to large hydroelectric projects in the regions with emerging economies.

Although the hydropower is clean and renewable, there are still some problems associated with it. First, the big dams and reservoirs cause a lot of environmental concerns and social disputes. Second, the repositioning of reservoir populations can be a big crisis. Moreover, hydropower is not as profuse as desired.

If at a certain point, the water falls through an appreciable vertical height, this energy can be converted into shaft work. As the water falls through a certain height, its potential energy is converted into kinetic energy and this kinetic energy is converted to the mechanical energy by allowing the water to flow through the hydraulic turbine runner. This mechanical energy is utilized to run an electric generator which is coupled to the turbine shaft.

Some Hydroelectric Power plants are located on rivers, streams, and canals, but for a reliable water supply, dams are needed. Dams store water for later release for such purposes as irrigation, domestic and industrial use, and power generation. The reservoir acts much like a battery, storing water to be released as needed to generate power.

3 Generators

3.1 AC Generators

AC generators are synchronous machines capable of generating AC electric power. The interactions between the multipole magnetic fields of the stators (armatures) and rotors of synchronous generators generate the electrical power. The interaction is called *synchronous* because when the generator is running, the stator and rotor magnetic fields turn at the same speed. A single small generator might have a rating of a few hundred watts, but the largest single machines have ratings that

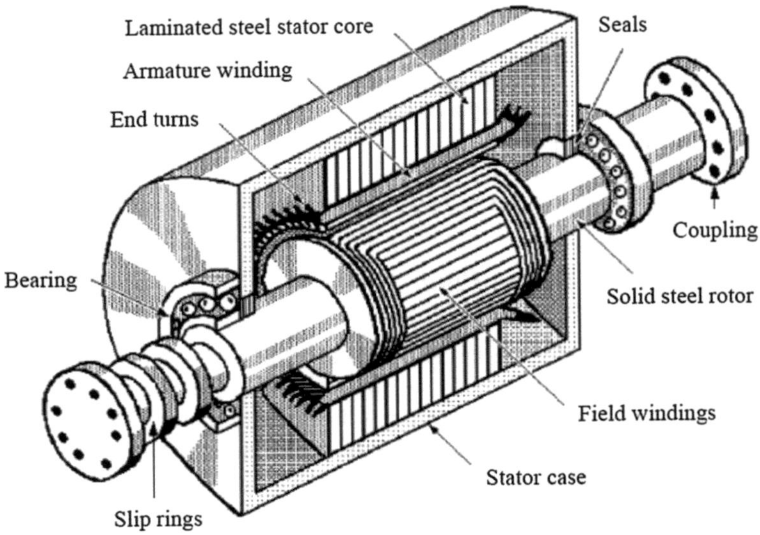

Fig. 3 A cutaway view of a synchronous generator [3]

exceed a billion watts. All synchronous generators have wound armatures and rotors, but the armature is wound on the stator rather than on the rotor, and the field winding is wound on the rotor. Figure 3 illustrates a cutaway view of a synchronous AC generator with a solid cylindrical wound rotor that permits it to turn at high speed without self-destructing.

The organization of a utility AC generator is opposite that of most DC generators and both AC and DC motors. If these machines have armatures, they are wound on their rotors; if they have field windings, they are wound on their stators. The construction of AC generators is reversed to eliminate the complexities of slip-ring mechanisms for obtaining the AC power and to provide more stable mechanical support for the stator windings. With more rigid support, the stators eliminate vibrations that create centrifugal forces which degrade the quality of the AC output. The armature windings are fitted tightly into slots on the inner surface of the stator formed by stacking magnetic sheet steel laminations. The field coils are then wound in axial slots in the outer surface of the solid cylindrical magnetic steel rotor. The rotor and stator together form the magnetic circuit. Most utility AC generators have a three-phase armature winding [3].

The insulation of the AC generator is simplified by having a revolving field and stationary armature. As the poles move under the armature conductors on the stator, the field flux cutting across the conductors induces an alternating voltage. It is alternating because poles of opposite polarity pass successively by a given stator conductor. The alternating voltage appears at the stator windings and is brought out directly through insulated leads from the stationary armature.

Because the most utility of AC generators run at constant speed, the voltage generated depends on field excitation. The rotating field is supplied with 120 or 240 V DC from a separate small DC generator called an *exciter* through two slip rings and brushes. This arrangement permits the generated voltage to be controlled by adjusting the amount of field excitation supplied to the exciter. The field excitation, in turn, is controlled by varying the excitation voltage applied to the alternator field.

Synchronous AC generators are fitted with one of two different rotor designs depending on their intended rotational speeds.

Round rotors are solid steel cylinders with the field winding inserted in slots milled into the surface or the rotor. They usually have two or four poles. Round rotors can withstand the stresses of high-speed rotation.

Salient-pole rotors have multiple pole pieces (typically six) mounted to the rotor structure, and the field winding is wound around the pole pieces. Because of their more complex construction and larger diameter-to-length ratios, salient-pole rotors cannot withstand the stresses of high-speed rotation.

Electric utility steam turbine-driven generators designed for 50- or 60-Hz AC output voltage have round rotors with two poles because they can withstand the stresses of speeds of 3000 and 3600 rpm. Hydroelectric, diesel, and natural-gas engines have far lower shaft speeds than steam turbines, so the generators they drive usually have six or more pole rotors, requirements usually met with more complex salient-pole rotors.

Three-phase AC generators have a winding that is made up of three separate stator windings, each displaced from the other two by 120 electrical degrees. The three windings can either be *wye* or *delta* connected. The wye connection is more common because it is better suited for direct high-voltage generation.

3.2 DC Generator

Although a far greater percentage of the electrical machines in service are A.C machines, the D.C machines are of considerable industrial importance. The main advantage of the D.C machine, specially the D.C motor, is that it provides a fine control of speed. Such an advantage is not claimed by any A.C motor. However, D.C generators are not as common as they used to be, because direct current, when required, is mainly obtained from an A.C supply by the use of rectifiers. Nevertheless, an understanding of D.C generator is important because it represents a logical introduction to the behavior of D.C motors. Indeed many D.C motors in industry actually operate as D.C generators for a brief period [4].

The D.C generators and D.C motors have the same general construction. On the other word, when the machine is being assembled, the workmen usually do not know whether it is a D.C generator or motor. Any D.C generator can be run as a

D.C motor and vice versa. All D.C machines have five main components namely, field system, armature core, armature winding, commutator, brushes.

The function of the field system is to produce uniform magnetic field in which the armature rotates. Field system consists of a number of salient poles (of course, even number) bolted to the inside of circular frame (generally called yoke). Whereas, the pole pieces are composed of stacked laminations the yoke is usually made of solid cast steel.

The armature core is keyed to the machine shaft and rotates between the field poles. It consists of slotted soft-iron laminations (about 0.4 to 0.6 mm thick) that are stacked to form a cylindrical. The laminations are separately encrusted with a thin insulating film so that they do not come in electrical contact with each other. The laminations are slotted to provide accommodation and provide mechanical security to the armature winding and to give shorter air gap for the flux to cross between the pole face and the armature "teeth".

The slots of the armature core hold insulated conductors that are connected in a suitable manner. This is known as armature winding. This is the winding in which "working" E.M.F. is induced. The armature conductors are connected in series–parallel; the conductors being connected in series so as to amplify the voltage and in parallel paths so as to increase the current. The armature winding of a D.C. machine is a closed-circuit winding; the conductors being connected in a symmetrical manner forming a closed loop or series of closed loops.

A mechanical rectifier which converts the alternating voltage generated in the armature winding into direct voltage across the brushes is commutator. It is made of copper segments insulated from each other by mica sheets and mounted on the shaft of the machine. The armature conductors are soldered to the commutator segments in a suitable manner to give rise to the armature winding. Depending upon the manner in which the armature conductors are connected to the commutator segments, there are two types of armature winding in a D.C. machine: lap winding and wave winding.

Great care is taken in building the commutator because any eccentricity will cause the brushes to bounce, producing unacceptable sparking. The sparks may bum the brushes and overheat and carbonize the commutator.

The principle of brushes is to guarantee electrical connections between the rotating commutator and stationary external load circuit. The brushes are made of carbon and rest on the commutator. The brush pressure is adjusted by means of adjustable springs. If the brush pressure is very large, the friction produces heating of the commutator and the brushes. On the other hand, the not good enough contact with the commutator may produce sparking if it is too weak.

Multipole machines have as many brushes as they have poles. For example, a 4-pole machine has 4 brushes. As we go round the commutator, the successive brushes have positive and negative polarities. Brushes having the same polarity are connected together.

3.3 Diesel Engines

The oil engines and gas engines are called Internal Combustion (IC) Engines. In IC engines, fuels burn inside the engine and the products of combustion form the working fluid that generates mechanical power. Whereas, in Gas Turbines the combustion occurs in another chamber and hot working fluid containing thermal energy is admitted in turbine. Reciprocating oil engines and gas engines are of the same family and have a strong resemblance in principle of operation and construction. The engines convert chemical energy in fuel into mechanical energy.

A typical oil engine has the following components [3]:

- Cylinder in which fuel and air are admitted and combustion occurs.
- Piston, which receives high pressure of expanding hot products of combustion and the piston, is forced to linear motion.
- Connecting rod, crankshaft linkage to convert reciprocating motion into rotary motion of shaft.
- Connected Load, mechanical drive, or electrical generator.
- Suitable valves (ports) for control of flow of fuel, air, exhaust gases, fuel injection, and ignition systems.
- Lubricating system, cooling system.

In an engine-generator set, the generator shaft is coupled to the Engine shaft. The main differences between the gasoline engine and the diesel engine are [4]:

- A gasoline engine intakes a mixture of gas and air, compresses it and ignites the mixture with a spark. A diesel engine takes in just air, compresses it and then injects fuel into the compressed air. The heat of the compressed air lights the fuel spontaneously.
- A gasoline engine compresses at a ratio of 8:1 to 12:1, while a diesel engine compresses at a ratio of 14:1 to as high as 25:1. The higher compression ratio of the diesel engine leads to better efficiency.

Gasoline engines generally use either carburetion, in which the air and fuel is mixed long before the air enters the cylinder, or port fuel injection, in which the fuel is injected just prior to the intake stroke (outside the cylinder). Diesel engines use direct fuel injection to the diesel fuel is injected directly into the cylinder.

The diesel engine is recommended due to their longevity think of an 18 wheeler capable of 1,000,000 miles of operation before major service), lower fuel costs (lower fuel consumption per kilowatt (kW) produced), and lower maintenance costs-no spark system, more rugged and more reliable engine.

Diesel engine power plants are installed where supply of coal and water is not available in desired quantity, power is to be generated in small quantity for emergency services, and standby sets are required for continuity of supply such as in hospital telephone exchange.

To conclude, the main advantages and disadvantages of diesel engines are as below:

- Very simple design also simple installation.
- Limited cooling water requirement.
- Standby losses are less as compared to other power plants.
- Low fuel cost.
- Quickly started and put on load.
- Smaller storage is needed for the fuel.
- Layout of power plant is quite simple.
- There is no problem of ash handling.
- Less supervision required.
- For small capacity, diesel power plant is more efficient as compared to steam power plant.
- They can respond to varying loads without any difficulty.
- High Maintenance and operating cost.
- The plant cost per kW is comparatively more.
- The life of diesel power plant is small due to high maintenance.
- Noise is a serious problem in diesel power plant.
- Diesel power plant cannot be constructed for large scale.

3.4 Transformers

A transformer is a static device consisting of a winding, or two or more coupled windings, with or without a magnetic core, for inducing mutual coupling between circuits. Transformers are exclusively used in electric power systems to transfer power by electromagnetic induction between circuits at the same frequency, usually with changed values of voltage and current. Power transformers are used extensively by traditional electric utility companies, power plants, and industrial plants. After transmission lines, transformers are the most important components of transmission and distribution systems.

The large transformers in power generation stations step up the output voltage of AC generators to higher values for more efficient transmission over transmission lines while also reducing the current values. Somewhat smaller transformers at electrical substations step the transmitted voltage down to the values more useful for regional and local distribution to customers while also stepping up the current.

Transformers can also isolate circuits, suppress harmonics, and regulate line voltage between distribution substations and consumers. Zigzag grounding transformers, for example, derive neutrals for grounding and a fourth wire from a three-phase neutral wire. They can be operated at voltages below their nameplate ratings, but they should not be operated at higher voltages unless they have taps intended for that purpose.

Transformers are classified in many different ways: dry- or liquid-insulated, single-phase or poly-phase, step-up or step-down, and single-winding or multi winding. In addition, they are classified by application. For example, there are voltages or potential transformers (VTs) and current transformers (CTs) that are used step high voltage and current down to safe levels for the measurement of voltage, current, and power with conventional instruments.

The efficiency of all power transformers is high, but efficiency is highest for large transformers operating at 50–100 % of full load. However, some losses are present in all transformers. They are classified as *copper* or *I2R losses* and *core losses.* Copper losses, also called *load losses,* are proportional to the load being supplied by the transformer. These losses can be calculated for a given load if the resistances of both windings are known. As in generators and motors, the core loss is due to *eddy-current induction loss* and *hysteresis* (molecular friction) *loss,* caused by the changing polarity of the applied AC. Provided that the cores are laminated from low-loss silicon steel, both eddy-current and hysteresis losses will be reduced. Nevertheless, well-designed transformers in all frequency and power ranges typically have efficiencies of 90 % or more.

4 Renewable Energy Resources

Generally, energy resources can be categorized into two groups: long term renewable resources, short-term renewable resources. Nonrenewable energy resources or long term renewable energy resources consist of types of energy whose appearance times are very long and provided that they are consumed gradually, they will be ended. Nonrenewable energy resources consist of all fossil fuel and nuclear fuel energies and the generation techniques based on this kind of energy are discussed previously.

However, short-term renewable energies are continuous and stable energy and can be found in nature as biomass, water energy, wind energy, solar energy, geothermal energy, etc.

As regards the high rate of energy consumption, fossil and nuclear fuels cannot responsible for energy demand of the world autonomously. Moreover, fossil fuels are running out and are rigorously against environment due to their emissions; hence, using of other alternative energy resources such as renewable energy and the development of their application is inevitable.

4.1 Wind Turbines

Wind Turbines are energy converters in which the kinetic energy of wind is converted to mechanical energy. Electricity generation is of several type of application which utilizes the generated mechanical energy.

Wind turbine based power plants have been categorized into two general groups namely; stand alone wind turbine and grid-connected wind turbine power plant

4.1.1 Standalone Wind Turbine Power Plant

Stand alone wind turbine power plants have different kind of applications such as electricity generation in far away points in which very reliable power supply is needed, battery charging, water heating, cooling, etc.

The turbine's rotor diameter of small wind turbine power plant whose main application is battery charging is less than 5 meters and their conventional capacity is between 400 and 1000 W. The heating and cooling application are usually utilized in residential. In this application, wind turbine power plants are directly connected to water heater or an electrical radiator. Moreover, small wind turbine power plants are usually used to generate electricity at far away points as an auxiliary power supply for agricultural affairs.

Standalone wind turbine power plants are usually utilized with batteries and they can be operated with photovoltaic system and diesel generator as a standalone hybrid power system.

4.1.2 Grid-Connected Wind Turbine Power Plant

Grid-connected wind turbine power plant can be classified into single wind turbine and wind farms.

Single wind turbine can be used to supply residential, commercial, industrial, and agricultural loads. The capacity of these turbines is between 10 and 100 kW and installed near the loads. The excess generate electricity may sell to global network and when a wind turbine cannot generate power, the loads should be supplied by energy purchased from network.

Moreover, the increase in capacity of wind turbine faces technical and economical problems; hence, a centralized wind farm with very large number of wind turbines should be used to construct a high capacity wind power plant. Today, wind farms with capacity of more than 100 MW are constructed. The conventional capacity of each wind turbine in a wind farm is between 50 and 500 kW. The amount of generated energy in wind farm depends on the number of wind turbine, the characteristic of each turbine, and wind speed. Furthermore, to prevent turbine's shadow on each other in a wind farm (maximum using of sweep area of all turbines), the turbines are installed with a minimum distance from each other.

4.1.3 Wind Turbine Model

The model of Wind conversion system is well reported in the literatures [5–8]. However, for simplification the wind turbine with doubly fed induction generator consists of three main components: turbine rotor, drive train, and generator. The power extracted from wind by the turbine rotor is given by:

$$P_w = \frac{1}{2}\rho A v^3 c_p(\lambda, \ \beta) \tag{1}$$

where P_w is the power extracted from the wind, ρ is the air density, A is the swept area, v is the wind speed and c_p is the power coefficient which is a function of the tip speed ratio (λ) and the pitch angle of the rotor (β).

The drive train transfers the power from the turbine rotor to the generator. The main modeling equations for the drive train are as below:

$$\frac{d\omega_M}{dt} = \frac{1}{2H_M}(T_M - K.\theta_{MG} - D_M.\omega_M) \tag{2}$$

$$T_M = \frac{P_w}{\omega_M} \tag{3}$$

where T_M is the accelerating torque, K is the effective shaft stiffness, θ_{MG} is the twist in the shaft system, ω_M is the speed of the wind turbine, and $D_M.\omega_M$ is the damping torque in the wind turbine.

A Park model approach is commonly used for the induction generator simulation. The stator is directly connected to the grid and the stator voltage ($\vec{v}_s$) is imposed by the grid. The rotor voltage ($\vec{v}_r$) is controlled by a converter and is used to perform the machine control. This model can be described as follows:

$$\vec{v}_s = R_s.\vec{i}_s + \frac{d}{dt}\vec{\psi}_s \tag{4}$$

$$\vec{v}_r = R_r.\vec{i}_r + \frac{d}{dt}\vec{\psi}_r - j\omega_r\vec{\psi}_r \tag{5}$$

where $\vec{i}$ is the current space vector, $\vec{\psi}_s$, $\vec{\psi}_r$ are the stator and rotor impedance, and R_s, R_r are the rotor and stator resistance.

4.1.4 Wind Turbines

The general categorization of wind turbines is based on the axis direction of turbine. The axis direction of wind turbines can be horizontal or vertical. Horizontal turbines are categorized based on the number of blades into three general categories of two-bladed, three-bladed, and multibladed. In a horizontal wind turbine, the more the wind speed is led to the less the number of blades. Moreover, the axis of vertical turbine is perpendicular to direction of wind. Most of vertical turbine has two blades and the most conventional types of vertical turbine are Darrieus Turbine, Evans Turbine, and Savonius Turbine. The blades of Darrieus turbine are made of aluminum and are in the form of curve-like blade while the Savonius turbines have hemispherical blades. In vertical turbines generator can be installed on earth which is led to simplification of maintenance and service. Figure 4 shows Darrieus, Evans, and Savonius turbines.

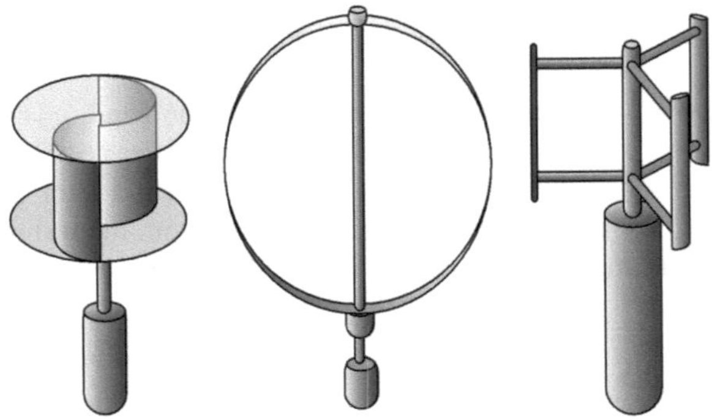

Fig. 4 The schematic of Savonius (*left*), Darrieus (*center*), and Evans (*right*) turbine

4.2 Photovoltaic Systems

PV panels turn sunlight directly into electricity, thanks to a property of their major component—silicon, the most abundant element on earth. Metals conduct electricity if the outer electrons on each atom are attached to the atom so lightly that they can drift away under the influence of a magnetic field. This electron drift is the electric current. Silicon atoms hold on to the electrons that surround them, but some are held less tightly than others and the right-sized hit of energy can knock them loose. Sunlight provides that energy hit, so when light shines on it some electrons are freed. Once the electrons are freed they can flow around a circuit—and that is an electric current. It is important to say that it is the *light*, not the *heat*, from the sun that enables the electricity to flow, so photovoltaics are just as effective in cold countries as in hot—provided there are long hours of sunlight.

4.2.1 Cell, Module, and Array

A solar cell is a semiconductor diode that when stimulated with light, produces an electron and a positive charge on opposite sides of the cell. Wires collect the charge off each side of the cell and take this electricity to the load circuit, such as a light bulb. Wiring cells in series increases the voltage, and in parallel increases the current output. Typically, it is a few square inches in size and produces about 1 W of power. To obtain high power, numerous such cells are connected in series and parallel circuits on a panel (module) area of several square feet. The solar array or panel is defined as a group of several modules electrically connected in a series–parallel combination to generate the required current and voltage.

Mounting of the modules can be in various configurations. In roof mounting, the modules are in a form that can be laid directly on the roof. In the newly developed amorphous silicon technology, the PV sheets are made in shingles that can replace the traditional roof shingles on a one-to-one basis, providing better economy in regard to building material and labor [9].

4.2.2 Types of Solar Cell

There are two ordinary types of solar cell panel [10]:

- Crystalline silicon solar cells have a solid silicon wafer as the semiconductor. The cells are sandwiched between tempered glass and a backing of tough ethylene vinyl acetate (EVA). These cells are protected from moisture. They need to remain cool as their output efficiency can drop by about 0.5 % for every degree Celsius above a standard test temperature of 25 °C. They typically incorporate a gap of approximately 150 mm behind the panels to allow for cooling.
- Amorphous silicon thin film solar cells have silicon in a thin film as the semiconductor. The silicon thin film is deposited on a low-cost substrate such as glass or a thin metal foil. The coating on top may be a flexible material (as opposed to glass), and they may use a flexible mounting system. This type of cell is generally cheaper. They are being developed for integration with materials so they can be part of the building fabric.

4.2.3 Application of Photovoltaic Cells

Some common applications of photovoltaic cells are listed below:

- Highway call boxes.
- Coast Guard buoys and navigation.
- Mountain top radio transmitting and repeating stations.
- Off-grid (rural) homes.
- Grid-tied homes.
- Utility interconnected for Demand Side Management (DSM).
- Utility grid support and bulk power generation.
- Railroad signaling.
- Satellites and space stations.
- Outdoor lighting.
- Calculators and watches.
- Telecommunications, mountaintop relay stations.
- Cathodic protection for metal exposed to the weather and earth.

4.2.4 Photovoltaic Cell Model

The simplest equivalent circuit of a solar cell is a current source in parallel with a diode. The output of the current source is directly proportional to the light falling on the cell.

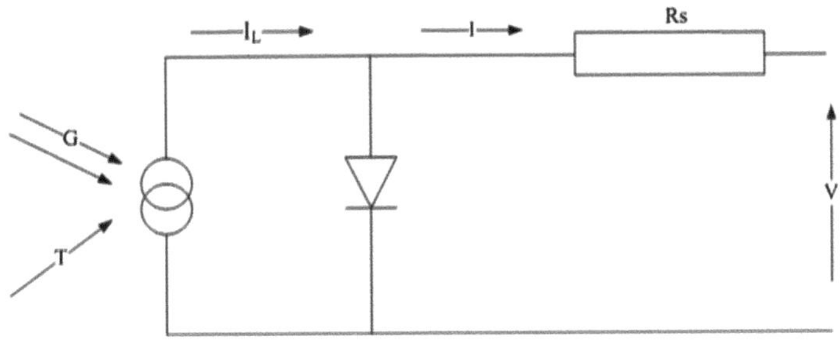

Fig. 5 A simple model of photovoltaic cell

The diode determines the I–V characteristics of the cell. Increasing sophistication, accuracy, and complexity can be introduced to the model by adding in turn.

The model includes temperature dependence of the photo-current I_L and the saturation current of the diode I_0. A series resistance R_S is also included, but not a shunt resistance. A single shunt diode is used with the diode quality factor set to achieve the best curve match [10]. The circuit diagram for the solar cell is shown in Fig. 5.

The equations which describe the I–V characteristics of the cell are

$$I_{PV} = I_{SC} - I_0(e^{q(V+IR_s)/AKT} - 1) \tag{6}$$

$$I_{sc} = I_{sc(T_1)}(1 + K_0(T - T_1)) \tag{7}$$

$$I_{sc(T_1)} = G \times I_{sc(T_1 \text{norm})}/G_{(\text{norm})} \tag{8}$$

$$I_0 = I_{0(T_1)} \times \left(\frac{T}{T_1}\right)^{3/A} \times e^{-qV_g/Ak\left(\frac{1}{T}-\frac{1}{T_1}\right)} \tag{9}$$

$$I_{0(T_1)} = I_{SC(T_1)}/(e^{qV_{OC(T_1)}/AkT_1} - 1) \tag{10}$$

$$R_s = -\left.\frac{dV}{dI}\right|_{V_{OC}} - \frac{1}{X_V} \tag{11}$$

where $X_V = I_{0(T_1)} \times \frac{q}{AKT_1} \times e^{qV_{OC(T_1)}/AKT_1}$; q is the electron charge; A is curve fitting constant; K is Boltzmann constant, T is temperature on absolute scale [10].

All of the constants in these equations can be determined by examining the manufacturer ratings of the PV array, and then the published or measured I–V curves of the array.

4.3 Geothermal Power Plant

Geothermal energy is of renewable energies and comes from the extractable heat due to heat of molten masses and decay of radioactive material existed in deep earth. Unlike other energy resources such as wind, solar, tidal, etc., this resource is continuously available. It means that it can be possible to generate electricity and thermal energy from geothermal resource in every hours of the day, while the other type of renewable resources are available seasonal or under special conditions.

4.3.1 Geothermal Resources

The geothermal resources can be classified as four main groups namely:

- *Hydrothermal*: The hot waters or steams which store in low-deep portion of earth (100–4500 m) is called hydrothermal resources. The hydrothermal resources which can generate electricity should have temperature between 90 and 350 °C; However, it is estimated that two third of hydrothermal resources have temperature between 150 and 200 °C. The richest type of hydrothermal resources contains dry steam or steam with low level of liquid. However, the number of dry steam hydrothermal resources is very few due to the special condition to be existed to form these resources.
- *Under high pressure layers*: These resources are hot springs containing methane which store in the deep of 3000–6000 m under high pressure in sedimentary layer. The temperature of these layers are estimated between 90 and 200 °C.
- *Hot dry rock*: The main idea of electricity generation using hot dry rock is to construct an artificial geothermal spring. To construct this spring, at first, two wells are drilled into hot rocks with 4000–5000 m deep. Then some cracks are made in rocks using nuclear explosion or hydraulic pressure; consequently, two wells are connected via the cracks. The cold water is injected to one well and hot water or hot steam is extracted from the other well. The Hot steam boils a secondary fluid with lower boiling point and finally, the electricity is generated by conducting the hot steam of secondary fluid to turbine. Nevertheless, the main problem of this type of electricity generation is that water does not circulate in a close loop circuit. Since the huge part of injected water penetrates into deep cracks.
- *Magma conduits*: Magmas have the temperature between 700 and 1200 °C, and they contain huge thermal energy. This resource is available in deep of 3000–10000 m; hence, operation of this resource is very difficult

Regardless of economical issues although these resources have different physical characteristics, all of these resources have the ability of electricity generation. Among all these resources, due to the competitive cost, hydrothermal energy is extended up to commercial generation much more than the other types while, the other resources are in the experiment stage. However, hot dry rocks resources have successfully passed the experimental stage and it can be possible to extract energy from these resources [1].

4.3.2 Types of Geothermal Power Plants

Geothermal resources can also be used for power generation. There are four methods to generate power from geothermal energy each can be described as:

- *Dry steam power generation*: In this type of power plant, hydrothermal resources are used to generate electricity. Hence, these power plants can be installed in fields in which there is the capability of dry steam extraction. The dry steam comes out using its natural pressure, then it is passed through a solid particle absorption filter and finally it is directly fed to turbine. The outflow steam of turbine enters a condenser and it loses its heat and is condensed; hence, a vacuum is created between turbine and condenser and consequently, the rotary force and efficiency of turbine is enhanced. The capacity of these power plants is considered about 15–20 MW.
- *Flash-steam power generation*: These power plants are used to utilize the sufficiently hot (more than 160 °C) liquid-dominated reservoirs and generally are categorized into two systems: one-stage and two-stage. In one-stage system, the geothermal hot water existing under high pressure flows over the well due to its very high temperature and pressure and is conducted into flash separator. Due to decrease of pressure, hot waters turn into steam and they are fed into turbine after passing special filter. In two-stage system, after steam generation in first separator, the hot water enters the second separator which have lower pressure. The generated steam of first and second separator flow to high and low-pressure turbine, respectively. The common capacity of these power plants is between 10 and 55 MW.
- *Binary cycle power generation*: These power plants are used to utilize the energy of liquid-dominated hydrothermal reservoirs whose temperature is not very high. The hot water whose temperature is about 65–200 °C is transferred to heat exchanger using pumps or its natural temperature. Then, the hot water transfers its heat to second fluid such as Isobutene or Freon whose boiling temperature is low. The steam generated by second fluid plays the role of working fluid of plant.
- *Combined/hybrid power generation*: In these power plants in addition to geothermal energy, other types of energy (such as fossil fuels) may be used to heat the working fluid. The geothermal fluid exchanges its heat with working fluid (pure water) in a heat exchanger leading to preheat of working fluid. The hot water is turned to steam in a steam generator using combustion of conventional fuels.
- *Power generation with Rotational Separation Turbine (RST)*: In these power plants, the geothermal hot water goes into the separator and parts of it turn into high pressure steam which can be fed into high pressure turbine. The remaining fluid in separator is conducted to RST which is coax with steam turbine. The

RST system has two advantages: not only does kinetic energy of RST help rotary power of steam turbine but also some low-pressure steam is generated in RTS which can be conducted to low-pressure turbine and enhance rotary power of steam turbine [11].

4.4 Biomass Power Plants

Generally, biomass resources refer to materials which are made up from planets and organisms. Unlike fossil fuels which are found in the centralized layers in the world, biomass resources are available in decentralized form. Biomass resources which are suitable for energy generation are namely as: wood fuels, agricultural waste, energetic shaw with short circulation period, herbal products such as sugarcane, herbals containing vegetable oil, herbals containing hydrocarbon, industrial and rural waste, etc.

Moreover, three technologies can be utilized to convert biomass energy. These technologies are known as: Direct combustion process, Thermochemical process, Biochemical process, each of which uses gasification reactors of type: fixed bed reactor and fluidized bed reactor [12].

Direct combustion process is the simplest technology to convert biomass energy. However, biomass material in this technology has low thermal energy contents; hence, the efficiency of this technology is less than efficiency of fossil fuel combustion. To enhance the thermal efficiency of biomass, the raw materials are processed using thermochemical process and different types of biofuels such as wood coal (at temperature less than 400 °C), ethanol, and methane (at temperature more than 1000 °C) are produced. The produced materials have more energy contents than biomass raw materials. However, in the biochemical process biomass material converts to biofuel using metabolic microbial organisms' actions. This process utilizes anaerobic fermentation to generate biofuel and biogas.

Several types of power plants are available to convert biomass energy to electricity. In big power plants, biomass is used as co-firing fuel; however, in small power plants biomass can be used as single fuel of plant. Plants with combustion engines, thermal biomass power plant, gas biomass power plant, combined cycle biomass power plants are the most commercial biomass which are used to electricity generation [13].

In power plants with combustion engines, different types of biofuel produced by biomass can be substituted with conventional fuel of combustion engines. However, in other types of power plants, biomass is used as the fuel of boiler furnace in thermal biomass or the fuel of combustion room of gas biomass power plant. Figure 6 illustrates a biomass combined cycle power plant.

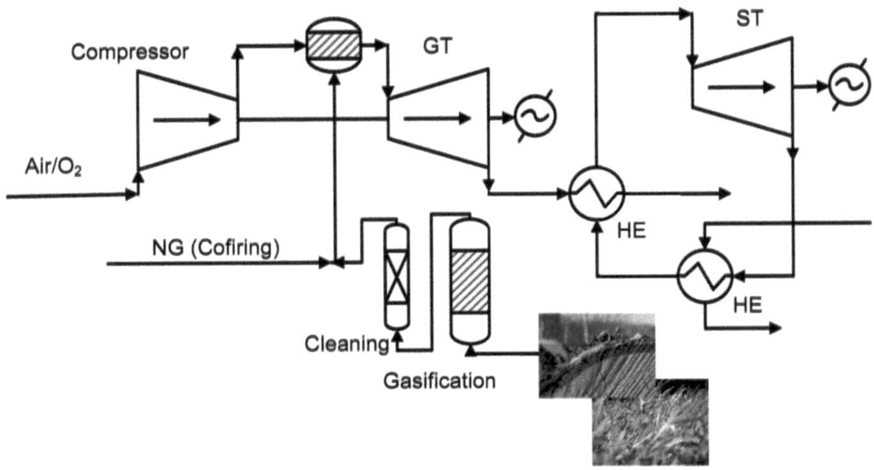

Fig. 6 Flow diagram of a combined cycle power plant using biomass as main fuel

4.5 Tidal Power Plants

The common technologies of tidal power plants are relatively full-fledged. Yet, the common methods to extract tidal energy require more economical performance. Generally, to generate electricity from tidal energy two different methods are suggested. First method needs to construct dam and water storage reservoir while in the second method tidal currents are used; hence, there is no need to dam construction.

Tidal power plants which need dam construction are categorized based on their reservoir combination. The categorization includes: a reservoir for ebb, a reservoir for flood, a two-way reservoir for ebb and flood, two reservoirs for ebb and flood, and two high and short reservoirs with one-way system [10, 14].

A reservoir for ebb: In this system, the water flows to reservoir via turbine in ebb situation and generator which is coupled with turbine generates electricity while the floodgates are opened in flood situation; consequently, the reservoir is evacuated.

A reservoir for flood: In this system, when floodgates are opened, at first, water storage reservoir will be fully filled. Then, if water level of reservoir is higher than water level of sea in flood situation the water of reservoir will be evacuated to sea via turbine and finally the generator which is coupled with turbine generates electricity.

A two-way reservoir for ebb and flood: In this system, a two-way single reservoir is used. However, comparing with other single reservoir, this scheme needs also a two-way turbine to generate electricity in both ebb and flood situations. Using this system, the pause time of electricity generation between ebb and flood situation is significantly reduced. Hence, more electricity can be generated comparing single reservoir schemes; however, generated energy still has fluctuations.

Two reservoirs for ebb and flood: The procedure of electricity generation in this system is that water flows from sea to one reservoir in the ebb situation and flows from another reservoir to sea in flood situation. Comparing with single reservoir systems previously described, the generated electricity is more continuously in this system; however, they need complicated scheme to generate energy.

Two high and low reservoirs with one-way system: In this system, two reservoirs are constructed in such way that one has a dam with high height while the other has a dam with low height. The turbine is placed in the wall between two reservoirs and spins with water flow from higher reservoir to lower reservoir; hence, the electricity is generated in generator which is coupled with turbine. Comparing to previous systems, the construction of the wall between two reservoirs is led to increasing in cost of constructions significantly. The mechanism of this system is that the higher reservoir is fully filled in the ebb situation by opening of floodgates of higher reservoir. When the water level of the sea decreases during flood situation, to prevent water flow from reservoir, the floodgates are closed. By continuously flowing of water from high reservoir to low reservoir via through turbine the electricity is generated. When the level of water in sea is less than the level of water in lower reservoir, the floodgates of lower reservoir will be opened to flow water to sea. This process continues until the water level of the sea equals to water level of lower reservoir. Then, the floodgates of this reservoir are closed to prevent flowing of water from sea to reservoir. Then lower reservoir is filled by higher reservoir through turbine. Finally to complete the loop, when the water level of sea is more than water level of higher reservoir, its floodgates will be opened and filled. Although the generation of electricity through this method is continuous, it has still fluctuations.

Tidal current power plants: These power plants are taken into consideration in recent years significantly due to no need for dam construction and high height tide and environmental issues. The power of tidal current power plants can be calculated as below:

$$P = 0.5\frac{M}{t}v^2 = 0.5\frac{\rho V}{t}v^2 = 0.5\rho Q v^2 = 0.5\rho A v^3 \qquad (12)$$

where ρ is water density (=1000 kg/m^3), Q is flow rate of tidal current (m^3/s), v is the speed of tidal current (m/s), M is the mass of flowing water (kg), A is the cross section of flowing water (m^2), and V is the volume of water (m^3).

As can be seen from Eq. (12), the most important condition for electricity generation of a tidal current power plant is tidal current with sufficient speed. Although the construction of these power plants eliminates the cost of dam construction, comparing with the other types which need dam, the power density of this method is very low.

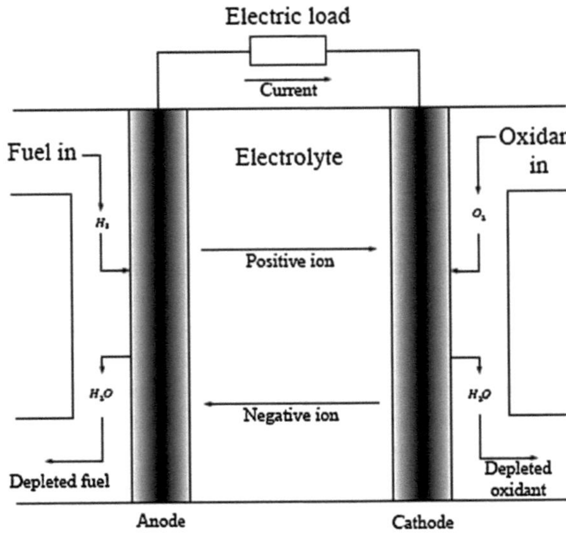

Fig. 7 Simple schematic diagram of fuel cell [17]

4.6 Fuel Cell Power Plant

A fuel cell is an electrochemical device that oxidizes fuel without combustion to directly convert the chemical energy of the fuel cell into electrical energy [15]. In simple terms, the fuel cell produces electric power by feeding a hydrogen-reach gaseous fuel to porous anode as an oxidant (air) is supplied to the cathode. The electrochemical reactions taking place at the electrodes result in electric current injected to the external circuit. Figure 7 schematically shows a simplified diagram of a fuel cell. In a fuel cell powered by hydrogen, the reverse reaction of water electrolysis takes place [16, 17]:

$$H_2 + 0.5O_2 \rightarrow H_2O \tag{13}$$

Except in Molten Carbonate Fuel Cell in which the general reaction is:

$$H_2O + CO_2 \rightarrow H_2 + CO_2 + 0.5O_2 \tag{14}$$

The chemical energy content of hydrogen is directly converted to electrical energy in the fuel cell. Therefore, a fuel cell can theoretically obtain higher electrical efficiency than thermal engines that are limited by the efficiency of the Carnot-cycle. However, kinetic over-voltages at the electrodes and electrical resistance cause relatively high losses in practical systems. Although several types of Fuel Cell technology are available, mostly important types of fuel cell technologies are [16]:

- Phosphoric Acid Fuel Cell (PAFC).
- Proton Exchange Membrane Fuel Cell (PEMFC).

- Alkaline Fuel Cells (AFC).
- Solid Oxide Fuel Cell (SOFC).
- Molten Carbonate Fuel Cell (MCFC).

PAFCs are currently being considered for use in heavy duty vehicles. Their major problem for use in vehicular application is their slow startup (since the cell has to be heated to over 200 °C), high costs, and excessive weight. Since PAFCs work best at a constant output, their application will be better in hybrid systems where a battery or other device meets the high-power demands of acceleration.

The basic components of a phosphoric acid fuel cell (PAFC) are the electrodes consisting of finely dispersed Pt catalyst on carbon paper, Sic matrix holding the phosphoric acid and a bipolar graphite plate with flow channels for fuel and oxidant. The operating temperature ranges between 35 and 200 °C and it can use either hydrogen or hydrogen produced from hydrocarbons (natural gas) or alcohols as the anodic reactant. In the case of hydrogen produced from a reformer with air as the anodic reactant, a temperature of 200 °C and a pressure of 8 atm are required for better performance. The heat generated by FC can be utilized in endothermic reaction of fuel process in fuel reformer or it can be used as heat supply to generate heat water. PAFCs are advantageous from a thermal management point of view. The rejection of waste heat and product water is very efficient in this system and the waste heat at 200 °C can be used efficiently for the endothermic steam reforming reaction. The system is extremely sensitive to CO and H_2S which are commonly present in the reformed fuels. A major challenge for using reformed fuel, therefore, lies in the removal of CO to a level of less than 200–300 ppm. CO tolerance is better at temperatures more than 180 °C. Removal of sulfur is still essential.

PEM fuel cells, also called solid polymer electrolyte fuel cells (SP(E)FC) use a proton (hydrogen ion) conducting membrane which stays sandwiched between two platinum-catalyzed porous electrodes728. Initially, these membranes were based on polystyrene, but at present a Teflon-based product "Nafion" is used. This offers high stability, high oxygen solubility, and high mechanical strength. The cell operating temperature is quite low (60–120 °C) and operating pressures can be in the range of 1–8 atmospheres. The fuel cell requires humidified hydrogen and oxygen for its operation. The pressures, in general, are maintained equal on either side of the membrane. Operation at high pressure is necessary to attain high-power densities, particularly when air is chosen as the anodic reactant. The major contaminant of the PEMFC system is carbon monoxide. Even a trace amount of CO drastically reduces the performance level. It is best to use a fuel which is virtually free of CO for PEMFC. On the other hand, it is tolerant toward CO_2 contamination [18].

AFCs offer high-power density and cold-start capabilities. Alkaline fuel cell represents the oldest and most widely used fuel cell systems in the U.S. space program and have gone onboard most of the manned space missions. AFCs use potassium hydroxide as the electrolyte and hydroxyl ions are the conducting species. Because of the alkaline electrolyte, no noble metal catalyst is required.

AFCs operate at 60–250 °C which is relatively low compared to other fuel cells. Operating pressure is normally atmospheric pressure. From a system point of view, removal of product water and heat is difficult at these low temperatures. In space shuttles, closed-loop hydrogen circulation as well as dielectric liquid circulation is used for heat management. Some of the terrestrial fuel cells are process air-cooled. Alkaline fuel cell can operate only with pure H_2 and pure O_2 Even a small level (less than 250 ppm) of CO_2 is sufficient: to carbonate the electrolyte and can spoil the electrode. Several processes for cleaning of the electrode after contamination are available (Physical adsorption-Selexol process, Fluor solvent process, pressure swing adsorption) but each is expensive and none are totally effective [18].

Solid oxide fuel cells (SOFCs) are solid-state power systems and at present use yttrium-stabilized zirconium as the electrolyte. The operating temperature is high, typically 1000 °C. SOFCs can be used as co-generators to supply both electricity and high quality waste heat. In this cell, a conversion efficiency of more than 50 % can easily be attained. Because of high temperature, the SOFCs can handle impurities in the incoming fuel better. SOFCs can operate equally well on dry or humidified hydrogen or carbon monoxide fuel or on mixture. The main poisoning factor for SOFC is H_2S. Though the sulfur tolerance level is approximately two orders of magnitude greater than other fuel cells, the level is below 80 ppm [19].

In the molten carbonate fuel cell, a molten alkali carbonate mixture, retained in a porous lithium aluminate matrix, is used as the electrolyte. The conducting species is carbonate ions. The operating temperature is in the range of 600–800 °C, high enough to produce suitable qualities of waste heat. This waste heat can be used for fuel processing and cogeneration, a bottoming cycle, or internal reforming of methane. MCFCs normally have 75 % fuel (hydrogen) utilization. The advanced form of MCFC referred to as internal reforming molten carbonate fuel cell (IRMCFC) may consume lower hydrocarbons (CH,) directlyi6. It is intrinsically efficient since the heat produced at the anode is used for reformation of hydrocarbons. Normally their efficiencies are 50 % or higher MCFCs do not suffer from CO poisoning and, in fact, can use CO in the anode gas as the fuel. They are extremely sensitive (1 ppm) to the presence of sulfur in the reformed fuel or oxidant gas stream. The presence of HCl, HF, HBr, etc., causes corrosion, while trace metals can spoil the electrodes. The presence of particulates of coayfine ash in the reformed fuel can clog the gas passages [15].

Table 1 summarizes the characteristics of different types of Fuel Cells. It should be noted that high rated power can be achieved by recovering of heat generated by FC in other thermal processes such as thermal power plants.

4.7 Storage Systems

Storage system is another component of hybrid power system. Due to the variable nature of wind and photovoltaic systems storage is required to supply demand continuously and reliably. Hence, when the sufficient wind and solar power are available, loads will directly supplied by these resources and the excess power can

Table 1 Characteristics of different types of fuel cell

	AFC	PAFC	MCFC	SOFC	PEMFC
Electrolyte	Potassium	Phosphoric acid	Molten alkali carbonate	Ceramic	Polymer
Operation temperature (°C)	60–250	35–200	600–800	700–1000	60–120
Fuel	Pure processed H_2	Processed H_2	Processed H_2/CO	Processed H_2/CO_2/CH_4	Processed H_2
Oxidant	O_2	O_2/air	CO_2/O_2/air	O_2/air	O_2/air
Electrical efficiency	>50	40–50	50–60	40–60	40–50
Applications	Space Industrial, transportation, military industrial, submarine	Industrial, commercial, transportation, airplane, power plants	Industrial, commercial, airplane, power plants	Industrial, commercial, airplane, power plants	Industrial, commercial, transportation, power plants(up to several kW)
Life time (hours)	>1000	>40000	>40000	>15000	>40000
Rated power (kW)	–	1–20000	50–50000	1–50000	1–3000

be stored in battery. There are three main goals for utilization of storage system in hybrid power system: to supply demand at a constant output, to compensate the instantaneous lack of renewable energies, to operate system as a dispatchable unit.

Different storage systems can be utilized in hybrid power system. The most important technologies are flywheel, compressed air energy storage (CAES), hydrogen production, ultra-capacitor, superconducting magnetic energy storage, thermal reservoir, etc., among whom battery is the most commercial technology to store energy in hybrid power system.

The classification of energy storage system is based on the energy, time, and transient response required for their application [12]. In medium and long term applications, storage system is classified based on energy density requirements and in short and very short-term applications, the classification is based on power density requirement.

In transients (milliseconds), the main features of storage is to compensate voltage sags, to improve harmonic distortion and power quality, to rid through disturbances, etc. In very short term (a few cycle of power frequency), storages cover load during startup, compensate transient response of renewable-based converters, increase system reliability during fault management. The short-term (minutes) capabilities of storages are to cover load during short-term load peak, to decrease needs of startup backup generator, and to improve maintenance needs of fossil fuel-based generators. In medium term (a few hours) they store the excess energy generated by renewable resources. In long term (e.g., several hours), storages provide reduction in fuel consumption and decrease waste of renewable energy, eliminate backup of conventional generators. Storage systems using in planning period are such as pumped hydro and compressed air systems. In this period, it can be possible to store hydrogen from biomass and renewable-based systems.

Figure 8 depicts storage technologies organized in categories of ride-through, power quality, and energy management. Storage systems include lead–acid batteries, advanced batteries, low- and high-energy flywheels, ultra-capacitors, superconducting magnetic energy storage (SMES) systems, heating systems, pumped hydro, geothermal underground, and compressed air energy storage (CAES).

5 Supervisory Control of Hybrid Power System

Supervisory controller is used to define a plant wide control system, and supervision refers to operation overview, planning & scheduling, co-ordination and execution of actions that improve performance, economy and reliability. The dynamic control is not a task of the supervisory controller.

A supervisory controller for hybrid power systems should continuously monitor the operating state of the system and keep it within the specified target domain. The control actions should ensure that the system's operating goals are achieved

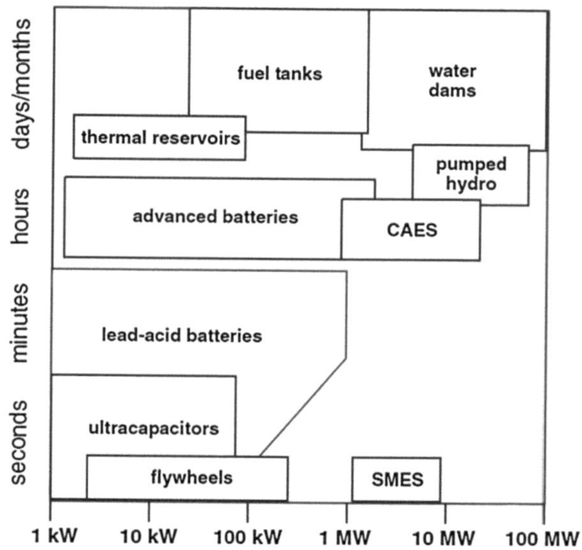

Fig. 8 Categorization of storages technologies based on energy [12]

and maintained in spite of uncertainties and resource constraints. It could also act to prevent safety critical system states or to alleviate consequences of failures in system components.

The primary goal of the supervisory controller is power supply optimization. Two basic tasks are necessary to accomplish this: (1) Improving the performance of the system (for example, increasing fuel savings or reducing generation cost); and (2) keeping the operational parameters (frequency and voltage, load at the diesel engines, etc.) within their limits. The control actions used to perform these tasks are (1) switching on/off components, i.e., changing the system's mode of operation, and (2) changing operation set points of some (controllable) components.

As is shown by Fig. 9, a supervisory control system includes three main parts namely Plant, Interface, and Controller. In this scheme to gain the benefits of discrete controllers, it is possible to design a discrete-event controller system for hybrid power systems. The explicit of identification of interfaces between continuous and discrete dynamics has been one of the characteristics of supervisory control of HPS. Hence, the supervisory controller supervises the operation of system while operation control is distributed between local controllers and regulators. The distributed control allows the system components to use their own control systems. For instance, a wind power generation system has a voltage regulator to control its output voltage and a rectifier-inverter system to maintain its output frequency at power frequency.

The supervisory controller selects the optimal operation mode for the system considering the specific operating goals, the system configuration and constraints. Moreover, the supervisory control can determine set points for different components

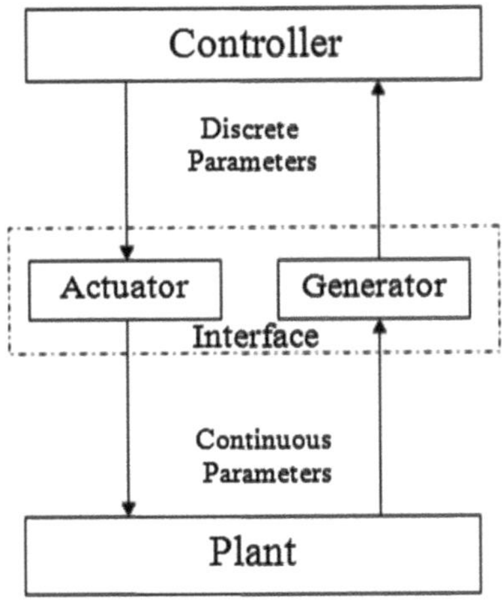

Fig. 9 Supervisory control system

at different modes of operation and send them to their respective local controllers. For wind generators the set points could be the power output value and the output voltage. Figure 10 illustrates a scheme of the supervisory control system [20]. The system consists of two main parts: *hardware* and *software*. The *hardware* is portrayed by the type of computer and its features. The interfaces with the plant which should be controlled (*process interface*) and the operators (*man–machine interface*) are major parts of the hardware and affect the *application programs* (*software*) related to the I/O operations of each interface. The *software* is divided into *operating system* (OS)—for each computer type there are several OS available in the market, and *application programs*, which are the algorithms of the control system. Normally, *development programs* develop offline, the control algorithms.

The aim of supervisory control is to enhance the economics of Hybrid Power Systems by ensuring efficient operation, reliability, and long life time. Furthermore, the controller should be robust enough to remain in operating mode efficiently after failures in components.

The difficulties related to operation of HPS, which are not present in large interconnected grids, are mainly caused by:

- The uncertainty and complication related to generation and conventional generators operation of HPS, respectively
- Lack of knowledge on power quality issues in HPS.

The stochastic nature of the renewable resources causes large variations of power generation. The main goal of HPS control is to supply load demand reliably regardless of generation variations while frequency and voltage maintain within acceptable limits.

Separation of different types of HPS control has a great importance. It should be noted that in order to control HPS dynamically, controllers performed in scales of

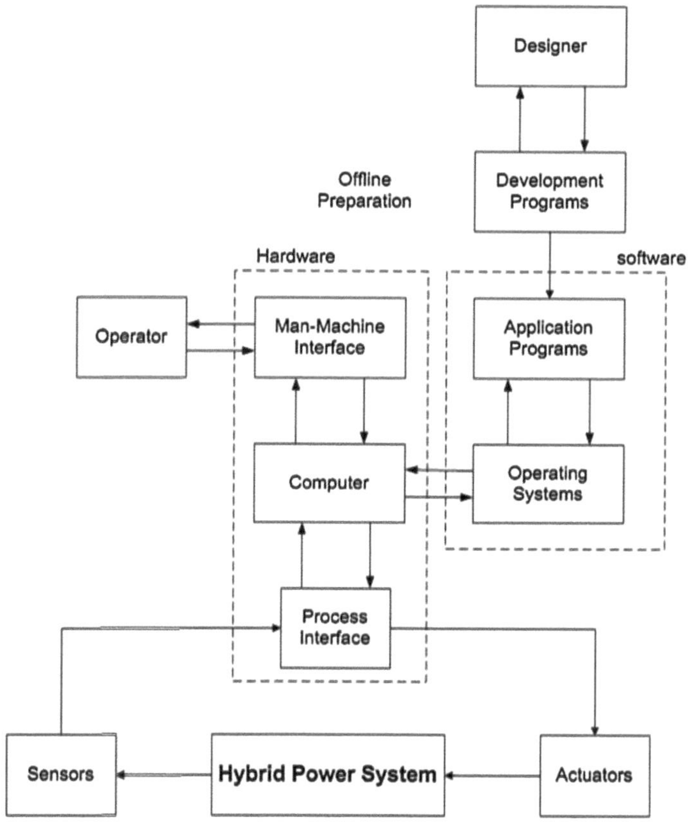

Fig. 10 Schematic of supervisory control problem [20]

seconds or milliseconds have been used. The dynamic control includes frequency and voltage control as well as stability issues and etc. Nevertheless, supervisory control makes decision about optimal operation of HPS including energy flow as well as actions to enhance system operation and the value of power generation of conventional generators. The time scale of operation strategy which is yielded by supervisory controller is about several seconds to hours.

In HPS with high renewable energy penetration, conventional generators whose fuel costs are significant can be switched off when the energy produced by renewable energy generation is sufficiently high to meet the demand. If accurate load and renewable power prediction are achieved, in order to significant enhance of system economically it would be possible to schedule service status of conventional generators.

Another problem to consider with conventional generator operation is that there can be relatively long time intervals necessary to start and stop the units. For example, a system collapse can happen during a diesel generator startup if the renewable power decreases faster than the startup time interval.

Normally, the procedure adopted by power plant operators is to keep always some conventional power capacity available in order to avoid system collapse in case of a sudden loss of renewable power or increase of load.

Nevertheless, this situation frequently conducts diesel generators to run under loaded and deteriorates the fuel savings.

5.1 Operation Strategies

An operation strategy should help the supervisory controller to make decision about generation, controllable loads, and storages.

The following parameters should be considered in the operation of HPS

- The characteristics of the demand, seasonal variations, peaks and valleys, etc.
- The characteristics of the renewable resources, such as mean values, standard deviations, frequency of occurrence, extreme values, diurnal and seasonal variations, etc.
- The characteristics of the conventional generators, such as specific fuel consumption, limits of operation, etc.
- The configuration of the system, i.e., number and type of components (renewable energy generators, conventional generators, controllable loads, type of storage, etc.).
- Power quality requirements—the required quality of the supply, in terms of variations of frequency and voltage and the probability of loss of load. Therefore, some operation strategies may be more effective in particular applications depending on the system goals and characteristics.

The operation strategies include spinning reserve, load management, minimum run-time of conventional generator, storage management, etc.

5.2 The Control System Architecture

Figure 11 illustrates the architecture of supervisory controller. It is worthwhile to note that the arrows in the Fig. 11 do not show the actual information flow but they depict the control flow of supervisory control process.

The data from sensors and messages from controller are sent to the *input interface*. The operator can be interfaced with control system to supply information, to change parameters, and to choose operation strategies.

The input data is stored in the *data base* module which is common database. The received data is filtered and analyzed in this module and the information in the

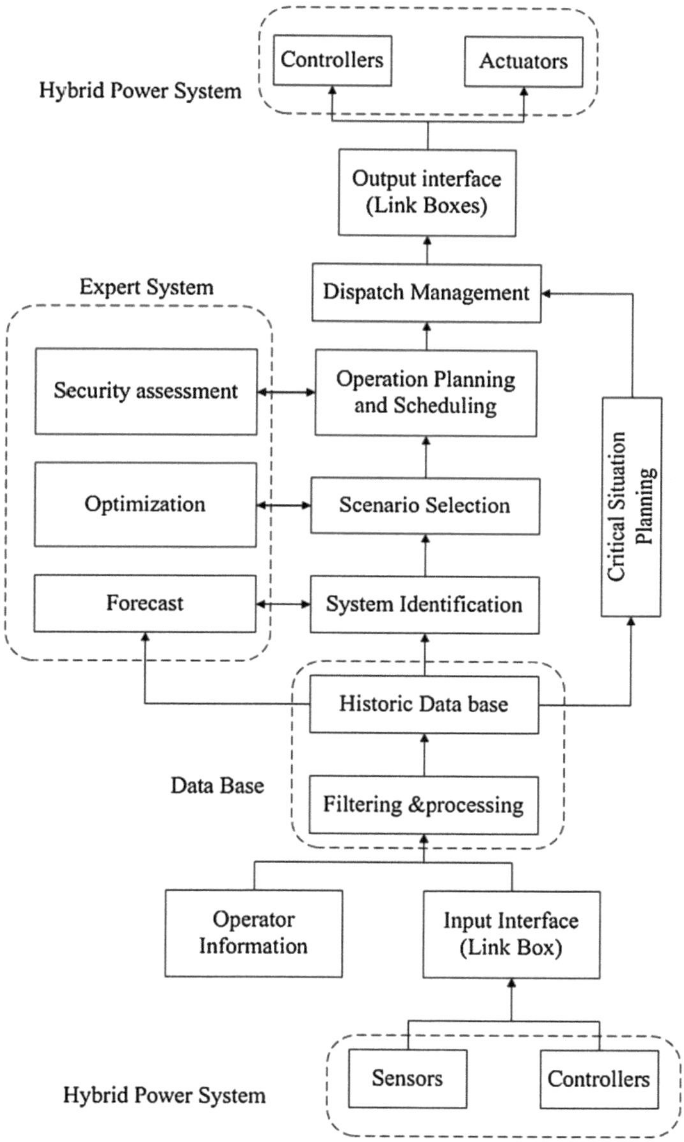

Fig. 11 Control flow of the supervisory controller architecture [20]

database can be accessed by the other parts of controller for further analysis. To run simulation and security assessment of system the information of historic data base are used by other parts of the system.

As far as dynamic control of power system is not a task of supervisory controller, the calculations, and information of abnormal situation are not included in database information.

Identification of current status of system based on values of state variables is interpreted by *system identification* module. The input of this module is the information of *data base* module. This information consists of grid parameters and operating status. This module identifies current operating system and predicts short period events of the system.

To optimize system operation via possible actions selection and to compensate changes in system parameters and to keep the operation conditions of conventional generators, *scenario selection* module is used. It should be noted that the list of possible actions are generated depending on the main goal, system configuration, and constraints. This module makes decisions about components' switching on/off and control parameters' changes.

The possible actions are sent to the *operation planning and scheduling* module and a plan for selected action is established. The planed action is executed in the *dispatch management* module. This module connects with interface of all components via *output interface* module. This module translates the commands and messages and sends signals and commands to actuators and controllers.

The *critical situations planning* module contains predefined rules and actions to compensate large disturbances in the system. It bypasses the basic decision model when safety critical system states are identified. This allows the supervisory controller to act very fast in case of emergency (or serious alarm) in order to protect the system.

The *expert systems* module comprises three distinct tasks: forecast, optimization, and security assessment. These expert systems are used to improve efficiency of the control algorithm. Several methods for forecast, optimization, and security assessment can be implemented and the structure of the program allows the user to select the algorithms (methods) to be included in the control system.

6 Hybrid Power System Modeling, Control, and Management

Hybrid power system (HPS) is any autonomous electricity generating system combining renewable energy sources and conventional generators. Wind-diesel, Fuel cell- Wind, Photovoltaic-Wind-Fuel cell systems, etc. are a subclass of HPS. The purpose of these systems is to supply reliable energy whose power quality is acceptable while the penetration of renewable resources is as much as possible. Furthermore, the lower generation costs due to low cost of fuel and the advantages of fuel saving should balance the high investment costs for renewable energy generators, controllers, dump loads, storage units, converters, etc.

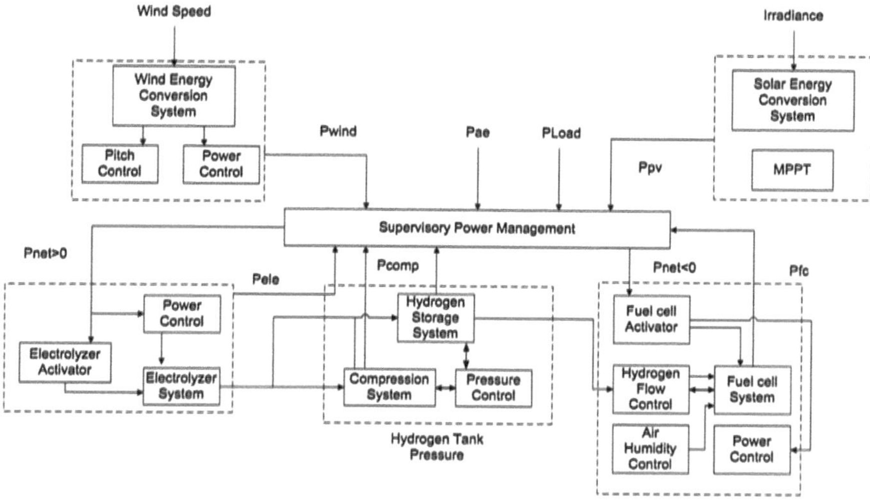

Fig. 12 Block diagram of management and control of a hybrid power system consisting of wind turbine, photovoltaic cell, fuel cell, storage system, and electrolyzer

The control system of Hybrid Power System consists of a supervisory controller for the overall power management, and secondary low-level controllers, which manage various parameters for the individual components.

The power flow in the hybrid system is shown in Fig. 12. The converted energy from the renewable sources can be either used directly to meet the load demand, or transferred to the hydrogen production process.

The logic of the decision process is based on Eq. (15) [21]:

$$P_{\text{net}} = (P_{\text{wind}} + P_{pv}) - (P_{\text{load}} + P_{ae}) \tag{15}$$

where P_{load} is the load demand and P_{ae} is the power consumed by auxiliary equipment including compressors, controllers, and safety equipment.

At each sampling interval, if the excess wind and Photovoltaic (PV) generated power is greater than the rated power of the electrolyzer ($P_{\text{net}} < P_{\text{ele}}$), the electrolyzer stack is activated to generate hydrogen, which is then delivered to the hydrogen storage tanks through the compressor unit. On the other hand, when there is a deficit in power generation ($P_{\text{net}} < 0$), the Fuel Cell (FC) is activated to consume the stored hydrogen and convert it to electricity. The fuel cell activation will occur if there is sufficient hydrogen in the storage tank. Otherwise, the hybrid system enters the "hydrogen storage" mode. This can occur as a consequence of either extreme operational conditions, such as low availability of renewable energy and very high load demand, or inappropriate unit sizing. An additional problem is operating the electrolyzer and fuel cell at their full capacities, i.e., without any local control. The amount of power required to run the electrolyzer depends on its capacity, and if it is operated at the rated capacity, at some point even if $P_{\text{net}} > 0$, the amount of power consumed by the electrolysis process would overcome the power generated by the RE sources. The solution to this problem is implementation of local controllers on the electrolyzer and fuel cell. Their primary objective is to ensure using the suitable extent of the electrolyzer and fuel cell's capacity in order to use the excess energy and stored hydrogen in the most efficient way. The local, or low level, controllers ensure maximum energy extraction of the RE side of the hybrid system, as well as proper hydrogen generation and utilization.

Despite all improvements, PV modules still have relatively low conversion efficiency. *V–I* and *V–P* characteristic curves for a PV array specify a unique operating point at which the maximum possible power is delivered. The Maximum Power Point Tracking (MPPT) algorithm is used for extracting the maximum available power from the PV module under certain voltage and current conditions. There are several MPPT techniques reported in the literature. The perturbation and observation method (P&O) is one of the common and effective ways of power tracking for PV arrays. The MPP tracker operates by periodically incrementing or decrementing the solar array current. If a given perturbation leads to an increase (decrease) of the output power of the PV, then the subsequent perturbation is generated in the same (opposite) direction.

The wind turbine power output varies with the wind speed. The control objective is dependent on the wind velocity range. Above the cut-in wind speed, the control system extracts maximum power according to the turbine specific maxima power trend. The control action is based on the difference between the actual turbine speed (ω_r) and the corresponding maxima power. This offset is sent to two PID controllers to adjust the current and voltage of the rotor converter in order to obtain the maximum power. Between the rated and cutout speed, the pitch angle controller takes action. In this velocity range, the turbine speed (ω_r) is compared to the desired turbine speed and the offset is sent to the pitch controller to manipulate the pitch angle and keep the output power constant. The pitch angle operational range and its rate of change are the constraints applied on this controller. In the case of wind speed higher than the cutout speed, the system is taken out of the operation for the protection of its components.

As previously mentioned, the electrolyzer and fuel cell are commonly operated at their maximum capacity and this can drastically decrease the overall efficiency of the system. In this study, separate model predictive controllers are designed and applied to control the electrolyzer and fuel cell performance.

For control design purposes, the nonlinear models of the electrolyzer and fuel cell were linearized around selected operating points to obtain a state space model in the following form:

$$\begin{cases} \dot{x} = Ax + Bu + B'w \\ y = Cx \end{cases} \tag{16}$$

where x, u, w and y are the model states, manipulated variables, disturbances, and model outputs, respectively.

These variables for the electrolyzer are [21]

$$\begin{cases} x_{ele} = \left[\delta N_{O_2,a}, \delta N_{H_2O,a}, \delta N_{H_2,c}, \delta N_{H_2O,c}\right]^T \\ y_{ele} = \left[\delta P_{ele}, \delta V_{ele}, \delta N_{H_2}, \delta p_{H_2}\right]^T \\ u_{ele} = \left[\delta I_{ele}\right] \\ w_{ele} = \left[\delta p_{ele}\right] \end{cases} \tag{17}$$

where the operator δ indicates the deviation from the operating point, I_{ele} is the electrolyzer's current, P_{ele} is the electrolyzer's consumed power, V_{ele} is the electrolyzer voltage, p_{ele} is the electrolyzer's operation pressure. P_{ele} is considered as the controlled variable and other outputs are measured outputs. The other parameters have been introduced in the electrolyzer modeling section.

The fuel cell state space model variables are:

$$\begin{cases} x_{fc} = \left[\delta p_{fc}, \delta N_{O_2,\,c}, \delta N_{H_2O,\,a}, \delta N_{H_2,\,a}, \delta N_{N_2,\,c}\right]^T \\ y_{fc} = \left[\delta P_{fc}, \delta V_{fc}\right]^T \\ u_{fc} = \left[\delta I_{fc}\right] \end{cases} \tag{18}$$

where P_{fc} is the fuel cell generated power, chosen as the controlled variable, and V_{fc} is fuel cell voltage, considered as the measured output.

For both systems, the control objective is to keep the power (P_{ele} and P_{fc}) at desired set points which are imposed by the P_{net} value from the power management controller. Constraints on upper and lower limits as well as the rate of change for power were implemented to avoid large and nonrealistic variations.

The model predictive controller is designed to minimize the following finite control and horizon performance index [21]:

$$\begin{aligned} &\min \sum_{i=1}^{n_y} \left|\alpha\left[y_i - y_{i,ref}\right]\right|^2 + \sum_{i=1}^{n_u} |\beta[\delta u_i]|^2 \\ &s.t. \begin{cases} y_{lb} \le y \le y_{ub} \\ \delta y_{lb} \le \delta y \le \delta y \\ u_{lb} \le u \le u_{ub} \end{cases} \end{aligned} \tag{19}$$

where α and β are input and output weight factors and n_y and n_u are the prediction and control horizons. The objective function was subjected to the set of constraints, the fuel cell and electrolyzer's operational limitations (y_{ub}, y_{lb}, u_{ub}, u_{lb}) and the rate of change in the electrolyzer and fuel cell power. Aside from power control, two PI controllers were implemented to minimize the pressure difference between the cathode and anode by manipulating the hydrogen flow, and keep the desired air humidity by injecting appropriate amount of water vapor into the air stream entering the cathode side of the fuel cell.

7 Conclusion

Hybrid power system is electricity generation system combining both renewable and conventional electricity generation beside energy storages. This power system may use any renewable energy resources such as wind, solar, water, geothermal, and biomass. These kinds of energies may either convert to hydrogen to store in energy storage system or generate electricity in fuel cells. To manage the wide range of energy resources and energy generation system, a control and management system is inevitable; hence, a supervisory control system is used to manage and control energy generation system.

In this Chapter, an overview of hybrid power system was presented. First, conventional fossil based power plants such as steam, gas, combined cycle, and nuclear and relevant generators such as AC and DC generators were introduced. Then, renewable-based power plants such as wind, solar, geothermal, tidal, hydro, and biomass power plant were introduced. Moreover, different kind of fuel cell as non-fossil based energy generation system and their generation mechanism was presented. Storage devices are other components of hybrid power system whose types were described in this chapter briefly. Any possible combination of these resources may be used in a hybrid power system. Finally, supervisory control and management system and its relevant operation strategies were described.

References

1. Tafreshi SMM (2010) Electrical energy systems in 21st century. Toosie University of Technology Publications, Tehran
2. Raja AK, Srivastava AP, Dwivedi M (2006) Power plant engineering. New Age International, New Delhi
3. Boldea I (2006) The electric generators handbook. Taylor & Francis Group, New York
4. Sclater N, Traister JE (2003) Handbook of electrical design details. McGraw-Hill, USA
5. Melicio R, Mendes VMF, Catalao JPS (2011) Transient analysis of variable-speed wind turbines at wind speed disturbances and a pitch control malfunction. Elsevier J Appl Energy 88:1322–1330
6. Dicorato M, Forte G, Trovato M (2012) Wind farm stability analysis in the presence of variable-speed generators. Elsevier J Energy 39:40–47. doi:10.1016/j.energy.2011.11.050
7. Vepa R (2011) Nonlinear, optimal control of a wind turbine generator. IEEE Trans Energy Convers. doi:10.1109/TEC.2010.2087380
8. Abedini A, Nikkhajoei H (2011) Dynamic model and control of a wind-turbine generator with energy storage. IET Renew Power Gener. doi:10.1049/iet-rpg.2009.0123
9. Hansen AD, Sorensen P, Hansen LH, Bindner H (2000) Models for a stand-alone PV system. Riso National Laboratory report, Roskilde, ISSN 0106–2840
10. Patel MR (2006) Wind and solar power systems: design, analysis and operation, 2nd edn. Taylor & Francis Group, New York
11. Hua CH (2008) Modeling of a renewable energy system: experimental innovation and technology centre. University of Waterloo, Dissertation
12. Farret FA, Simoes MG (2006) Integration of alternative sources of energy. John Wiley & Sons Publication, New Jersey

13. Jurado F, Cano A, Carpio J (2003) Modeling of combined cycle power plants using biomass. Eur J Renew Energy 28:743–753
14. Wood J (2008) Local energy distributed generation of heat and power. IET power energy series 55. The Institution of Engineering and Technology, London
15. International Corporation (2000) Fuel cell handbook, 5th edn. EG&G Services Parsons, Inc, Morgantown
16. Borbely AM, Kreider JF (2001) Distributed Generation: the power paradigm for the new millennium. CRC Press, New York
17. Knazkins V (2004) Stability of power system with large amounts of distributed generation. Dissertation, KTH Institution for Elektrotekniska System
18. Masters GM (2004) Renewable and efficient electric power systems. John Wiley & Sons Publication, New Jersey
19. Grew KN, Chiu WKS (2012) A review of modeling and simulation techniques across the length scales for the solid oxide fuel cell. Elsevier J Power Sources 199:1–13. doi:10.1016/j.jpowsour.2011.10.010
20. Pereira AL (2000) Modular supervisory controller for hybrid power systems. Dissertation, University of Denmark
21. Wang C (2006) Modeling and control of hybrid wind/photovoltaic/Fuel cell distributed generation system. Dissertation, Montana state university

Revisiting and Generalizing Barkhausen's Equality

Horia-Nicolai Teodorescu

Abstract A study of the conditions for sustaining signals in a loop shows that loop equations are essentially fixed-point equations over a space of functions, with the loop performing a mapping on that space of functions. When the space of functions is specified, one can derive particular conditions for the loop has a solution. Barkhausen conditions fall in this category. Loops composed of two subsystems are in the first place analyzed. The purpose of the chapter is to put into a general perspective the problems of loops, showings the general conditions that must be satisfied. The analysis aims to clarify several perspectives on and the framework of loops operation.

1 Introduction

This chapter is motivated in the first place by the intuitive nature and generality of the concept of loop, and by the numerous applications of this concept. In the second place, the motivation relates to the richness of results that may be derived from the basic condition of signals sustained by a loop, which is essentially an identity.

Probably the most favored method of approaching today loop-like problems is to write, for the subsystems in the loop, their differential equations, taking into account that the output of a system is the input to the next one. Then, the corresponding system of differential equations is solved, deriving the loop solutions.

H.-N. Teodorescu (✉)
GheorgheAsachi Technical University of Iasi, Iasi, Romania
e-mail: hteodor@etti.tuiasi.ro

H.-N. Teodorescu
Institute for Computer Science of the Romanian Academy, Iasi, Romania

N. Bizon et al. (eds.), *Analysis, Control and Optimal Operations in Hybrid Power Systems*, Green Energy and Technology, DOI: 10.1007/978-1-4471-5538-6_2,

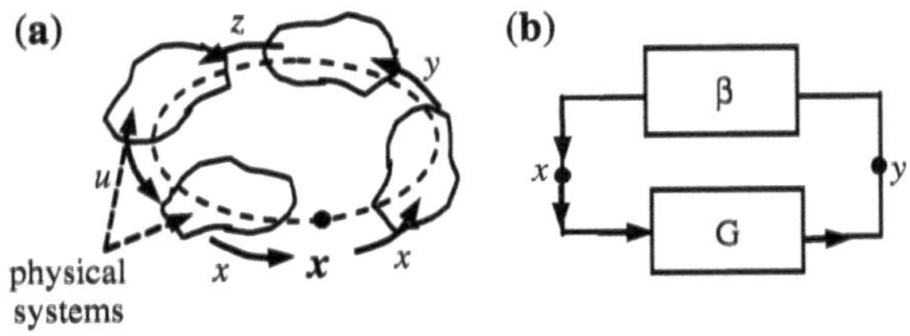

Fig. 1 **a** Loop with a serried set of physical systems; each system output is the input to the subsequent one. **b** A two-system standard loop

We are interested in this chapter in another point of view. Namely, for a given loop composed of several subsystems, supposing that the loop has a non-trivial solution, we ask what conditions the subsystems do necessarily obey. While this problem is more general than that partly addressed by Barkhausen, it seems to be in line with Barkhausen's one.

Interestingly, it seems that sometimes the problem of existence of solution of a loop is confused with that of the stability of the solution; the confusion translates into the name often given to Barkhausen criterion indicating a stability criterion, not an existence one.

The concept of loop emerges in the physical and structural representation of sets of coupled equations whose variables are produced by individualized interdependent systems (Fig. 1a). Loops are system configurations used in numerous engineering fields. Control loops also appear in all natural sciences as a result of the coupling of equations in subsystems and are recently used to explain cellular and molecular "clocks". In economy and social sciences, loops are used to represent regulatory systems.

In many engineering applications, the equations representing systems take a simple multiplicative form, $y = G_1 \cdot x$, $z = G_2 \cdot y$, ... For this type of systems, Barkhausen established a condition on the systems in a loop for the loop "accepts" a specified signal on it. The condition is a necessary one, not caring about the solution of the loop, i.e., on the signals in the loop. Barkhausen's conditions had reverberations in circuit theory in electronics and in control for several decades. They preserve some interest in relation with large systems, biologic models, and similar topics where loops occur. A longstanding academic tradition includes Barkhausen's condition in the introductory lectures on oscillators, but the interest in these conditions waned in scientific research during the recent decades.

Recently, Barkhausen's condition has attracted revived interest, along with some criticism [1–6]. He and coauthors wrote [1] "For too long a time, the Barkhausen criterion has been widely used as a condition of oscillation for the harmonic feedback oscillator. It is intuitive and simple to use. However it is wrong, and cannot give the real characteristics of an oscillator such as the startup condition and the oscillation frequency." In another direction, Singh [3, 4] shows that the supra-unitary Barkhausen condition, $G\beta > 1$ is not, contrary to the popular belief reflected in some textbooks, the condition for the starting up of oscillations; instead, Singh proves on a system example that the condition $G\beta < 1$ must hold. In a later paper, he proved for harmonic oscillators that the setting up of oscillations in a loop can occur in some specific cases if $G\beta < 1$, while in some other cases it occurs when $G\beta > 1$. Singh discuss the issue in the framework of Nyquist stability condition.

While Barkhausen's condition(s) of oscillation is a strong instrument, these conditions are offering only a partial answer to the problem of sustained signals in a loop. They do not give enough information on the signals in nonlinear loops, on the stability of the solutions, and on the starting up of such signals. For a discussion on the original Barkhausen identity, see the Appendix.

Simple feedback loops essentially represent an identity equation in the form $x \equiv x$, where $\equiv$ denotes the identity relation. The identity refers to any point in the loop, where literally the left part of the identity refers to the loop physical path from that point leftward, while the right part "sees" the loop rightward (Fig. 1a).

Sometimes, Barkhausen equality (frequently written as $G\beta = 1$or $A\beta = 1$) is confused with a criterion of stability of oscillations, or with a criterion for starting oscillations (see [6]), which Barkhausen criterion is not. Even worse, in some technical documentation, Barkhausen equality is replaced by an inequality. For example, in an older application note [7], it is written "For oscillation to occur, the Barkhausen criteria must be met: (1) The loop gain must be greater than one. …." Also, the above condition is not valid except for a subset of loops, including linear loops. Two more general criteria for loop solution existence are discussed, namely the loop identity condition (LIC) and the loop differential condition (LDC).

The chapter structure is as follows. In the next section, we present the general settings and introduce several notations and term definitions. In Sect. 2, the general setting is presented. Section 3 is devoted to loops depending on the derivatives of time-dependent functions. Conclusions and a short annex on the work of Barkhausen end the chapter.

2 General Setting and Non-Differential Loops

In this section, we set the loop problem framework. Two concepts are necessary for the discussion. The first is the concept of loop. The second is the notion of loop solution. In the first part of this section we will be less formal and precise, to connect to the general intuitive understanding of the topic in applications.

The loops are formally described by a system of (coupled) equations in the form $\{y = f_1(x);\ z = f_2(y);\ \ldots;\ x = f_n(u)\}$, or, more generally, $\{y = f_1(x, \dot{x}, \ddot{x}, \cdots), \cdots\}$, when the systems operate in continuous time and depend on the derivatives of the input. It is essential to recall that in this chapter $x, y, z, \ldots$ denote functions, not necessarily real valued, from a given space of functions, while f_1, …, the subsystems are seen as mappings from a space of functions to itself. This point of view is justified by the fact that we are interested in signals supported by the loops, not by constant values (equilibrium points) of the loops.

Throughout this chapter, one assumes that all the functions of the subsystems and the signals are derivable in the domain of interest, when time is continuous. The states of the system are not considered.

2.1 Terminology

A solution of the loop is defined as follows. First, reduce the loop to a single function, the function overall performed by all the systems in the loop, by combining all systems in one, f_{loop}. Assume that f_{loop} has a single variable, x. Then, a ν variable dependent solution of the loop is a ν-dependent function that is a fixed point of f_{loop},

$$x(\nu) = f_{loop}(x(\nu)). \tag{1}$$

A time-dependent solution of the loop is a time-dependent function, named also signal (in the time domain), $x(t)$, which is a fixed point of f_{loop}. Assuming that the loop function does not depend on the derivatives of $x(t)$, then, a solution of the loop satisfies

$$x(t) = f_{loop}(x(t)). \tag{1'}$$

Equation (1) should be valid whatever is the variable of x, that is, whatever is the representation of x. For example, if x is a Fourier representation of a time-domain signal, then the loop is represented by a mapping F_{loop} such that $x(\omega) = F_{loop}(x(\omega))$, where $x(\omega)$ is the complex spectrum. Consider that the loop is composed of two systems performing multiplicatively, $y(x) = x(\omega)G(\omega)$, $x(y) = y(\omega)H(\omega)$, with $F_{loop}(x) = x(\omega)G(\omega)H(\omega)$, that is, F_{loop} is linear in the input x, while H, G are independent of x. Then, from $x(\omega) = F_{loop}(x(\omega))$ it results that, for all $\omega$where $x(\omega) \neq 0$, $G(\omega)H(\omega) = 1$. One may view the subsystems in the loop having the functions $\frac{dy(x)}{dx} = G, \frac{dx(y)}{dy} = H$, with $HG = 1$ ($y'_x x'_y = 1$.)

One can say about (1) that a loop is a mapping from the set of ν-dependent functions to itself, and a solution of the loop is a fixed point of the mapping. A solution will be named trivial when $x(\nu) = constant$. Following Barkhausen and the literature, we are interested in the loops that can be written, for practical reasons, as $f_{loop} = f \circ g$, and in the conditions that relate the two mappings, f and g, under specified general conditions, for the loop to have non-trivial solutions. For practical reasons, having physical loops in mind, a loop that allows the trivial solution $x(t) \equiv 0$ is named passive loop, or equilibrium-able loop. For example, a loop consisting of amplifiers and filters is an equilibrium-able loop when the amplifiers have no offset.

While the loop condition $x = L(x)$ is nothing but a fixed point definition for the "signal" in the loop, it resides on the physical trivial identity $x(t) = x(t)$. We know the general conditions for a fixed-point problem to have solutions, according to theorems in the class of theorems developed along Brouwer, Kakutani, and Schauder theorems. However, in this chapter we are interested in less general problems, related to loops composed of one or two subsystems. The focus is on the conditions satisfied by the subsystems, as implied by the hypothesis of non-trivial solutions of the loop general equations.

Throughout this chapter, we will say that a function $x(v)$ is accepted (supported) by the loop when it is a solution of the loop equation, that is, the fixed point condition (1). We will consider loops with at most two serried systems. In general, we will assume that the systems are intrinsically time-invariant.

In the case of loops consisting of two systems, assume $x = g(f(x)) = g(y)$ and $y = f(g(y)) = f(x)$. This formally identical (interchangeable) role of f and $g$explains many of the symmetric results derived in this chapter.

The number of elements (subsystems) in the loop is irrelevant, because the string of elements in the loop can be seen as a single element. Also, the loop identity must hold on any loop in a graph, in a multi-loop configuration system where appropriate (Kirchhoff-like) conditions are added. The simplest case of loops we deal with is that of linear systems in the loop, $y = kx$.

The systems in the loop may produce an output depending on one or several variables, including the input, the derivatives of the input, and one or several parameters, $y = y(x, \dot{x}, \ddot{x}, \lambda_1, \cdots, \lambda_p)$. We need to consider these cases one by one.

Notice that loop Eq. (1) does not refer to steady state or transitory regimes specifically—it is just a translation for a loop of an identity condition. As such, it can be applied to steady state oscillations in nonlinear systems.

2.2 Basic Conditions

Consider a loop with the overall function denoted by L and dependent only on the input, not on the input derivatives with respect to time, $L = L(x(v))$, where v represents some variable in the representation space of x. For example, v can stand for ω in the Fourier representation, or it may be some other suitable variable in another representation. Assume that both L and x have derivatives. By differentiation of the fixed point condition, we obtain that, at any fixed point x^* of the loop,

$$\frac{\partial L}{\partial x}(x^*) \cdot \frac{dx^*}{dv} = \frac{dx^*}{dv}. \tag{2}$$

When $\frac{dx^*}{dv} \neq 0$, the above implies that

$$\frac{\partial L}{\partial x}(x^*) = 1 \tag{3}$$

for any function $x^*(v)$ that is a loop solution; in technical terms, this means that the "amplification" (slope) of the loop must be equal to 1 at x^*, for x^* is a solution. A stability condition must be added, for the solution to be stable.

The condition (1), $x = L(x)$ will be named *loop identity condition* (LIC), while the equality (3) $1 = \partial L(x)/\partial x|_{x^*}$ will be named the *loop differential condition* (LDC).

Subsequently, we provide the formal general setting of the loop problem.

Consider a space of functions over the set U, with values in $\boldsymbol{R}$ or $\boldsymbol{C}$. Denote these functions as $x(v), y(v), \ldots, v \in U$. The set of these functions is denoted by Ξ. We assume that U, Ξ are endowed with metrics and that derivation is defined for mappings from Ξ to itself. Consider three operators, $L, G, H : \Xi \to \Xi$, all of them derivable at the fixed points of L. Assume that L has at least one fixed point, $x^*(v)$. Also assume that on Ξ, $\frac{dx}{dv} = 0$ only when $x(v) \neq c \in \mathbf{R}(\boldsymbol{C})\forall v$. Then,

Definition. An operator L that satisfies the condition $L = HG$ is named loop. A fixed point of the loop is named loop solution.

A loop solution is non-trivial when $x(v) \neq c \in \mathbf{R}(\boldsymbol{C})\forall v$. In the next Property, one assumes that derivatives are defined in some way (e.g., Fréchet differentiable, or in the sense of Alexandroff, or as in Sobolev spaces); for various definitions of derivatives, see [8–11]. One also assumes that the typical properties of the derivatives, as the derivative of a composed function and the chain rule hold. In the next property, x is a function, for example a time-domain function, or the Fourier spectrum of a time domain function; x^* is a solution of the loop.

Property 1 *If a loop with two systems has a non-trivial solution* x^*, *then*

(1) $HoG(x^*) = \mathrm{Id}(x^*)$, where Id is the identity application on Ξ.
(2) When x is real valued, $\frac{dL}{dx}(x^*) = 1$, where 1 denotes the unity in $\mathbf{R}$.
(3) $\frac{dG}{dy}(y)\frac{dH}{dx}(x^*) = 1$.
(4) If the loop is twice derivable, then $\frac{d^2L}{dx^2}(x^*) = 0$.
(5) If G is invertible, then, $H(x^*) = G^{-1}(x^*)$.
(6) If x is a derivable function in two variables, $x(v, w)$, then the loops solution satisfies the conditions $\frac{dH}{dy}\frac{dG}{dx}\frac{\partial x}{\partial v} = \frac{\partial x}{\partial v}$ and $\frac{dH}{dy}\frac{dG}{dx}\frac{\partial x}{\partial w} = \frac{\partial x}{\partial w}$ at the point (function) $x = x^*$.
(7) If the variable can be written as a linear decomposition $x(v) = \sum_k a_k x_k(v)$, x real valued, with all functions x_k derivable with respect to all x_j, $a_k \in \mathbf{R}$, then the solution of the loop satisfies the set of conditions $\frac{dH}{dy}\frac{dG}{dx}\frac{\partial x}{\partial x_j} = a_j + \sum_{k \neq j} a_k \frac{\partial x_k}{\partial x_j}$.

The above property directly results from the definitions. Conditions (1) and (2) resembles the typical form of Barkhausen condition written in the frequency domain, $H(\omega)G(\omega) = 1$; in fact, (1) is the general form of that condition.

We next give a negative result. Consider that "signal" means a real-valued, time-dependent function, and $x(t)$ is a solution of a loop, then for all values of $x(t) \in \mathbf{R}$, the loop identity $x(t) = L(x(t))$ holds. Suppose that $L(x(t))$ is a function only in x, not in its derivatives. But this implies that, in the range of $x(t)$, L coincides with the identity operator $I(c) = c$, $c \in \mathbf{R}$, which satisfies (3). Then,

Property 2 *A loop that depends exclusively on the value of the real-valued function $x(t)$, not on its derivatives, either can support only trivial solutions, $x(t) = constant$ ($\frac{dx}{dt} = 0$), or is an identity loop.*

Example 1 Consider a loop with the function $L(x) = x^2$. It supports only two constant signals, $x = 0$ and $x = 1$.

Example 2 Assume that a transform, for example Fourier transform, is applicable to (1), and that (1) becomes

$$X(v) = F(v)X(v),$$

where $X(v)$ is the transform of x and $F(v)$ is the transform of L. When in (2) the loop corresponds to two serried systems with functions G and H, from the above we obtain $F(v) = G(v)H(v) = 1$, one of the first Barkhausen condition for linear systems. This example refers to linear loops, which allow us to apply the Fourier transform.

The LIC criterion also implies that: If a steady-state signal appears in the loop, and the loop has a delay, τ, independent on the value of the signal and on its derivatives, then the solution satisfies the condition $x(t) = x(t + \tau)$, meaning that the loop can support only periodical signals, moreover, the delay must be a multiple of the signal period. When the delay is dependent on the signal and its derivatives, we obtain the condition

$$x(t) = x(t + \tau(x, \dot{x}, \ddot{x}, \ldots)) \tag{4}$$

which is a supplementary condition (differential equation) to be satisfied by the fixed points of the loop.

Assuming that the loop is composed of two systems, characterized by two function $y(v) = f(x(v))$, $x(v) = g(y(v))$, then $x = (g \circ f)(x) = g(f(x))$. Further assuming that all functions are differentiable, we obtain,

$$\frac{dx}{dv} = \frac{dg}{dy}\frac{df}{dx}\frac{dx}{dv}.$$

Whenever $dx/dv \neq 0$, it follows that

$$\frac{dg}{dy}\frac{df}{dx} = 1. \tag{5}$$

This condition represents, for loops with two serried systems with differentiable functions f, g, both depending only in the input variable, not on its derivatives, the condition that the functions are one the inverse of the other. This could be named a *loop differential identity condition* (LDIC).

When the two functions are linear in the respective variables, the first Barkhausen condition is obtained, $A\beta = 1$, where $A = \partial f/\partial x$, $\beta = \partial g/\partial x$. We assumed above that x is real and f, g are real-valued functions of real variables. Whenever the function g is invertible on the range of $f(x)$, the loop has a solution if the system satisfies the condition $f = g^{-1}$, which is an obvious generalization of the Barkhausen equality $A = 1/\beta$.

The differential equation $\frac{\partial g}{\partial y} \cdot \frac{\partial f}{\partial x} = 1$ does not need to be satisfied when the solution of the loop is trivial, that is when x is a constant value with respect to the variable v; in that case, because in that case $dx/dv = 0$ maintains the differential equality above always true, independent of g and f.

Recall that a solution of the loop fixed point equation may be stable or unstable according to the absolute value of the derivative of $g(f(.))$ in a vicinity of the solution, with a smaller than 1 absolute value of the derivative ensuring the stability. In the next section, we deal with general conditions for time variable solutions of the loop.

Assume that we look for solutions of a loop that are functions of time (such functions are named here “signals”). The next remark provides a sufficient condition for solutions of this type of loop.

Remark 1 Denote by f, g the two functions and assume g is invertible for some range of y. Then, if the function g is invertible, the loop identity $x = (g \circ f)(x)$ is satisfied by $f(x) = g^{-1}(y)$ for all y in the given range.

A loop with two systems, both with functions depending only on the values of the signals (not of their derivatives) has a solution if (i) one of the two systems composing the loop has an invertible function on a domain of values of the function of the other system and (ii) the inverse function is equal to the function of the first system.

The above remark provides the principle for a method to build loops having a given solution, $x^*(t)$. Choose a function g which is invertible on the range of $x^*(t)$. Then complete the loop with a function $f = g^{-1}$ on $\{g(x^*(t)) | t \in \mathbf{R}\}$.

3 Systems Depending on the Derivatives of the Signal

Consider the loop $L = L(x(v))$ and the loop condition $x(v) = L\big(x(t), x^{(1)}(t), \cdots, x^{(n)}(t)\big)$ for a time-dependent function $x(t)$. An equivalent, n-th order differential equation is readily obtained as

$$L\Big(x(t), x^{(1)}(t), \cdots, x^{(n)}(t)\Big) - x(t) = 0.$$

Assuming that all involved functions are derivable, taking once again the derivative as in (2), one obtains the equation representing the loop as

$$\frac{dx}{dt} = \frac{d}{dt}(L(x(t))) = \frac{dL}{dx^{(n)}}\frac{dx^{(n)}}{dt} + \cdots + \frac{dL}{dx}\frac{dx}{dt}.$$

Conversely, any differential equation $E\left(x^{(n)}, \cdots, x\right) = 0$, involving the derivatives of a function $x(t)$ can be seen as a loop; for that, take $L(x(t)) = E + x(t)$. The interesting situation is when the loop is not trivial, in the sense that it includes two subsystems, $L = H(G(x))$. Then, the equation writes $E = H\left(G\left(x, x^{(1)}(t), \cdots\right)\right) - x = 0$.

Example 1 Consider the second order differential equation $\ddot{x} + \gamma\dot{x} + \lambda x = 0$. We wish to model it using a loop conveniently designed. Let $y = f(\dot{x}, x) = a\dot{x} + bx$, $g(y) = \alpha\dot{y} + \beta y$. The loop equation becomes then $g(y) = \alpha(a\dot{x} + bx)' + \beta(a\dot{x} + bx) = x$, or $\alpha a\ddot{x} + (\alpha b + \beta a)\dot{x} + \beta bx = x$, with $\alpha a = 1$, $\alpha b + \beta a = \gamma$, and $\beta b - 1 = \lambda$.

Consider a loop described by the function depending on the input derivatives, $x = L(x, \dot{x}, \ddot{x})$, assuming that $x(t)$ is derivable. We assume here that the loop systems have as variables the input signals and their first derivative, with $G(x, \dot{x})$ and $H(y, \dot{y})$ functions in the input variable and its derivative, where $y = G(x, \dot{x})$. It is easy to see that the loop depends on the second derivative of x, as in the Example 1.

The use of two systems is equivalent to using the d'Alembert decomposition of second order differential equations into first-order equations (see [8], p. 5). The loop identity (Barkhausen equality) has the form $x = \beta(G(x, \dot{x}), \dot{G}(x, \dot{x}))$.

Differentiating the equality

$$\frac{dx}{dt} = \frac{\partial\beta}{\partial u} \cdot \left(\frac{\partial G}{\partial x}\frac{dx}{dt} + \frac{\partial G}{\partial\dot{x}}\frac{d\dot{x}}{dt}\right) + \frac{\partial\beta}{\partial v} \cdot \left(\frac{\partial\dot{G}}{\partial x}\frac{dx}{dt} + \frac{\partial\dot{G}}{\partial\dot{x}}\frac{d\dot{x}}{dt}\right) \tag{6}$$

we obtain, assuming that all functions are derivable in x, and x has its $n+1$ derivative,

$$\frac{dx}{dt} = \left(\frac{\partial\beta}{\partial u}\frac{\partial G}{\partial x} + \frac{\partial\beta}{\partial v}\frac{\partial\dot{G}}{\partial x}\right)\frac{dx}{dt} + \left(\frac{\partial\beta}{\partial u}\frac{\partial G}{\partial\dot{x}} + \frac{\partial\beta}{\partial v} \cdot \frac{\partial\dot{G}}{\partial\dot{x}}\right)\frac{d\dot{x}}{dt}. \tag{7}$$

This is a $(2 + r)$ order differential equation $x''\left(\frac{\partial H}{\partial y}\frac{\partial G}{\partial x} + \frac{\partial H}{\partial y'}\frac{\partial G'}{\partial x'}\right) + x'\left(\frac{\partial H}{\partial y}\frac{\partial G}{\partial x} + \frac{\partial H}{\partial y'}\frac{\partial G'}{\partial x} - 1\right) = 0$, where r is the highest order of the derivative of x in the parentheses.

Suppose that we require a solution $x(t) = \cos(\omega t)$. Then, x' and x'' are independent and the above equality can hold only if

$$\frac{\partial H}{\partial y}\frac{\partial G}{\partial x} + \frac{\partial H}{\partial y'}\frac{\partial G'}{\partial x} = 1, \tag{8}$$

$$\frac{\partial H}{\partial y}\frac{\partial G}{\partial x} + \frac{\partial H}{\partial y'}\frac{\partial G'}{\partial x'} = 0. \tag{9}$$

The above are generalizations of the Barkhausen conditions—compare the first with $\frac{\partial H}{\partial y}\frac{\partial G}{\partial x} = 1$. We conclude:

Property 3 *When $\dot{x}$ and $\ddot{x}$ are independent in a second or larger order differential loop, (8) and (9) hold.*

The conditions (8) and (9) extend Barkhausen equality for differential loops of order at least 2.

Example 2 Consider the second order differential equation $\ddot{x} + \gamma\dot{x} + \lambda x = 0$. It can be modeled using a loop conveniently designed. Let $y = G(x, \dot{x}) = a\dot{x} + by$, $x = H(y, \dot{y}) = \alpha\dot{y} + \beta y$. The loop equation becomes then $g(y) = \alpha(a\dot{x} + bx)' + \beta(a\dot{x} + bx) = x$, or $\alpha a\ddot{x} + (\alpha b + \beta a)\dot{x} + \beta bx = x$.

Replacing $a = \partial G/\partial\dot{x}$, $b = \partial G/\partial x$, $\alpha = \partial H/\partial\dot{y}$, $\beta = \partial H/\partial y$, and assuming that $\dot{x}$ and $\ddot{x}$ are independent, the condition (9) is satisfied when $\alpha b + \beta a = 0$. The last condition is known to mean that the loop is without attenuation—allowing the loop to sustain the oscillation.

4 Conclusions

The broad applicability of loops in technical systems ranging from oscillators to control applications requires a solid understanding of loops in the broader context of mappings on sets of functions and of fixed point theorems for such mappings. According to the application domain, one is interested to have trivial solutions of the loop, that is, non-oscillations, as in control, or non-trivial solutions, electronic as in oscillators. Barkhausen, and this chapter, adopted the latter point of view, emphasizing the "oscillation conditions."

The requirement for the loop to consist of at least two subsystems (two mappings and the related mapping composition) brings specificities to the properties of the solutions. Whenever the mappings are derivable (in a weak, general sense), assuming the derivatives preserve some properties like chain property (which is true for the weak derivative in some spaces), the contour of stronger properties of the solutions appear, as shown in the properties in this chapter. Some of these properties can be seen as general conditions for a non-trivial fixed point exists, extending the type of conditions currently called with Barkhausen's name. A few examples completed the general perspective proposed in this chapter.

While there is no true novelty in the results presented in this chapter, because all are only facets and generalizations of well-known facts like the condition of fixed point, the differential conditions for accepted signals on loops clarify some cases of loops of practical importance. We analyzed differential loops and determined conditions of fixed point of such loops. The cases of differential loops described by equations as $x = L(x, \dot{x}, \ddot{x})$ show the entire power of the loop concept and of fixed point conditions. We believe that the terms "inverse function" and "fixed point" should not lack from any textbook dealing with loops and oscillations.

N.B. Several serious errors appearing in the brief paper (H.N. Teodorescu, Revisiting Barkhausen Conditions, ECAI 2011 Conf., Pitesti, (Rev. Univ. Pitesti, Series Electronics, Computers, and Artificial Intelligence, vol. 4, nr. 1, 2011, pp. 7–10) have been corrected in this chapter.

Acknowledgments I thank colleagues and referees who made constructive critics on preliminary forms of this chapter.

Appendix

Historical note. Barkhausen has proposed what is better named but less known as Barkhausen *equation*, written by him as SDRi = 1, which translates to $G\beta = 1$ or $HG = 1$ in the notations used in this chapter. Barkhausen was also interested in the starting of oscillations (we name the topic today "stability of oscillations"), as proved by the manuscript pictured in [Eugen-Georg Woschni, The Life-Work of Heinrich Barkhausen http://www2.mst.ei.tum.de/ahmt/publ/symp/2004/2004_075.pdf]. The oscillation problem has been the subject of Brakhausen's doctoral thesis, Das Problem Der Schwingungserzeugung: MitBesondererBerucksichtigungSchnellerElektrischerSchwingungen, published in 1907. According to the above quoted article on his life, Barkhausen equation first appeared in a book he published in 1917. More on Barkhausen's life at http://de.wikipedia.org/wiki/Heinrich_Barkhausen. Publications by Barkhausen are listed in the Katalog der DeutschenNationalbibliothek, at https://portal.d-nb.de/opac.htm?query=Woe%3D118657240&method=simpleSearch.

References

1. He F, Ribas R, Lahuec C, Jezequel M (2009) Discussion on the general oscillation startup condition and the Barkhausen criterion. Analog Integr Circ Sig Process 59:215–221
2. Wang H-Y, Huang C-Y, Liu Y-C (2007) Comment: a note on determination of oscillation startup condition. Analog Integr Circ Sig Process 51:57–58
3. Singh V (2007) Failure of Barkhausen oscillation building up criterion: further evidence. Analog Integr Circ Sig Process 50:127–132
4. Singh V (2010) Discussion on Barkhausen and Nyquist stability criteria. Analog Integr Circ Sig Process 62:327–332
5. von Wangenheim L (2010) On the Barkhausen and Nyquist stability criteria. Analog Integr Circ Sig Process. doi 10.1007/s10470-010-9506-4. Published on-line 15 July 2010
6. Bible S (2002) Application Note AN826 Microchip™, Crystal oscillator basics and crystal selection, p 2. 2002 Microchip Technology Inc
7. Aleaf A (1986) A study of the crystal oscillator for CMOS-COPS™, National Semiconductor, Application Note 400, August National Semiconductors

8. Ginoux J-M (2009) Differential geometry applied to dynamical systems, chapter 1. World Scientific Publishing Co. Pte. Ltd., p 5
9. Nirenberg L (1981) Variational and topological methods in nonlinear problems. Bull (New Series) Am Math Soc 4(3):267–302
10. Istratescu VI (1981,2001) Fixed point theory: an introduction (mathematics and its applications series), 1st edn. D. Reidel Publishing Company, Kluwer Group, Dordrechht
11. Skalland K (1975) Differentiation on metric spaces. Proc SD Acad Sci 54:75–77

State Energy-Based Approach as a Tool for Design and Simulation of Linear and Nonlinear Systems

Milan Stork, Josef Hrusak and Daniel Mayer

Abstract This chapter deals with a new problem of physical correctness detection in the area of strictly causal system representations. The starting point is energy and the assumption that a system can be represented by a proper interconnection. The interconnection or, better, the interaction between physical systems can be described in terms of power exchange through power ports. The proposed approach to the problem solution is based on generalization of Tellegen's theorem well known from electrical engineering. Consequently, mathematically as well as physically correct results are obtained. The contribution is mainly concerned with presentation of a new structural approach to analysis and synthesis of linear and nonlinear causal systems. It has been proven that complete analysis of system behavior reduces to two independent tests: the monotonicity test of abstract state space energy and that of complete state observability, eventually of its dual, i.e., complete state controllability property. For comparison, the example of port-Hamiltonian approach is also presented.

Keywords Chaoticity · Conservativity · Convergence · Dissipativity · Port Hamiltonian · Stability · State space energy · Tellegen's principle

M. Stork (✉) · J. Hrusak
Department of Applied Electronics and Telecommunications, University of West Bohemia, Plzen, Czech Republic
e-mail: stork@kae.zcu.cz

J. Hrusak
e-mail: hrusak@kae.zcu.cz

D. Mayer
Department of Theory of Electrical Engineering, University of West Bohemia, Plzen, Czech Republic
e-mail: mayer@kte.zcu.cz

N. Bizon et al. (eds.), *Analysis, Control and Optimal Operations in Hybrid Power Systems*, Green Energy and Technology, DOI: 10.1007/978-1-4471-5538-6_3,

1 Introduction

It is well known that a large class of real-world systems admits modeling based on Hamilton's principle of classical mechanics either in the form of Euler–Lagrange equations or in the form of Hamilton canonical equations [1]. In deriving both the sets of equations a crucial role play is the representation of the energy storage in the system. Although the Hamilton principle was originally formulated for mechanical systems only, from a general point of view the specific physical nature of the system is immaterial. Despite the almost generally accepted opinion that both the energy-based paradigm and system conservativity are necessary for construction of physically correct system representations, there are also some known approaches in which a power-based paradigm plays the key role. Most of them are based on the general concept of state [2, 3]. One of the resulting nonlinear electrical network models is known as the Brayton–Moser equations [4]. The well-known topological Brayton–Moser approach [5], leading to a set of the Brayton–Moser equations, is based on a special choice of physically motivated state variables frequently used in electrical circuits theory. It represents one of the most powerful systematic tools for construction of state space representations for nonlinear electrical networks.

It is familiar that there are two basic approaches to system modeling. The first consists in using mathematical formulas and physical principles (a causality principle, different forms of conservation laws, power balance relations, etc.) in order to describe appropriate system behavior. It has been successfully used in many fields of science and engineering so far. However, there are also situations where physical laws are not known or cannot be expressed in a proper mathematically exact form. In that case the second basic so-called cybernetical approach to system modeling can be turned to. It is based on identification methods working in terms of experimentally gained data [6, 7]. It is possible to divide the identification methods into two groups: parametric and non-parametric, respectively. If any prior information about a system structure is not available then one of the non-parametric procedures has to be chosen for system identification. On the other hand, assume that a physical structure of an investigated system is known. In such cases some of the available parametric methods can be used and consequently, more adequate results from the physical correctness point of view should be obtained [8–14]. Unfortunately, any reliable explicit knowledge about a physical system structure is more likely an exception than a rule. Therefore, a system structure is mostly chosen ad hoc on behalf of heuristic arguments. Subsequently, it has to be verified whether obtained quantitative results are not in conflict with obvious qualitative expectations concerning regular system behavior and/or results of additional experiments performed on a real system. In Fig. 1, the different energy-based systems description and their relations are shown [7, 15–21].

The chapter is organized as follows: Section 1 is devoted to the problem of physical correctness of electrical circuit models. In Sect. 2, the state space energy is introduced and a generalized form of the Tellegen's theorem called the Lyapunov–Tellegen's principle is investigated. In Sect. 3, the power and energy in

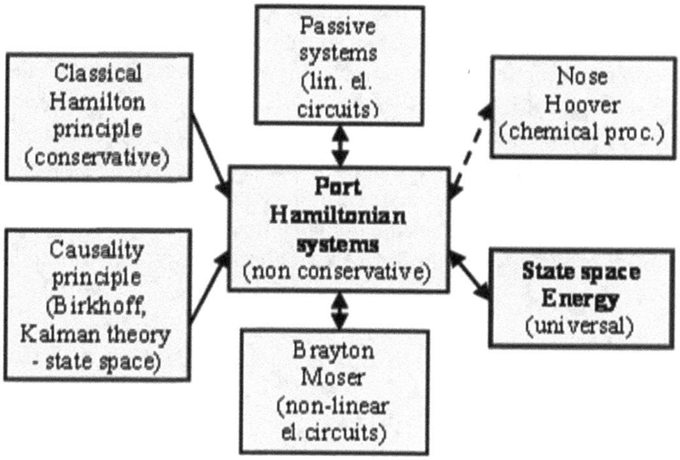

Fig. 1 The different energy-based approach simulations and their relations

systems and topological structures are discussed. Section 4 summarizes some knowledge concerning observability, controllability and state minimality of linear system representations, as well as state energy additivity property. In Sect. 5, the problem of stability analysis is used for illustration. In Sect. 6, the abstract form of the state space energy conservation principle is developed and illustrated by nonlinear systems. Section 7 demonstrates the application of state space energy principle for switching system. In Sect. 8 the energy approach based on port Hamiltonian is used for switching system analysis.

1.1 Physical Correctness in Electrical Circuits Theory

Certainly, any realizable system has to fulfill some causality and energy conservation requirements. Recall that existence of a state space representation is necessary and sufficient for a system to be causal. On the other hand causality does not imply energy conservation. In electrical engineering, validity of Kirchhoff's laws is standardly used as necessary and sufficient condition for energy conservation. Thus, for the correctness of a mathematical model of an electrical network we have:

Definition 1 (*Physical correctness of an abstract electrical circuit*) Electrical circuit is physically correct if it is not in contradiction with both the voltage and the current Kirchhoff's laws.

There arises a natural question of how the concept of physical correctness could be generalized beyond the electrical circuit theory. Tellegen's theorem seems to be a proper starting point, because it has proven to be a very general and extraordinarily elegant abstract form of the energy conservation principle for a class of physically correct system representations, in which voltages and currents have been chosen as state variables [20– 21].

At first, let us briefly summarize the essential features of the classical version of Tellegen's theorem and assume that an arbitrarily connected electrical network of b components is given. Let us disregard the specific nature of the individual network components and represent the *network structure* by an oriented graph with n vertices and b branches. Let the set of Kirchhoff's laws constraints be given as

$$Ai(t) = 0 \quad Bv(t) = 0 \tag{1}$$

where A is a *node incidence matrix*, B is *loop incidence matrix*, and i and v are:

$$i = [i_1, i_2, \ldots, i_n]^T, \ v = [v_1, v_2, \ldots, v_n]^T \tag{2}$$

Let J be the set of all current vectors i and V be the set of all voltage vectors v such that i and v satisfy (1). Then the classical Tellegen's principle can be expressed in the form as follows:

Theorem 1 (Classical Tellegen's theorem) *If $i \in J$ and $v \in V$ then for the inner product of both the vectors it holds*:

$$\langle i(t), v(t) \rangle = 0 \tag{3}$$

The most important observation is that the *orthogonality condition* (3) is basically *not a physical statement,* but rather of *geometrical nature.*

In fact, from abstract system theory point of view, it expresses a *neutrality of interactions* between a *gradient field* of an abstract system *state space energy* and the *vector field* of an abstract *state space velocity.* Thus for a *dual product* of a *row gradient vector* with a *column state velocity vector* we could write equivalently:

$$\langle \mathrm{grad}_x E(x), f(x) \rangle = 0, \ \mathrm{grad}_x E(x) = \psi(x) \tag{4}$$

where $E(x)$ is the *total electromagnetic energy accumulated* in all the electrical circuit components and $f(x)$ is the vector field of *state space velocity* defined by the corresponding *state space representation:*

$$\begin{aligned} \frac{\mathrm{d}x}{\mathrm{d}t} = f(x), \quad & x(t_0) \in X \subset R^n, \quad 1 \leq n < \infty \\ y = h(x), \quad & y(t) \in R^n, \quad 1 \leq p < n \end{aligned} \tag{5}$$

The idea of a generalized Lyapunov-Tellegen's principle (4) will be further developed in Part 2.

2 State Space Energy Approach and Generalized Lypuunov–Tellegen's Principle

As an alternative to the well-known physical energy motivated method of Lyapunov functions, a new conceptually different approach to stability problems, has recently been proposed in [23–26] and called the signal energy-metric approach. The *crucial idea* is that, in fact, *it is not the energy by itself*, but only a *measure of distance from the system equilibrium to the actual state* $x(t)$, which is needed for stability and instability analysis.

Thus a *state space metric* $\rho[x(t), x^*]$, where x^* denotes the *equilibrium state*, has been *defined*, and the basic idea of a new *state energy-metric approach* has then formally been expressed by the following *state space energy definition:*

$$E(x) = \frac{1}{2}\rho^2[x(t), x^*] \tag{6}$$

Let us consider a linearly observed *nonlinear* abstract system representation $R\{S\}$ defined by the *state equation in semilinear form*

$$\begin{aligned} \dot{x}(t) &= f[x(t)] + B\,u(t), \; x(t_0) = x^0, \\ y(t) &= C\,x(t) \end{aligned} \tag{7}$$

with an uncontrolled *state velocity vector field* given by

$$f(x) = A(x)x \tag{8}$$

1. Let $X \subset R^n$ be a class C_∞*-manifold,* representing the *state space* of the representation $\Re\{S\}$ containing the *unique zero input equilibrium state vector*

$$x^* = \left[x_1^*, x_2^*, \ldots, x_n^*\right]^T, \quad x^* \in X \tag{9}$$

defined by the condition

$$u(t) = 0, \quad t \in [t_0, \infty) : f(x^*) = 0 \Leftrightarrow x^* = 0, \quad \dot{x}(t)|_{x=x^*} = 0 \tag{10}$$

and $x(t)$ be an instantaneous value of the state vector x in a time instant t:

$$x(t) \in X \subset R^n \tag{11}$$

2. Let ψ be a continuous sufficiently *smooth vector mapping* of the class $C_k, k > 1$

$$\psi : X \to R^n \tag{12}$$

representing the *state space gradient vector*

$$\psi(x) = \mathrm{grad}_x E(x) \tag{13}$$

of a *state space energy* scalar field $E(x)$, defined *independently of the integration path* by

$$E(x) = \int_{x_1^*=0}^{x_1} \psi_1(\xi_1, 0, \ldots 0)\mathrm{d}\xi_1 + \int_{x_2^*=0}^{x_2} \psi_1(x_1, \xi_1, 0, \ldots 0)\mathrm{d}\xi_2 + \cdots$$
$$+ \int_{x_n^*=0}^{x_n} \psi_n(x_1, x_2, \ldots x_{n-1}, \xi_n)\mathrm{d}\xi_n \tag{14}$$

For the sake of mathematical correctness we have to require the field ψ to be *nonsingular,* fulfilling the *condition*:

$$\forall x \in X, \quad x \neq x^* : \det J_x(\psi) = \det\left[\frac{\partial\psi}{\partial x}\right] \neq 0 \tag{15}$$

and irrotational, i.e., fulfilling the *condition*:

$$\forall x \subset X : \ \mathrm{curl}(\psi(x)) = 0 \tag{16}$$

or equivalently, satisfying the *Jacobian symmetry condition*

$$[J_x(\psi)]^T - J_x(\psi) = 0 \tag{17}$$

3. Let f^o be a continuous sufficiently *smooth vector field* of the class $C_k, k > 1$

$$f^o : X \to V, \ V \subset R^n \tag{18}$$

representing the *zero-input state space velocity vector* defined by

$$v(t) = f^o[x(t)] = \left.\frac{\mathrm{d}x}{\mathrm{d}t}\right|_{u(t)=0} \qquad t \in [t_o, \infty) \tag{19}$$

4. Let h be a continuous sufficiently *smooth vector mapping* of the class C_n

$$h : X \to Y, \quad Y \subset R^p \tag{20}$$

representing the *output observation vector of the internal state of a strictly causal system representation*

$$y(t) = h[x(t)], \quad t \in [t_o, \infty) \tag{21}$$

Let D be a continuous *scalar mapping* of the class C_2

$$D : X \to R \tag{22}$$

defined by the dual product

$$\langle \psi(x), f^o(x) \rangle = 0 \tag{23}$$

where $\psi(x)$ is the *gradient field of the state space energy* defined by

$$\psi(x) = \mathrm{grad}_x E(x) \tag{24}$$

and $f^o(x)$ is the *zero input state velocity vector field* defined by

$$f^o(x) = \frac{\mathrm{d}x}{\mathrm{d}t}\Big|_{u(t)=0} \tag{25}$$

Theorem 2 (Generalized Tellegen's theorem) *If* $\psi(x)$ satisfies the relation (13) and $f^o(x)$ is defined by (18) then for the *dual product* of both the vectors it holds:

$$\forall x \in X, \quad \forall t, t \in [t_o, \infty) : D(x) = \langle \psi(x), f^o(x) \rangle = 0 \tag{26}$$

The proof is obvious by inspection: Because the gradient vector $\Psi(x)$ is at any fixed state $x(t)$ *orthogonal* to the state velocity vector $f^o(x)$, the state space energy $E(x)$, defined by integration of the dissipation function $D(x)$ has to be constant along any trajectory of the uncontrolled system representation $\Re\{S\}$ iff $D(x) = 0$ holds identically.

Remark 1 Theorem 2 should be considered rather as a principle than a provable statement. It represents an *abstract form of power balance relation* in the case of *zero input power.* Equivalently, we may interpret it as a *state space energy conservation principle in the differential form.*

In fact, the proposed approach is closely related to the well-known *Direct method of A. M. Lyapunov*, in particular to the *Schulz's variable gradient method* for solving the nontrivial *problem of Lyapunov functions generation* [2, 7, 8]. Recall that in spite of obvious conceptual similarity, there are at least *two fundamental diversities* of the proposed approach:

- At first, in contrast to the *concept of the state space energy* which is implicitly assumed to be *unique* it is known that *unicity of the Lyapunov function* is not warranted by Lyapunov's axioms.
- Second, in contrast to the Lyapunov function concept, the *concept of the state space energy does not depend* on the *chosen structure and parameterization* of the *given system representation.*

In order to apply the direct Lyapunov‘s method a *proper choice* of the *unknown Lyapunov function structure and parameters* (or that of its gradient), has to be made in such a way that a set of axioms required by Lyapunov method will be fulfilled. It is not an easy task, because no general constructive method for Lyapunov functions generation has been found, yet. Recall that the *class of the Lyapunov functions* is characterized axiomatically under the implicit assumption that the information about the *dynamics of the system* is contained in the given vector field $f^o(x)$, which is considered as the *only acceptable fixed model* of the system. Consequently, the gradient field $\psi(x)$ is considered as the *completely unknown element of a certain class of vector mappings.*

It is obvious from the Generalized Tellegen's principle (26) that the *operation of the dual product is symmetric.* It follows that it can be *dualized* in such a sense that instead of the fixed vector field $f^o(x)$ an *equivalence class of vector fields,* generated by the given field will be considered, and instead of the *unknown gradient field representation* one specific, but *universally* acceptable *fixed form* of the gradient vector $\psi(x)$ will be chosen within the *class of vector mappings*, satisfying the *nonsingularity condition* (15) and *set Jacobi symmetry conditions* (17).

The following natural questions arise: Should some forms of the mapping $\psi(x)$ be preferred? If so, what are the most *relevant criteria* for such a decision? The answer to the second question is plausible: besides *maximal universality the requirement of minimal complexity* should be the dominant features of the proposed paradigm. The *structure symmetry* requirement of the Jacobian $J_x(\psi)$ in the form

$$\left.\begin{array}{c} \dfrac{\partial\psi_2}{\partial x_1}-\dfrac{\partial\psi_1}{\partial x_2}=0, \\ \dfrac{\partial\psi_3}{\partial x_1}-\dfrac{\partial\psi_1}{\partial x_3}=0, \dfrac{\partial\psi_3}{\partial x_2}-\dfrac{\partial\psi_2}{\partial x_1}=0, \\ \text{-------------------------------} \\ \dfrac{\partial\psi_{n-1}}{\partial x_1}-\dfrac{\partial\psi_1}{\partial x_{n-1}}=0, \dfrac{\partial\psi_{n-1}}{\partial x_2}-\dfrac{\partial\psi_2}{\partial x_{n-1}}=0, \ldots, \dfrac{\partial\psi_{n-1}}{\partial x_{n-2}}-\dfrac{\partial\psi_{n-2}}{\partial x_{n-1}}=0, \\ \dfrac{\partial\psi_n}{\partial x_1}-\dfrac{\partial\psi_1}{\partial x_n}=0, \dfrac{\partial\psi_n}{\partial x_2}-\dfrac{\partial\psi_2}{\partial x_n}=0, \ldots, \dfrac{\partial\psi_n}{\partial x_{n-1}}-\dfrac{\partial\psi_{n-1}}{\partial x_n}=0, \end{array}\right\} \qquad (27)$$

reduces the complexity measure given by the number $N = n^2$ of its elements from n^2 to the total number of *independent elements*

$$N = n + \frac{1}{2}n(n-1) \qquad (28)$$

where the second term determines the total number of the *off-diagonal elements.*

Using a proper *state space equivalence transformation* and putting

$$\left.\begin{array}{c}\frac{\partial\psi_1}{\partial x_2}=\frac{\partial\psi_1}{\partial x_3}=\ldots=\frac{\partial\psi_1}{\partial x_{n-1}}=\frac{\partial\psi_1}{\partial x_n}=0\\ \frac{\partial\psi_2}{\partial x_3}=\frac{\partial\psi_2}{\partial x_4}=\ldots=\frac{\partial\psi_2}{\partial x_n}=0\\ ------------------\\ \frac{\partial\psi_{n-2}}{\partial x_{n-1}}=\frac{\partial\psi_{n-2}}{\partial x_n}=0\\ \frac{\partial\psi_{n-1}}{\partial x_n}=0,\end{array}\right\} \tag{29}$$

we can, without any restriction of generality, *eliminate* all the *off-diagonal elements* and further reduce the complexity of the gradient $\psi(x)$ to its *minimal structural complexity,* corresponding to $N = n$ *nonvanishing elements:*

$$\left.\begin{array}{l}\psi_1(x_1,x_2,\ldots,x_n)=\psi_1(x_1)\\ \psi_2(x_1,x_2,\ldots,x_n)=\psi_2(x_2)\\ \psi_3(x_1,x_2,\ldots,x_n)=\psi_3(x_2)\\ \qquad\qquad\vdots\\ \psi_n(x_1,x_2,\ldots,x_n)=\psi_n(x_1)\end{array}\right\}\Leftrightarrow$$

$$J_x(\psi)=\begin{bmatrix}\frac{\partial\psi_1}{\partial x_1} & 0 & \ldots & 0 & 0\\ 0 & \frac{\partial\psi_2}{\partial x_2} & 0 & \ldots & 0\\ \vdots & \vdots & \vdots & \vdots & \vdots\\ \vdots & \vdots & \vdots & \vdots & \vdots\\ 0 & 0 & \ldots & 0 & \frac{\partial\psi_n}{\partial x_n}\end{bmatrix} \tag{30}$$

with easy to check *nonsingularity condition*

$$\det J_x(\psi)=\prod_{i=1}^{n}\frac{\partial\psi_i}{\partial x_i}=\prod_{i=1}^{n}\frac{\mathrm{d}}{\mathrm{d}x_i}\psi_i(x_i)\neq 0 \tag{31}$$

As a consequence, the *nested additive form* of the *total state space energy* is

$$E(x)=E_1(x_1)+E_2(x_1,x_2)+E_3(x_1,x_2,x_3)+\cdots+E_n(x_1,x_2,x_3,\ldots,x_n) \tag{32}$$

where the *k*th term is defined by

$$E_k(x_1,x_2,x_3,\ldots,x_n)=\int_{x_k^*}^{x_k}\psi_k(x_1,x_2,\ldots,x_{n-1},\xi_k)\mathrm{d}\xi_k,\quad k\in\{1,2,\ldots,n\} \tag{33}$$

reduces to the *sum of state space energies* for a *complete set* of the *individual first order system components*

$$E(x) = E(x_1, x_2, x_3, \ldots, x_n) = \sum_{k=1}^{n} E_k(x_k) \tag{34}$$

where kth term defined by

$$E_k(x_k) = \int_{x_k^*}^{x_k} \psi_k(\xi_k) \mathrm{d}\xi_k, \quad k \in \{1, 2, \ldots, n\} \tag{35}$$

determines *the same amount* of the state space energy corresponding to the kth *interacting subsystem* as *it would be isolated.*

Remark 2 The last observation is of crucial importance from the structure of state space representation point of view, because the class of allowed state space representations is now significantly restricted: any state space representation of this class must namely satisfy the so-called *interaction neutrality condition*, which could also be deduced from the generalized Lyapunov–Tellegen's principle.

Remark 3 It is worthwhile to notice that as a by-product *of the Jacobian diagonalization* the n-dimensional mapping $\psi(x)$ reduces to n equivalent independent *one-dimensional nonsingular* mappings $\psi_i(x_i), i \in \{1, 2, \ldots, n\}$.

Let us now make the last step in the direction of *complexity minimization* requiring an additional property of *a structural nonsingularity*

$$\forall k, k \in \{1, 2, \ldots, n\} : \frac{\mathrm{d}}{\mathrm{d}x_i} \psi_i(x_i) = 1 \Rightarrow \forall x \in X \subset R^n : \det J_x(\psi) = 1 \tag{36}$$

as expressed by *gradient renormalization* in the form

$$\left.\begin{aligned} \psi_1(x) &= x_1 \\ \psi_2(x) &= x_2 \\ \psi_3(x) &= x_3 \\ &\vdots \\ \psi_n(x) &= x_n \end{aligned}\right\} \tag{37}$$

corresponding to the most natural choice of gradient mapping $\psi(x)$ as the *identity mapping*, i.e.,:

$$\psi(x) = x^T = [x_1, x_2, \ldots, x_n] \tag{38}$$

Thus the *total state space energy* $E(x)$ after integration (35) is now expressed as follows:

$$E(x) - E(x^*) = \frac{1}{2} \|x(t)\|^2 = \frac{1}{2} \rho_2^2(x, x^*) \Big|_{x^*=0} \tag{39}$$

and from the Lyapunov–Tellegen's principle the *conservativity of the state velocity vector field* follows, or equivalently the *state space energy conservation principle* holds:

$$\begin{aligned} &\forall x \in X : \langle x^T, f^0(x) \rangle = 0 \Leftrightarrow \\ &\forall t, t \in [t_0, \infty) : E[x(t)] = E[x(t_0)] \end{aligned} \tag{40}$$

3 Power and Energy in Systems and Topological Structures

Let us now return to more *general dissipativity problems:*

$$\begin{aligned} &\dot{x}(t) = f[x(t)] + Bu(t), x(t_0) = x^0, \\ &y(t) = Cx(t) \end{aligned} \tag{41}$$

and with an uncontrolled *state velocity vector field* given by

$$f(x) = A(x)x \tag{43}$$

Let the *immediate value* of the *output signal power* be defined:

$$P(t) \ = \ \|y(t)\|^2, \tag{44}$$

and the *total value of the state space energy*, accumulated in a *time instance t* in the *state* $x(t)$ be given by

$$E(t) \ = \frac{1}{2} \ \|x(t)\|^2 \tag{45}$$

Putting $u(t) = 0,\ \forall t \geq t_0$ and assuming validity of the state space energy conservation principle in the following differential form

$$\frac{\mathrm{d}E(x)}{\mathrm{d}t} = -P(t) \tag{46}$$

we get by computing the *derivative of the state space energy* $E[x(t)]$ along the state representation of the given system a *signal power balance relation* in the following explicit *Lyapunov equation-like* matrix form

$$\frac{\mathrm{d}E(x)}{\mathrm{d}t} = \frac{1}{2} x^{\mathrm{T}}(t)[A(\cdot) + A^{\mathrm{T}}(\cdot)]x(t) = -\|y(t)\|^2 \tag{47}$$

Remark 4 In contrast to the *standard physical* concept *of system energy*, the proposed *state space energy concept* represented by a class C_2 mapping $E: X \rightarrow R^+$ is *defined independently of any physical variables* or *parameters having any* specific *physical meaning*. In contrast, it depends *exclusively* on the *abstract system structure* and on the corresponding *state space topology*.

It has been shown in Refs. [8–10] that a special form of a *structurally dissipative* state equivalent system representation called *dissipation normal form exists* and its internal algebraic structure can be described by the following matrix:

$$A = \begin{bmatrix} -\alpha_{11} & \alpha_2 & 0 & 0 & 0 & 0 \\ -\alpha_2 & -\alpha_{22} & \alpha_3 & 0 & 0 & 0 \\ 0 & -\alpha_3 & -\alpha_{33} & \ddots & 0 & 0 \\ 0 & 0 & \ddots & \ddots & \alpha_{n-1} & 0 \\ 0 & 0 & 0 & -\alpha_{n-1} & -\alpha_{n-1,n-1} & \alpha_n \\ 0 & 0 & 0 & 0 & -\alpha_n & -\alpha_{n,n} \end{bmatrix} \tag{48}$$

Remark 5 It is important to emphasize that any of the internal or external power-informational interactions *may be time-varying* or *nonlinear* with respect to state or input variables.

Remark 6 It follows directly from expression (47) that the *output dissipation power* $P[x(t)]$ is unambiguously *determined by the symmetric part* $A_S(x)$ defined by

$$A_s(x) = \frac{1}{2}\left[A(x) + A^{\mathrm{T}}(x)\right] \tag{49}$$

and is *completely independent* of the *skew-symmetric part:*

$$A_A(x) = \frac{1}{2}\left[A(x) - A^{\mathrm{T}}(x)\right] \tag{50}$$

of the matrix $A(x)$ defined by

$$A(x) = A_s(x) + A_A(x) \tag{51}$$

Remark 7 Modification of the *power balance relation* and the *state space energy conservation principle* for the general case of non-vanishing input signals is straightforward.

Remark 8 An *abstract form of state space energy conservation principle* for a proper chosen *equivalent* state space representation follows by integration. Hence, in case of zero input:

$$u(t) = 0, \quad t \geq t_0 \tag{52}$$

the *total energy accumulated in the system* in time t_0 must be equal to an *amount of the energy dissipated* on the interval $[t_0, \infty)$ *by the vector output signal given by the expression*:

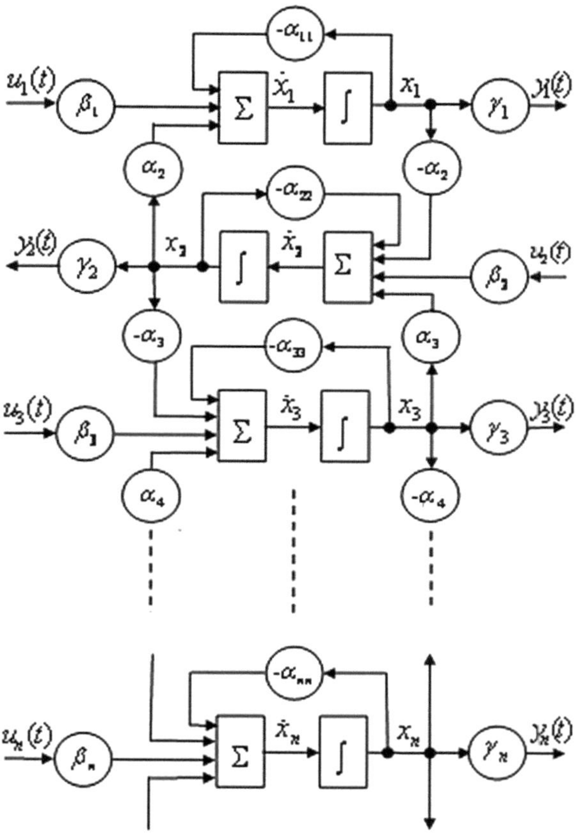

Fig. 2 Topological structure of the dissipation normal form with n inputs and with n output channels

$$E(t_0) = \int_{t_0}^{\infty} \|y(t)\|^2 \mathrm{d}t \tag{53}$$

or, *in case of scalar output signal*, alternatively, *by the expression*:

$$E(t_0) = \int_{t_0}^{\infty} [y(t)]^2 \mathrm{d}t \tag{54}$$

The topological structure corresponding to the algebraic structure defined by the matrix (48) is displayed in Fig. 2.

It is important to realize that the topological structure of the system representation derived from the algebraic structure (48) is *strongly over-parameterized.* In many cases it may be reasonable to remove all the redundant diagonal elements of the matrix $A_S(x)$, and *concentrate the total dissipation power* into a *single* (scalar) *observed output signal* $y(t)$.

In such a special case of *parameter minimal system representation* arises, and the corresponding *internal algebraic structure* of the state space representation *reduces* to:

$$\boldsymbol{A} = \begin{bmatrix} -\alpha_1 & \alpha_2 & 0 & 0 & 0 & 0 \\ -\alpha_2 & 0 & \alpha_3 & 0 & 0 & 0 \\ 0 & -\alpha_3 & 0 & \ddots & 0 & 0 \\ 0 & 0 & \ddots & \ddots & \alpha_{n-1} & 0 \\ 0 & 0 & 0 & -\alpha_{n-1} & 0 & \alpha_n \\ 0 & 0 & 0 & 0 & -\alpha_n & 0 \end{bmatrix} \tag{55}$$

The correspondingly *modified topological structure* of the *parametrically minimal state space representation* is displayed in Fig. 3.

Thus any parametrically minimal system representation defined by a *matrix* (55) is *dissipative* if and only if the matrix A *has negative trace, i.e., if and only if* the *single parameter* α_1 has a positive value, $\alpha_1 > 0$. The time evolution of state space energy $E[x(t)]$ for some typical cases is shown in Fig. 4.

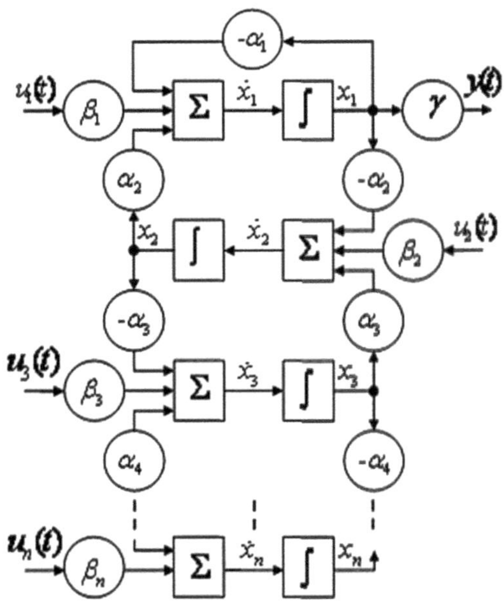

Fig. 3 Parametrically minimal structure of the dissipation normal form with single output dissipation channel

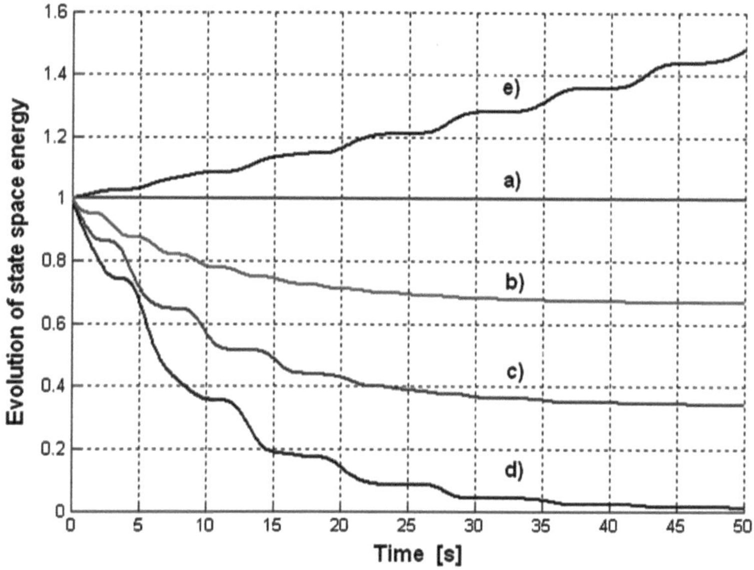

Fig. 4 Time evolution of state space energy $E[x(t)]$ for $n = 6$: a conservative case $\alpha_1 = 0$; b dissipativity $\alpha_1 > 0,\ \alpha_3 = 0$; c dissipativity $\alpha_1 > 0,\ \alpha_5 = 0$; d asymptotic stability $\alpha_1 > 0$, $\alpha_k \neq 0,\ k = 2, 3, \ldots, n$; e instability $\alpha_1 < 0$

4 State Dimension Minimality and Additivity of State Energy

Let us consider *continuous-time linear time-varying strictly causal state space system representation:*

$$\Re\{S\}: \left.\begin{array}{ll} \frac{\mathrm{d}x(t)}{\mathrm{d}t} = A(t)x(t) + B(t)u(t), & \dim x(\cdot) = n \\ y(t) = C(t)x(t), & \dim u(t) = r \dim y(\cdot) = p \end{array}\right\} \tag{56}$$

where $1 \leq r \leq n$, $1 \leqq p \leq n$, and the matrices $A(\cdot)$, $B(\cdot)$, $C(\cdot)$ are supposed to be known. Assume that the given system S has the *asymptotic stability* property, and that its representation $\Re\{S\}$ has the *minimal order n*, i.e., it is *controllable and observable.* In such a case the *controllability and observability Grammian matrices* $W_c(t)$ and $W_o(t)$ exist, are symmetric and positive definite, and satisfy the following two linear matrix *Lyapunov-like ordinary differential equations* [7]:

$$\left.\begin{array}{l} A(t)W_c(t) + W_c(t)A^{\mathrm{T}}(t) + \dfrac{\mathrm{d}W_c(t)}{\mathrm{d}t} = -B(t)B^T(t) \\ A^T(t)W_o(t) + W_o(t)A(t) + \dfrac{\mathrm{d}W_o(t)}{\mathrm{d}t} = -C^{\mathrm{T}}(t)C(t) \end{array}\right\} \tag{57}$$

Remark 9 Any such representation induces an equivalence class of minimal, controllable, observable, and asymptotic stable state equivalent system representations given by the *conditions of state equivalence* in the form

$$\bar{A}(t) = \left[T(t)A(t) + \frac{\mathrm{d}T(t)}{\mathrm{d}t}\right]T^{-1}(t) \tag{58}$$

$$\bar{B}(t) = T(t)B(t) \tag{59}$$

$$\bar{C}(t) = C(t)T^{-1}(t) \tag{60}$$

generated by the group of *linear time-varying state equivalence transformations:*

$$\bar{x}(t) = T(t)x(t), \quad x(t) = T^{-1}(t)\bar{x}(t) \tag{61}$$

Remark 10 It is a well-known fact that in the linear time-invariant case the Grammian matrices $W_c(t)$ and $W_o(t)$ are closely related to the input and output signal energy functions, respectively [8, 9]. In the previous part it was shown that the state minimality property of any linear strictly causal system representation is equivalent to the property of positive definiteness of both the observability and controllability Grammian matrices if these are well-defined, i.e., if the representation has the property of asymptotic stability.

From the mathematical correctness point of view there is no reason to require such a property in the general case. Hence there is a need to separate the minimality conditions from the asymptotic stability conditions. It is well known that the minimality of linear time-invariant system representations can instead of Grammians be tested independently of stability property using rank conditions of the controllability and observability matrices H_c and H_o defined by the expressions

$$H_c = \left[B, AB, A^2B, \ldots, A^{n-1}B\right] \tag{62}$$

$$H_o = \left[C^T,\ A^TC^T, (A^T)^2C^T, \ldots, (A^T)^{n-1}C^T\right] \tag{63}$$

It is easy to show that for any two state equivalent state representations the following relations hold:

$$\overline{H}_c = TH_c \tag{64}$$

$$\overline{H}_o = (T^{-1})^T H_o \tag{65}$$

and can be used for *determination of the state transformation* (61)

$$T = \overline{H}_c H_c^{-1} \tag{66}$$

or alternatively

$$T = (\overline{H}_o^{-1})^T H_o^T \tag{67}$$

if the structure and parameters of both the equivalent representations are known.

Consider the *time-invariant* version of the system representations given by Eq. (56). The finite controllability Grammian matrix $W_c(t)$ at time t is defined as follows:

$$W_C(t) = \int_0^t e^{A\tau} \cdot B \cdot B^T \cdot e^{A^T\tau} \mathrm{d}\tau \tag{68}$$

and has two important properties [8].

First, the controllability Grammian matrix $W_c(t)$ is *symmetric and nonnegative definite*

$$W_c^T(t) = W_c \geq 0 \tag{69}$$

and secondly, the columns of $W_c(t)$ span the controllable space, i.e., the image of $W_c(t)$ equals to the image of the controllability H_c:

$$\mathrm{Im}[W_c(t)] = \mathrm{Im}[H_c(A, B)] \tag{70}$$

It is easy to show that the state vector defined by the chosen structure of matrices A, B is controllable if and only if $W_c(t)$ is positive definite for some $t > 0$.

If the controllability Grammian matrix is invertible, then the minimum energy input signal $u(\cdot)$ exists and the *minimum input signal energy* corresponding to the state transfer from the zero state to $x(t)$ is known to be given by

$$\mathrm{Energy}\{u(.)\} = x^T(t) W_c^{-1} x(t), \quad t > 0 \tag{71}$$

Similarly, using the observability Grammian matrix $W_O(t)$, the *output signal energy* at time t caused by the initial state $x(t_0)$ is known to be given by the quadratic form:

$$\mathrm{Energy}\{y(.)\} = x^T(t_0) W_o(t) x(t_0), \quad t_0 < t \tag{72}$$

where the finite observability Grammian $W_o(t)$ at time $t < +\infty$ is defined by the expression [8]

$$W_\mathrm{o}(t) = \int_{t_0=0}^t e^{A^T\tau} \cdot C^T \cdot C \cdot e^{A\tau} \mathrm{d}\tau \tag{73}$$

and has two important properties.

First, the observability Grammian matrix $W_c(t)$ is *symmetric and nonnegative definite*

$$W_o^T(t) = W_o(t) \geq 0 \tag{74}$$

Secondly, for positive values of t, the kernel of finite observability Grammian matrix is equal to the kernel of observability matrix H_o

$$\mathrm{K}er[W_o(t)] = \mathrm{K}er[H_o(A, C)] \tag{75}$$

In the case of standard infinite time observability Grammian, i.e., for $t \to \infty$ *the largest observation energy* produced by any given value of initial state $x(t_0)$ is given by

$$E_o[x(t_0)] = x^T(t_0) W_o x(t_0) \tag{76}$$

assuming the couple (A, C) is observable and A is such that asymptotic stability conditions are satisfied. In such case the well-known special form of Lyapunov's equation

$$A^T W_o + A W_o = -C^T C \tag{77}$$

which in fact expresses a form of the state-output energy transfer balance relation, can be used for determination of the unknown output energy function $E_o(t)$.

In the dual case of standard infinite time controllability Grammian, i.e., for $t \to \infty$ the *minimal value of the input energy* required to transfer the zero initial state to any state $x(t_1)$ is given by

$$E_I[x(t_0)] = x^T(t_1) \cdot W_c^{-1} \cdot x(t_1) \tag{78}$$

assuming the couple (A, B) is controllable and A is such that asymptotic stability conditions are satisfied. In such a case the well-known special form of Lyapunov's equation

$$A W_c + A^T W_c = -B B^T \tag{79}$$

which in fact expresses a form of *the input-state energy transfer balance relation, can be used as a tool for determination of the unknown input energy function* $E_I(t)$.

Remark 11 In general, both the energy functions above are generated by symmetric positive semidefinite matrices, which depend heavily on the chosen system representation structure.

From physically motivated *energy additivity principle* (EAP):

$$E[x(\cdot)] = \sum_{i=1}^{n} E_i[x_i(\cdot)], \quad x_i(\cdot) \in R \tag{80}$$

it follows that only such system representation structures can be *accepted as physically correct*, which are not in contradiction with energy additivity principle, or equivalently for which the *Grammian matrices, generated by state equivalent triplets of matrices {A, B, C}, are diagonal and nonsingular, i.e., for which it holds:*

$$W = W_o \text{ or } W_c \Leftrightarrow W = \mathrm{diag}\{\delta_1, \delta_2, \ldots, \delta_n\}, \delta_i \neq 0 \tag{81}$$

5 Stability Analysis Based on State Space Energy

In this part mainly some *elementary consequences* of *dissipativity* and the *state space energy-metric approach* for stability analysis will be presented. We prove that the set of *real parameters* satisfy the following *conditions:*

1. $\alpha_1 > 0$ is *necessary and sufficient for dissipativity*
2. $\alpha_1 < 0$ is *sufficient for structural instability*
3. $\alpha_1 = 0$ is *necessary and sufficient for conservativity*
4. $\alpha_1 > 0$ is *necessary for asymptotic stability*
5. $\forall i, i \in \{2, 3, \ldots, n\}$: $\alpha_i \neq 0$, $\alpha_1 \neq 0$ is equivalent to *parametric and state minimality*
6. $\forall i, i \in \{2, 3, \ldots, n\}$: $\alpha_i \neq 0$, $\beta_1 \neq 0$ is equivalent to *structural controllability*
7. $\forall i, i \in \{2, 3, \ldots, n\}$: $\alpha_i \neq 0$, $\gamma \neq 0$ is equivalent to *structural observability*
8. $\forall i, i \in \{2, 3, \ldots, n\}$: $\alpha_i \neq 0$, $\alpha_1 > 0$ is equivalent to *structural asymptotic stability.*

Example 1 In order to demonstrate the way the proposed concept may be used, let us consider a simple linear time-invariant system for $n = 6$, and for the input signal $u(t) = 0$, $t > t_0$, represented by the linear differential equation with constant coefficients

$$y^{(6)} + a_1 y^{(5)} + \cdots + a_4 \ddot{y}(t) + a_5 \dot{y}(t) + a_6 y(t) = 0 \tag{82}$$

with *characteristic polynomial*

$$A(s) = s^6 + a_1 s^5 + a_2 s^4 + \cdots + a_4 s^2 + a_5 s + a_6 \tag{83}$$

Now, let the parameters $a_1, a_2, \ldots, a_n$ be considered as unknown, and *specify the region of asymptotic stability in the parameter space.* The state space representation having the required structure (55) is described by

$$\begin{aligned}
\Re(S) : \dot{x}_1(t) &= -\alpha_1 x_1(t) + \alpha_2 x_2(t) \\
\dot{x}_2(t) &= -\alpha_2 x_1(t) + \alpha_3 x_3(t) \\
\dot{x}_3(t) &= -\alpha_3 x_2(t) + \alpha_4 x_4(t) \\
\dot{x}_4(t) &= -\alpha_4 x_3(t) + \alpha_5 x_5(t) \\
\dot{x}_5(t) &= -\alpha_5 x_4(t) + \alpha_6 x_6(t) \\
\dot{x}_6(t) &= -\alpha_6 x_5(t) \\
y(t) &= \gamma_1 x_1(t)
\end{aligned} \tag{84}$$

Before we use the expression for the *state space energy function in the reduced form*

$$E(t) = \frac{1}{2}\|x(t)\|^2, \tag{85}$$

i.e., for the *zero input and zero equilibrium state* defined by the conditions

$$\begin{aligned} u(t) = 0, \quad t \in [t_0, \infty) : f(x^*) = 0 \Leftrightarrow \\ x^* = 0, \quad \dot{x}(t)|_{x=x^*} = 0 \end{aligned} \tag{86}$$

Thus, we have to check *the necessary and sufficient conditions for existence and uniqueness* of the equilibrium state $x^* = 0$. From Eq. (84) the following set of linear algebraic equations is obtained:

$$\begin{aligned} 0 &= -\alpha_1 x_1(t) + \alpha_2 x_2(t) \\ 0 &= -\alpha_2 x_1(t) + \alpha_3 x_3(t) \\ 0 &= -\alpha_3 x_2(t) + \alpha_4 x_4(t) \\ 0 &= -\alpha_4 x_3(t) + \alpha_5 x_5(t) \\ 0 &= -\alpha_5 x_4(t) + \alpha_6 x_6(t) \\ 0 &= -\alpha_6 x_5(t) \end{aligned} \tag{87}$$

The *necessary and sufficient condition for existence and unique equilibrium state* $x^* = 0$, resulting from the solution of the set of algebraic equations above read as

$$\alpha_2 \neq 0, \quad \alpha_4 \neq 0, \quad \alpha_6 \neq 0 \tag{88}$$

Remark 12 It means that from the existence of a unique equilibrium state point of view, the *dissipation parameter* α_1, as well as *interaction parameters* α_3, α_5 can be *arbitrary.*

Alternatively, using the *algebraic structure* of the matrix A *in the dissipation normal form* for $n = 6$

$$A = A_6 = \begin{bmatrix} -\alpha_1 & \alpha_2 & 0 & 0 & 0 & 0 \\ -\alpha_2 & 0 & \alpha_3 & 0 & 0 & 0 \\ 0 & -\alpha_3 & 0 & \alpha_4 & 0 & 0 \\ 0 & 0 & -\alpha_4 & 0 & \alpha_5 & 0 \\ 0 & 0 & 0 & -\alpha_5 & 0 & \alpha_6 \\ 0 & 0 & 0 & 0 & -\alpha_6 & 0 \end{bmatrix} \tag{89}$$

we could get conditions (88) from the equivalent *invertibility condition* of the matrix A:

$$\det A_6 = a_6 = \alpha_2^2 \alpha_4^2 \alpha_6^2 \neq 0 \tag{90}$$

Remark 13 In the general case of *any finite order n* the *invertibility condition* of matrix A_n can be checked *recursively*

$$a_n = (-1)^n \det A_n = \alpha_n^2 \det A_{n-2} \neq 0 \tag{91}$$

with the *initialization*

$$\det A_1 = -\alpha_1, \quad \text{for}\, n = 3, 5, 7, \ldots, \tag{92}$$

for the class of system representations of any *odd* order, and with the *initialization*

$$\det A_2 = \alpha_2^2, \quad \text{for}\, n = 4, 6, 8, \ldots, \tag{93}$$

for the class of the *conservative system* representations.

Remark 14 It is important to emphasize that as a *consequence of unique zero equilibrium* requirement the order of any *physically correct* state space representation of a *conservative system must be even.* On the other hand, for the class of *dissipative* and/or *anti-dissipative systems* no such restriction is needed.

Now, let all the *parameters* a_1, a_2,..., a_n of the characteristic polynomial $A(s) = A_n(s)$ be considered as *unknown,* and let us specify the *region of asymptotic stability in the parameter space.* We can compute the *instantaneous value of the state space energy* using the expression

$$E[x(t)] = \frac{1}{2}\rho^2[x(t), 0] = \frac{1}{2}\|x(t)\|^2 = \frac{1}{2}\sum_{i=1}^{n} x_i^2(t) \tag{94}$$

For the *derivative of the state space energy function E(x) along the system representation* (84) we get

$$\left.\frac{\mathrm{d}E(t)}{\mathrm{d}t}\right|_{\Re\{s\}} = -\alpha_1 x_1^2(t) = -\frac{\alpha_1}{\gamma_1^2} \cdot y^2(t) \tag{95}$$

where γ_1 is a *real output scaling parameter* $0 < \gamma_1 < \infty$.

Remark 15 It is of crucial importance that the *dissipation parameter* α_1 is *the only element* of the matrix *A,* whose *sign separates* the *system dissipativity* from its *anti-dissipativity.* The *critical value* of $\alpha_1 = 0$ corresponds to the *system conservativity* and *separates stability* of the *zero equilibrium state* from its *instability.*

Thus, for *real output dissipation power* $y^2(t)$ the *state space energy conservation principle* holds if and only if:

$$P_o(t) = y^2(t) \Leftrightarrow \alpha_1 = \gamma_1^2 \geq 0 \tag{96}$$

or equivalently if and only if *the state space energy conservation principle* holds

$$\left.\frac{\mathrm{d}E(t)}{\mathrm{d}t}\right|_{\Re(s)} = -P_0(t), \quad P_0(t) \geq 0 \tag{97}$$

It can be concluded from (95) that the *total state space energy* is *constant*, i.e., the system is *conservative*, if and only if the *observed output vanishes*

$$P_o(t) = y^2(t) = 0 \Leftrightarrow \alpha_1 = \gamma_1^2 = 0 \tag{98}$$

and the system is *dissipative* if and only if the *state space energy monotonously decreases*

$$P_o(t) = y^2(t) > 0 \Leftrightarrow \alpha_1 = \gamma_1^2 > 0 \tag{99}$$

Formally, there is no problem to show that the total *state space energy monotonously increases,* i.e., the system is *anti-dissipative,* if it holds $\alpha_1 < 0$, or equivalently

$$\left.\frac{\mathrm{d}E(t)}{\mathrm{d}t}\right|_{\Re(s)} = +P_0(t), \quad P_0(t) \geq 0 \tag{100}$$

but we would have to *change the physical meaning of the power* $P_o(t) = y^2(t)$and to interpret it as *an input power.*

Remark 16 Notice that if we put $\alpha_5 = 0$, then the state variables x_5 and x_6 become *unobservable by the output y*; thus *only the first isolated subsystem* with the state variables x_i, $i = 1, 2, 3, 4$, which are *observable,* will be *asymptotic stable*, while the *second one will oscillate at a constant level of the state space energy*. The situation is illustrated by case (c) in Fig. 4.

Similarly, if we put $\alpha_3 = 0$, then the state variables x_i, $i = 3, 4, 5, 6$ become *unobservable by the output y,* and *only the observable subsystem*

$$\begin{aligned} \dot{x}_1(t) &= -\alpha_1 x_1(t) + \alpha_2 x_2(t) \\ \dot{x}_2(t) &= -\alpha_2 x_2(t) \\ y(t) &= \gamma_1 x_1(t) \end{aligned} \tag{101}$$

will be asymptotic stable. The situation is illustrated by case (b) in Fig. 4.

In order to be able to interpret the obtained stability results for any given *n*th order ordinary differential equation, we need an effective tool to *transform any given state space representation,* such as (84), but *of arbitrary finite order*, into the form like (81), and back. One straightforward technique suggests to compute *successive derivatives of the output equation* up to the order *n,* combined with *successive elimination all of the state variables* as follows:

$$y(t) = \gamma_1 x_1(t) \tag{102}$$

$$\dot{y} + \alpha_1 y = \gamma_1 \alpha_2 x_2 \tag{103}$$

$$\ddot{y} + \alpha_1 \dot{y} + \alpha_2^2 y = \gamma_1 \alpha_2 \alpha_3 x_3 \tag{104}$$

$$\dddot{y} + \alpha_1 \ddot{y} + (\alpha_2^2 + \alpha_3^2)\dot{y} + \alpha_1 \alpha_3^2 y = \gamma_1 \alpha_2 \alpha_3 \alpha_4 x_4 \tag{105}$$

$$y^{(4)} + \alpha_1 \dddot{y} + (\alpha_2^2 + \alpha_3^2 + \alpha_4^2)\ddot{y} + \alpha_1 (\alpha_3^2 + \alpha_4^2)\dot{y} + \alpha_2^2 \alpha_4^2 y = \gamma_1 \alpha_2 \alpha_3 \alpha_4 \alpha_5 x_5 \tag{106}$$

$$y^{(5)} + \alpha_1 y^{(4)} + (\alpha_2^2 + \alpha_3^2 + \alpha_4^2 + \alpha_5^2)\dddot{y} + \alpha_1(\alpha_3^2 + \alpha_4^2 + \alpha_5^2)\ddot{y} \\ +[\alpha_2^2\alpha_4^2 + \alpha_2^2\alpha_5^2 + \alpha_3^2\alpha_6^2]\dot{y} + \alpha_1\alpha_3^2\alpha_5^2 y = \gamma_1\alpha_2\alpha_3\alpha_4\alpha_5\alpha_6 x_6 \tag{107}$$

$$y^{(6)} + \alpha_1 y^{(5)} + (\alpha_2^2 + \alpha_3^2 + \alpha_4^2 + \alpha_5^2 + \alpha_6^2) y^{(4)} \\ + \alpha_1(\alpha_3^2 + \alpha_4^2 + \alpha_5^2 + \alpha_6^2)\dddot{y} \\ + (\alpha_2^2\alpha_4^2 + \alpha_2^2\alpha_5^2 + \alpha_3^2\alpha_6^2 + \alpha_2^2\alpha_6^2 + \alpha_3^2\alpha_6^2 + \alpha_4^2\alpha_6^2)\ddot{y} \\ + \alpha_1(\alpha_3^2\alpha_5^2 + \alpha_3^2\alpha_6^2 + \alpha_4^2\alpha_6^2)\dot{y} + \alpha_2^2\alpha_4^2\alpha_6^2 y = 0 \tag{108}$$

As a by-product of the *state elimination procedure* above the following *state transformation relations* have been obtained:

$$x_1 = \frac{1}{\gamma_1} y, \quad \gamma_1 \neq 0 \tag{109}$$

$$x_2 = \frac{1}{\gamma_1\alpha_2}(\alpha_1 y + \dot{y}), \quad \gamma_1 \neq 0, \alpha_2 \neq 0 \tag{110}$$

$$x_3 = \frac{1}{\gamma_1\alpha_2\alpha_3}(\alpha_2^2 y + \alpha_1\dot{y} + \ddot{y}), \quad \gamma_1 \neq 0, \alpha_2 \neq 0, \alpha_3 \neq 0 \tag{111}$$

$$x_4 = \frac{1}{\gamma_1\alpha_2\alpha_3\alpha_4}\left[\alpha_1\alpha_3^2 y + (\alpha_2^2 + \alpha_3^2)\dot{y} + \alpha_1\ddot{y} + \dddot{y}\right], \\ \gamma_1 \neq 0, \alpha_2 \neq 0, \alpha_3 \neq 0, \alpha_4 \neq 0 \tag{112}$$

$$x_5 = \frac{1}{\gamma_1\alpha_2\alpha_3\alpha_4\alpha_5} \\ \left[\alpha_2^2\alpha_4^2 y + \alpha_1(\alpha_3^2 + \alpha_4^2)\dot{y} + (\alpha_2^2 + \alpha_3^2 + \alpha_4^2)\ddot{y} + \alpha_1\dddot{y} + y^{(4)}\right], \\ \gamma_1 \neq 0, \alpha_2 \neq 0, \ldots, \alpha_5 \neq 0 \tag{113}$$

$$x_6 = \frac{1}{\gamma_1\alpha_2\alpha_3\alpha_4\alpha_5\alpha_6} \\ [\alpha_1\alpha_3^2\alpha_5^2 y + (\alpha_2^2\alpha_4^2 + \alpha_2^2\alpha_5^2 + \alpha_3^2\alpha_6^2)\dot{y} + \alpha_1(\alpha_3^2 + \alpha_4^2 + \alpha_5^2)\ddot{y}, \\ + (\alpha_2^2 + \alpha_3^2 + \alpha_4^2 + \alpha_5^2)\dddot{y} + \alpha_1 y^{(4)} + y^{(5)}], \\ \gamma_1 \neq 0, \alpha_2 \neq 0, \alpha_3 \neq 0, \ldots, \alpha_6 \neq 0 \tag{114}$$

which can be expressed in a vector–matrix form

$$x = T\bar{x}, \quad \bar{x} = T^{-1}x \tag{115}$$

with the corresponding *invertible transformation matrix T*

$$T = \begin{bmatrix} \frac{1}{\gamma_1} & 0 & 0 & 0 & \cdots \\ \frac{\alpha_1}{\gamma_1\alpha_2} & \frac{1}{\gamma_1\alpha_2} & 0 & 0 & \cdots \\ \frac{\alpha_2}{\gamma_1\alpha_3} & \frac{\alpha_1}{\gamma_1\alpha_2\alpha_3} & \frac{1}{\gamma_1\alpha_2\alpha_3} & 0 & \cdots \\ \frac{\alpha_1\alpha_3}{\gamma_1\alpha_2\alpha_4} & \frac{\alpha_2^2+\alpha_3^2}{\gamma_1\alpha_2\alpha_3} & \frac{\alpha_1}{\gamma_1\alpha_2\alpha_3\alpha_4} & \frac{1}{\gamma_1\alpha_2\alpha_3\alpha_4} & \cdots \\ \vdots & \vdots & \vdots & \vdots & \vdots \end{bmatrix} \tag{116}$$

Remark 17 Notice, that the technique above has the advantage that, in principle, it works also for a broad *class of nonlinear state representations.*

In the linear and time-invariant situations various Laplace transform-based techniques may be more appropriate. A simple method based on computation of the *characteristic polynomial* $A_n(s)$ of the matrix A like that used in (83) will be presented in the sequel.

Let us consider matrix A_n as follows:

$$A_n = \begin{bmatrix} -\alpha_1 & \alpha_2 & 0 & 0 & 0 & 0 \\ -\alpha_2 & 0 & \alpha_3 & 0 & 0 & 0 \\ 0 & -\alpha_3 & 0 & \ddots & 0 & 0 \\ 0 & 0 & \ddots & \ddots & \alpha_{n-1} & 0 \\ 0 & 0 & 0 & -\alpha_{n-1} & 0 & \alpha_n \\ 0 & 0 & 0 & 0 & -\alpha_n & 0 \end{bmatrix} \tag{117}$$

where the conditions

$$\gamma_1 \neq 0, \alpha_2 \neq 0, \alpha_3 \neq 0, \ldots, \alpha_k \neq 0, \quad k \in \{2, 3, 4, \ldots n\}, \alpha_1 = \gamma_1^2 \tag{118}$$

are the *observability conditions of the dissipation normal form* (117), and *simultaneously* represent the set of *necessary and sufficient* conditions for *existence and invertibility* of the derived *state space transformation.*

The nth degree *characteristic polynomial* $A_n(s)$ of any matrix A_n is defined by the determinant

$$A_n(s) = \det[sI - A_n], \quad s = \sigma + j\omega \tag{119}$$

and can be explicitly expressed in the following standard form:

$$A_n(s) = s^n + a_1 s^{n-1} + a_2 s^{n-2} + a_3 s^{n-3} + \cdots + a_{n-2} s^2 + a_{n-1} s + a_n \tag{120}$$

It is easy to prove that for the *class of matrices A_n of the given structure* (117) the characteristic polynomial $A_n(s)$ can be computed *recursively*, using the following *recursive relation:*

$$A_k(s) = sA_{k-1}(s) + \alpha_k^2 A_{k-2}(s), \quad k \in \{3, 4, \ldots, n\} \tag{121}$$

with the *initialization* for $n = 1$ and $n = 2$:

$$A_1(s) = s + \alpha_1, \quad A_2(s) = s^2 + \alpha_1 s + \alpha_2^2 \tag{122}$$

Using it, for instance, for matrix A_6 in the form (89) we obtain:

$$A_6(s) = s^6 + \underbrace{\alpha_1}_{a_1} s^5 + \underbrace{[\alpha_2^2 + \alpha_3^2 + \cdots + \alpha_6^2]}_{a_2} s^4 + \underbrace{\alpha_1 [\alpha_3^2 + \alpha_4^2 + \cdots + \alpha_6^2]}_{a_3} s^4 + \underbrace{\cdots + \alpha_2^2\alpha_4^2\alpha_6^2}_{a_n} \tag{123}$$

and by comparing the result (123) with the standard form (120), we obtain a *set of algebraic equations* from which the *unknown elements* α_k of matrix A_n can be computed for given a_i.

For example for the system order $n = 6$, the set of parameters a_i, $i \in \{1, 2, \ldots, 6\}$ is obviously given by

$$\begin{aligned} a_1 &= \alpha_1 \\ a_2 &= \alpha_2^2 + \alpha_3^2 + \alpha_4^2 + \alpha_5^2 + \alpha_6^2 \\ a_3 &= \alpha_1(\alpha_3^2 + \alpha_4^2 + \alpha_5^2 + \alpha_6^2) \\ a_4 &= \alpha_2^2(\alpha_4^2 + \alpha_5^2 + \alpha_6^2) + \alpha_3^2(\alpha_5^2 + \alpha_6^2) + \alpha_4^2\alpha_6^2 \\ a_5 &= \alpha_1\alpha_3^2(\alpha_5^2 + \alpha_6^2) + \alpha_1\alpha_4^2\alpha_6^2 \\ a_6 &= \alpha_2^2\alpha_4^2\alpha_6^2 \end{aligned} \tag{124}$$

and can easily be *arbitrary reduced or extended for any finite order.*

For instance for $n = 5$ it reduces to

$$\begin{aligned} a_1 &= \alpha_1 \\ a_2 &= \alpha_2^2 + \alpha_3^2 + \alpha_4^2 + \alpha_5^2 \\ a_3 &= \alpha_1(\alpha_3^2 + \alpha_4^2 + \alpha_5^2) \\ a_4 &= \alpha_2^2(\alpha_4^2 + \alpha_5^2) + \alpha_3^2\alpha_5^2 \\ a_5 &= \alpha_1\alpha_3^2\alpha_5^2 \end{aligned} \tag{125}$$

for $n = 4$ we get:

$$\begin{aligned} a_1 &= \alpha_1 \\ a_2 &= \alpha_2^2 + \alpha_3^2 + \alpha_4^2 \\ a_3 &= \alpha_1(\alpha_3^2 + \alpha_4^2) \\ a_4 &= \alpha_2^2\alpha_4^2 \end{aligned} \tag{126}$$

and for $n = 3$ it follows:

$$\begin{aligned} a_1 &= \alpha_1 \\ a_2 &= \alpha_2^2 + \alpha_3^2 \\ a_3 &= \alpha_1 \alpha_3^2 \end{aligned} \tag{127}$$

For any finite order the analytical solution has been found easily, and can be expressed as follows

$$\begin{aligned} \alpha_1 &= a_1 = \Delta_1, \\ \alpha_2 &= \sqrt{\frac{a_1 a_2 - a_3}{a_1}} = \sqrt{\frac{\Delta_2}{\Delta_1}} \\ \alpha_3 &= \sqrt{\frac{a_1 a_2 a_3 - a_3^2 - a_1^2 a_4}{(a_1 a_2 - a_3) a_1}} = \sqrt{\frac{\Delta_3}{\Delta_2 \Delta_1}} \\ \alpha_k &= \sqrt{\frac{\Delta_k \Delta_{k-3}}{\Delta_{k-2} \Delta_{k-1}}}, \quad k = 4, 5, 6, \ldots, n \end{aligned} \tag{128}$$

where the set of new parameters Δ_k $k = 1, 2, \ldots, n$ correspond to the *set of diagonal minors of the Hurwitz determinant.* Using the *state space energy monotonicity* condition (97), together with the *state observability conditions* (118), it is easy to prove the resulting *asymptotic stability conditions*:

$$\begin{aligned} &\alpha_1 > 0, \alpha_2 \neq 0, \alpha_3 \neq 0, \ldots, \alpha_k \neq 0, \ k \in \{2, 3, 4, \ldots, n\} \\ &\Leftrightarrow \Delta_k > 0, \ k = 1, 2, \ldots, n \end{aligned} \tag{129}$$

which are obviously *equivalent to the set of necessary and sufficient stability* conditions representing the well-known *Hurwitz stability criterion.*

Example 2 (Linear asymptotic stability analysis) For $n = 6$, the *necessary and sufficient observability conditions* are given by the condition:

$$\det H_o \neq 0 \tag{130}$$

where H_o is the *observability matrix* defined by matrices A and C as follows:

$$H_o = \left[C^T, A^T C^T, (A^T)^2 C^T, \ldots, (A^T)^{n-1} C^T \right] \tag{131}$$

and the calculated *observability conditions* for the order $n = 6$ are:

$$\alpha_2 \neq 0, \alpha_3 \neq 0, \alpha_4 \neq 0, \alpha_5 \neq 0, \alpha_6 \neq 0, \gamma_1 \neq 0 \tag{132}$$

According to (99) the *dissipativity appears* if and only if the *state space energy monotonously decreases*

$$P_o(t) = y^2(t) > 0 \Leftrightarrow \alpha_1 = \gamma_1^2 > 0 \tag{133}$$

Thus the derived *set of necessary and sufficient conditions of asymptotic stability* reads as

$$\alpha_1 > 0, \alpha_k \neq 0, k \in \{2,3,\ldots,6\}, \Delta_k > 0, k = 1,2,\ldots,6 \tag{134}$$

It has been clearly demonstrated that linear algebraic methods for stability analysis can be seen as a *special case* of methods based on the proposed state space energy approach. It will be shown in the next section that even nonlinear stability problems can be effectively solved by the same technique. Moreover, the state space energy interpretation makes it possible to gain a better insight into many classical results.

6 State Space Energy Approach for Nonlinear Systems

Example 3 (Nonlinear asymptotic stability analysis—Lewis servo system) Let us consider a well-known *nonlinear system* called *Lewis servomechanism* given by equation:

$$\ddot{y}(t) + a_1[1 - \beta y(t)\text{sign}[y(t)]]\dot{y}(t) + a_2 y(t) = 0 \tag{135}$$

with $\alpha_2 > 0$, and α_1, β as arbitrary real design parameters. Using the state space energy approach for:

$$A(x_1, x_2) = \begin{bmatrix} -a_1[1 - \beta x_1(t)\text{sign}[x_1(t)]], & \sqrt{a_2} \\ -\sqrt{a_2}, & 0 \end{bmatrix} \tag{136}$$

the *unique Lyapunov function* $V(x)$ can be determined by *isometric transformations as the state space energy* $E(x)$. For simplicity we assume $a_2 = 1$ and get:

$$E = \frac{1}{2}\left[\frac{1}{4}a_1^2\beta^2x_1^4 - a_1^2\beta x_1^3\text{sign}[x_1(t)] - a_1\beta x_1^2x_2\text{sign}[x_1(t)] + (a_1^2+1)x_1^2 + 2a_1x_1x_2 + x_2^2\right] \tag{137}$$

The dissipation power $P(t)$ is given by

$$\forall a_2 \geq 0 : \frac{dE}{dt} = -2a_1x_1^2\left[1 - \frac{1}{2}\beta x_1\text{sign}[x_1(t)]\right] \tag{138}$$

i.e., the *state* $x^* = 0$ is *asymptotically stable in the region* $D \subset X \subset R^2$, as shown in Fig. 5. Phase portraits for different initial conditions are shown in Fig. 5 and

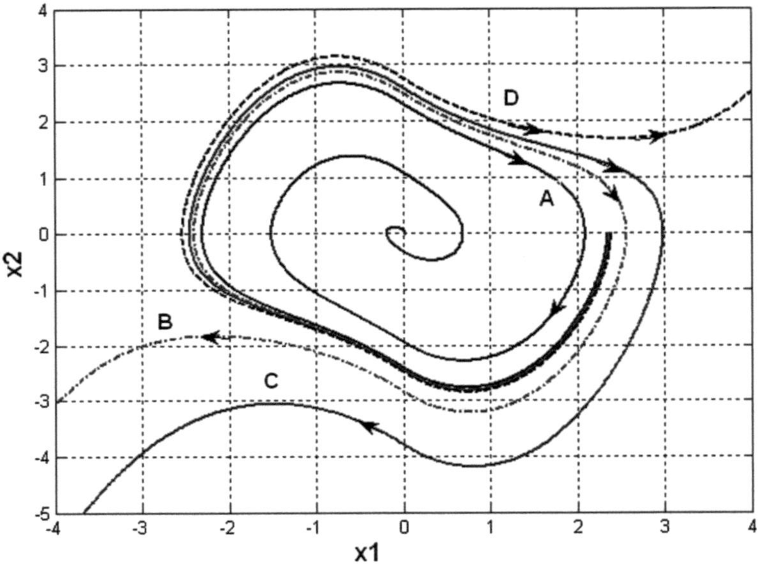

Fig. 5 Phase portrait and the typical state space topology generated by the state space energy function of Lewis system. Initial conditions are: *a* $x_0 \in [2.35;\ 0]^T$, *b* $x_0 \in [2.369;\ 0]^T$, *c* $x_0 \in [2.378;\ 0]^T$, *d* $x_0 \in [2.392;\ 0]^T$

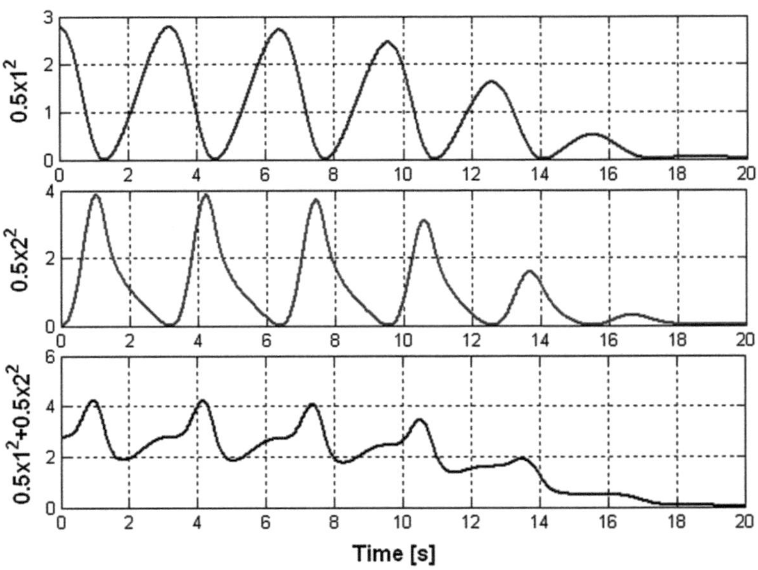

Fig. 6 State space energy of x_1, x_2 and total energy for Initial conditions $x_0 \in [2.35;\ 0]^T$

state space energy evolution for asymptotically stable system is displayed in Fig. 6.

For a *special case* ($\beta = 0$), i.e. *linear* (and for $\alpha_2 = 1$), the *total state space energy reduces to*

$$E[x_1, x_2] = 0.5(a_1^2 + 1)x_1^2(t) + a_1 x_1(t) x_2(t) + 0.5 x_2^2(t) \tag{139}$$

Example 4 (Nonlinear asymptotic stability analysis—modified Van der Pol equation). Let us consider a simple *nonlinear system given* by the representation:

$$\ddot{y}(t) - \varepsilon\left[\alpha - \beta y^2(t)\right]\dot{y}(t) + a_2 y(t) = 0 \tag{140}$$

Let matrix C be defined as $C = [\gamma, 0]$, and the *structure of the matrix* $A(x)$ be

$$A(x_1, x_2) = \begin{bmatrix} \varepsilon\left(\alpha - \frac{1}{3}\beta x_1^2\right) & \sqrt{a_2} \\ -\sqrt{a_2} & 0 \end{bmatrix} \tag{141}$$

The system representation is *locally observable if and only if* $\gamma \neq 0$, $a_2 > 0$, and dissipation power is

$$\left.\frac{\mathrm{d}E(t)}{\mathrm{d}t}\right|_{\Re(s)} = x_1\dot{x}_1 + x_2\dot{x}_2 = \varepsilon\left(\alpha - \frac{1}{3}\beta x_1^2\right)x_1^2 \tag{142}$$

Example 5 (Nonlinear asymptotic stability analysis—generation of the Lyapunov function) Let us consider the same *nonlinear system,* but instead of the *structure of* $A(x)$, the state $x(t)$ is defined by the standard way, i.e.,: $x_1 = y$, $x_2 = \mathrm{d}y/\mathrm{d}t$. Then the *unique Lyapunov function* $V(x)$ can be determined by *isometric transformations of the state space energy function* $E(x)$, and for $\alpha = \beta = a_2 = 1$ we get:

$$V(x) = E(x) = \frac{1}{2}\left[\frac{1}{9}\varepsilon^2 x_1^6 - \frac{2}{3}\varepsilon^2 x_1^4 + (1 + \varepsilon^2)x_1^2 - \cdots \frac{2}{3}\varepsilon x_1^3 x_2 + 2\varepsilon x_1 x_2 + x_2^2\right] \tag{143}$$

and again for the *linear conservative* case ($\varepsilon = 0$) it reduces to:

$$V(x) = E(x) = \frac{1}{2}(x_1^2 + x_2^2) = \frac{1}{2}\rho_2^2[x, 0] \tag{144}$$

Evaluation of state space variables is shown in Fig. 7, phase projection in Fig. 8, and state space energy in Fig. 9.

Example 6 In this example the *chaotic system based on 3rd order of nonlinear system* is simulated. The equation's are:

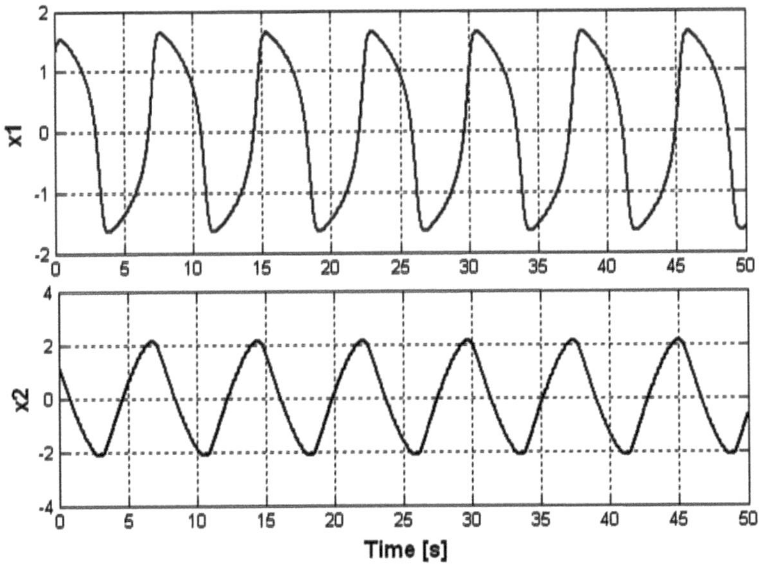

Fig. 7 Van der Pol state space variable x_1 and x_2 time evaluation. For $\varepsilon = 1$; $\alpha = 2$; $\beta = 1$; $a_2 = 1$ and initial conditions $x_0 \in [1.2;\ 1.2]^T$

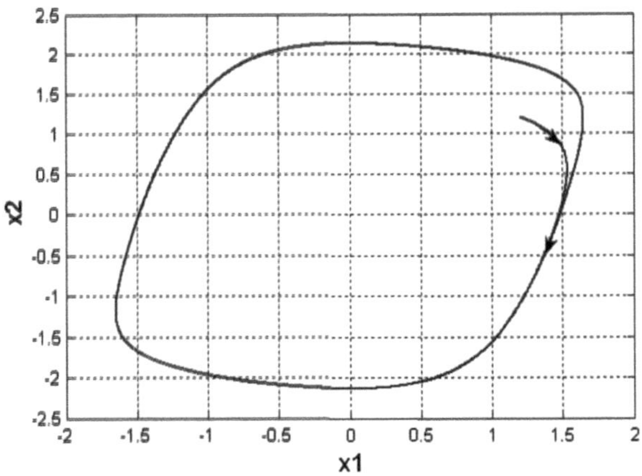

Fig. 8 Phase projection of x_1 and x_2 for initial conditions $x_0 \in [1.2;\ 1.2]^T$

$$
\begin{aligned}
\dot{x}_1 &= (1 - k_1 x_2^2 + k_2 x_3^2) x_1 + \alpha_2 x_2 \\
\dot{x}_2 &= -\alpha_2 x_1 + \alpha_3 x_3 \\
\dot{x}_3 &= -\alpha_3 x_2
\end{aligned}
\tag{145}
$$

where $k_1 = 0.1$; $k_2 = 6$; $\alpha_2 = 1$; $\alpha_3 = 0.89$ and initial conditions are $x_0 \in [0;\ 0.1;\ 0]^{\mathrm{T}}$.

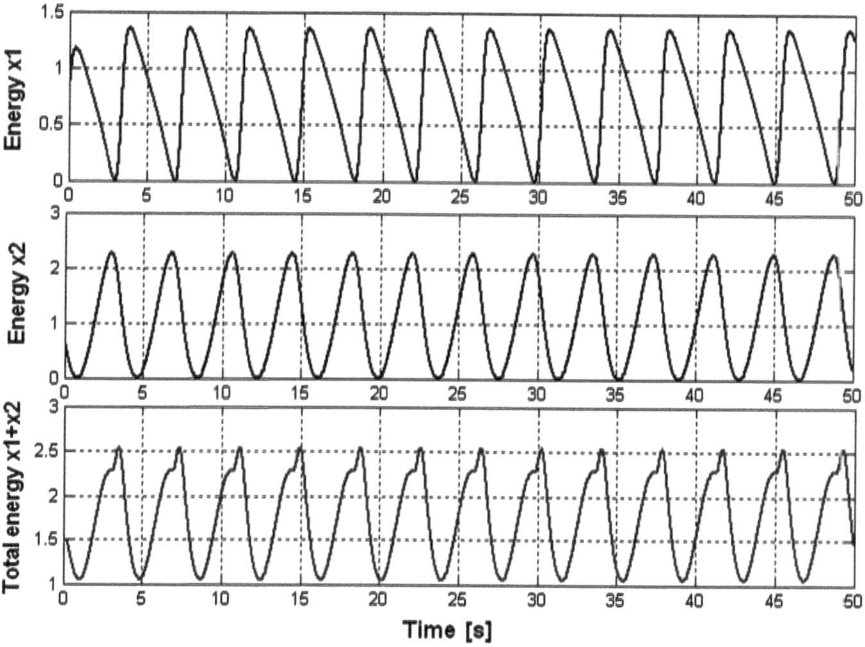

Fig. 9 Van der Pol space state energy. Mean values of energy $E_1 = 0.73$, $E_2 = 1.03$, $E_1 + E_2 = 1.762$; initial conditions $x_0 \in [1.2; 1.2]^T$

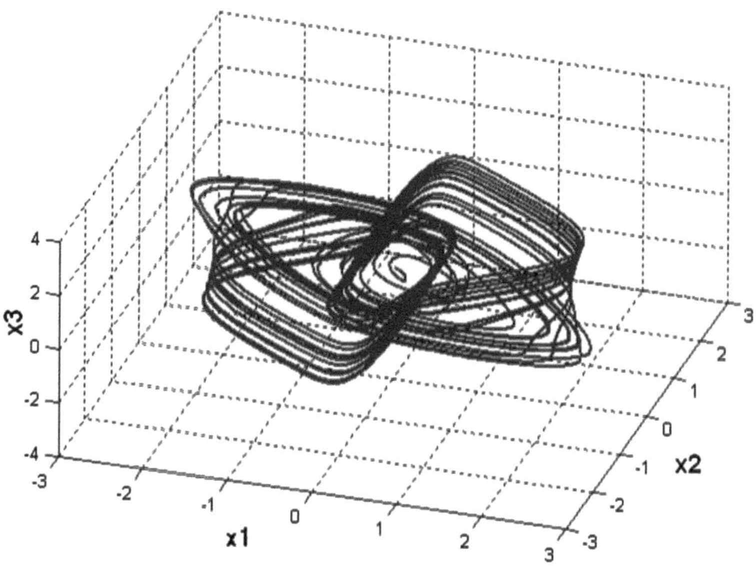

Fig. 10 3D phase portrait of chaotic system

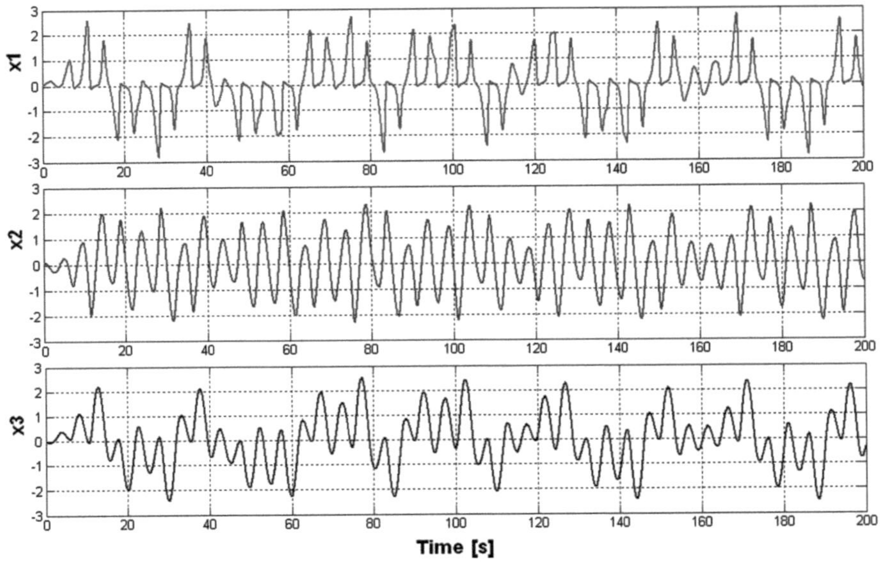

Fig.. 11 Time evolution of state space variables

The state space energy is given as

$$E = 0.5\left(x_1^2 + x_2^2 + x_3^2\right) \tag{146}$$

The dV/dt is

$$\begin{aligned}\frac{\mathrm{d}E}{\mathrm{d}t} &= \dot{x}_1 x_1 + \dot{x}_2 x_2 + \dot{x}_3 x_3 = \left[\left(1 - k_1 x_2^2 + k_2 x_3^2\right)x_1 + \alpha_2 x_2\right]x_1 \\ &+ (-\alpha_2 x_1 + \alpha_3 x_3)x_2 + (-\alpha_3 x_2)x_3 = \left(1 - k_1 x_2^2 + k_2 x_3^2\right)x_1^2\end{aligned} \tag{147}$$

The three-dimensional phase portrait is shown in Fig. 10, time evolution of state variables in Fig. 11, and state space energy of x_1, x_2 and x_3 variables and total state space energy are shown in Fig. 12.

7 Switching System Example: State Space Energy Approach

Example 7 In this part the more complex example of state space energy approach is presented. At the beginning, the *non-switched linear LRC circuit* (Fig. 13) is analyzed.

The circuit, according to Fig. 13, can be described as

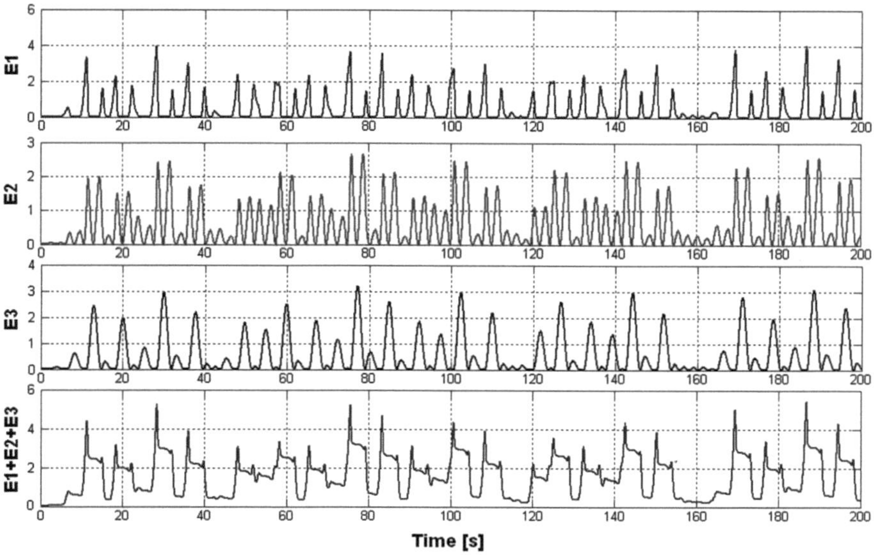

Fig. 12 Time evolution of state space energy, where $E_1 = 0.5x_1^2$; $E_2 = 0.5x_2^2$; $E_3 = 0.5x_3^2$; and bottom $E_1 + E_2 + E_3$

$$\left.\begin{aligned} \underbrace{Ri_{L1}}_{u_{R1}} + \underbrace{L_1\frac{\mathrm{d}i_{L1}}{\mathrm{d}t}}_{u_{L1}} + u_{C1} = v \\ -u_{C1} + \underbrace{L_2\frac{\mathrm{d}i_{L2}}{\mathrm{d}t}}_{u_{L2}} + u_{C2} = 0 \\ -i_{L1} + i_{L2} + \underbrace{C_1\frac{\mathrm{d}u_{C1}}{\mathrm{d}t}}_{i_{C1}} = 0 \\ -i_{L2} + \underbrace{C_2\frac{\mathrm{d}u_{C2}}{\mathrm{d}t}}_{i_{C2}} = 0 \end{aligned}\right\} \tag{148}$$

where state space variables are:

Fig. 13 The RLC circuit

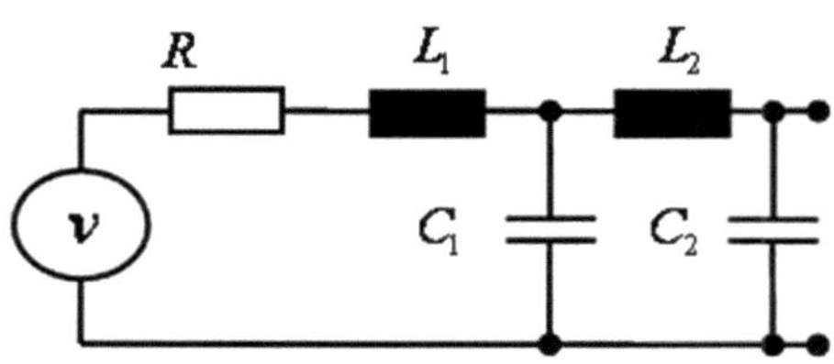

$$\begin{aligned} x_1 = i_{L1}, \quad x_2 = i_{L2} \\ x_3 = u_{C1}, \quad x_4 = u_{C2} \end{aligned} \tag{149}$$

and after substitution into (148)

$$\begin{aligned} \underbrace{R_1 x_1}_{u_{R1}} + \underbrace{L_1 \dot{x}_1}_{u_{L1}} + \underbrace{x_3}_{u_{C1}} &= v \\ \underbrace{-x_3}_{u_{C1}} + \underbrace{L_2 \dot{x}_2}_{u_{L2}} + \underbrace{x_4}_{u_{C2}} &= 0 \\ \underbrace{-x_1}_{i_{L1}} + \underbrace{x_2}_{i_{L2}} + \underbrace{C_1 \dot{x}_3}_{i_{C1}} &= 0 \\ \underbrace{-x_2}_{i_{L2}} + \underbrace{C_2 \dot{x}_4}_{i_{C2}} &= 0 \end{aligned} \tag{150}$$

Energy in the system is given as

$$\begin{aligned} E(t) &= \frac{1}{2} L_1 i_{L1}^2(t) + \frac{1}{2} L_2 i_{L2}^2(t) + \frac{1}{2} C_1 u_{C1}^2(t) + \frac{1}{2} C_2 u_{C2}^2(t) \\ &= \frac{1}{2} L_1 x_1^2(t) + \frac{1}{2} L_2 x_2^2(t) + \frac{1}{2} C_1 x_3^2(t) + \frac{1}{2} C_2 x_4^2(t) \end{aligned} \tag{151}$$

and state space equations of system S_1

$$\begin{aligned} \begin{bmatrix} \dot{x}_1 \\ \dot{x}_2 \\ \dot{x}_3 \\ \dot{x}_4 \end{bmatrix}_{S1} &= \begin{bmatrix} -R_1/L_1 & 0 & -1/L_1 & 0 \\ 0 & 0 & 1/L_2 & -1/L_2 \\ 1/C_1 & -1/C_1 & 0 & 0 \\ 0 & 1/C_2 & 0 & 0 \end{bmatrix} \begin{bmatrix} x_1 \\ x_2 \\ x_3 \\ x_4 \end{bmatrix} + \begin{bmatrix} v/L_1 \\ 0 \\ 0 \\ 0 \end{bmatrix}; \\ y &= [1 \quad 0 \quad 0 \quad 0] \begin{bmatrix} x_1 \\ x_2 \\ x_3 \\ x_4 \end{bmatrix} \end{aligned} \tag{152}$$

For values of $R = 1$; $C_1 = 0.4545$; $C_2 = 0.3454$; $L_1 = 1$; $L_2 = 6.368$ the numerical values of S_1 are

$$\begin{bmatrix} \dot{x}_1 \\ \dot{x}_2 \\ \dot{x}_3 \\ \dot{x}_4 \end{bmatrix}_{S_1} = \underbrace{\begin{bmatrix} -1 & 0 & -1 & 0 \\ 0 & 0 & 0.157 & -0.157 \\ 2.2 & -2.2 & 0 & 0 \\ 0 & 2.89 & 0 & 0 \end{bmatrix}}_{A} \begin{bmatrix} x_1 \\ x_2 \\ x_3 \\ x_4 \end{bmatrix} + \underbrace{\begin{bmatrix} 1 \\ 0 \\ 0 \\ 0 \end{bmatrix}}_{B} v, \tag{153}$$

$$y = \underbrace{\begin{bmatrix} 1 & 0 & 0 & 0 \end{bmatrix}}_{C} \begin{bmatrix} x_1 \\ x_2 \\ x_3 \\ x_4 \end{bmatrix}$$

The state space representation (152); (153) must be converted into an appropriate version for state space energy approach. Therefore, α_1; α_2; α_3; α_4 are calculated from characteristic polynomial of matrix A

$$x^4 + \underbrace{1.0}_{a1} x^3 + \underbrace{3.0}_{a2} x^2 + \underbrace{0.8}_{a3} x + \underbrace{1}_{a4} = 0 \tag{154}$$

Values are: $\alpha_1 = 1$; $\alpha_2 = 1.4832$; $\alpha_3 = 0.5878$; $\alpha_4 = 0.6742$ and matrix A_H (system S_2) is

$$A_H = \begin{bmatrix} -1 & 1.48 & 0 & 0 \\ -1.48 & 0 & 0.59 & 0 \\ 0 & -0.59 & 0 & 0.67 \\ 0 & 0 & -0.67 & 0 \end{bmatrix} \tag{155}$$

Matrix A_H was determined, but now transformation matrix must be computed to find also matrix B_H and C_H. They can be evaluated by observability.

The system S_1 observability matrix O_B is:

$$O_B = \begin{bmatrix} C & CA & CA^2 & CA^3 \end{bmatrix}^T \tag{156}$$

and numerical values of observability matrix of system S_1 are

$$O_B = \begin{bmatrix} 1 & 0 & 0 & 0 \\ -1 & 0 & -1 & 0 \\ -1.2 & 2.2 & 1 & 0 \\ 3.4 & -2.2 & 1.546 & -0.346 \end{bmatrix} \tag{157}$$

The desired matrix C_H of system S_2 is given as

$$C_H = \begin{bmatrix} 1 & 0 & 0 & 0 \end{bmatrix} \tag{158}$$

and the system S_2 observability matrix O_{BH} is:

$$O_{BH} = \begin{bmatrix} C_H & C_H A_H & C_H A_H^2 & C_H A_H^3 \end{bmatrix}^T \tag{159}$$

and numerical values are

$$O_{BH} = \begin{bmatrix} 1 & 0 & 0 & 0 \\ -1 & -1.483 & 0 & 0 \\ -1.2 & -1.483 & 0.872 & 0 \\ 3.4 & -2.29 & -0.872 & 0.588 \end{bmatrix} \tag{160}$$

Transformation matrix T is determined by:

$$\mathrm{T} = O_{BH}^{-1} O_B \tag{161}$$

the transformation matrix is

$$\mathrm{T} = \begin{bmatrix} 1 & 0 & 0 & 0 \\ 0 & 0 & -0.674 & 0 \\ 0 & 2.52 & 0 & 0 \\ 0 & 0 & 0 & -0.588 \end{bmatrix} \tag{162}$$

The transformed system S_2 has the form:

$$\begin{bmatrix} \dot{x}_{1H} \\ \dot{x}_{2H} \\ \dot{x}_{3H} \\ \dot{x}_{4H} \end{bmatrix} = \underbrace{\begin{bmatrix} -1 & 1.48 & 0 & 0 \\ -1.48 & 0 & 0.59 & 0 \\ 0 & -0.59 & 0 & 0.67 \\ 0 & 0 & -0.67 & 0 \end{bmatrix}}_{A_H} \begin{bmatrix} x_{1H} \\ x_{2H} \\ x_{3H} \\ x_{4H} \end{bmatrix} + \underbrace{\begin{bmatrix} 1 \\ 0 \\ 0 \\ 0 \end{bmatrix}}_{B_H} v$$
$$y_H = \underbrace{\begin{bmatrix} 1 & 0 & 0 & 0 \end{bmatrix}}_{C_H} \begin{bmatrix} x_{1H} & x_{2H} & x_{3H} & x_{4H} \end{bmatrix}^T \tag{163}$$

For $v = 0$ and initial conditions:

$$i_{L1} = 0, \quad i_{L2} = 0, \quad u_{C1} = 0, \quad u_{C2} = 1 \tag{164}$$

Initial energy of the electric circuit is

$$E = 0.5C_2u_{C2}^2 = 0.5 \cdot 0.3454 \cdot 1^2 = 0.1727 \tag{165}$$

The system initial conditions $x_0 \in [0\ 0\ 0\ 1]^{\mathrm{T}}$ must also be transformed

$$x_{0H} = \mathrm{T}x_0 = \mathrm{T}[0 \quad 0 \quad 0 \quad 1]^T = [0 \quad 0 \quad 0 \quad -0.5879]^T \tag{166}$$

The energy in system S_2 is calculated by equation

$$E_H(t) = \frac{1}{2}\sum_{i=1}^{4} x_{iH}^2(t) \tag{167}$$

The time evolution of energy in *RLC* system according to Fig. 13 (response on nonzero initial conditions and zero input voltage) was simulated by SPICE (Fig. 14) and energy evolution calculated by state space energy approach is shown

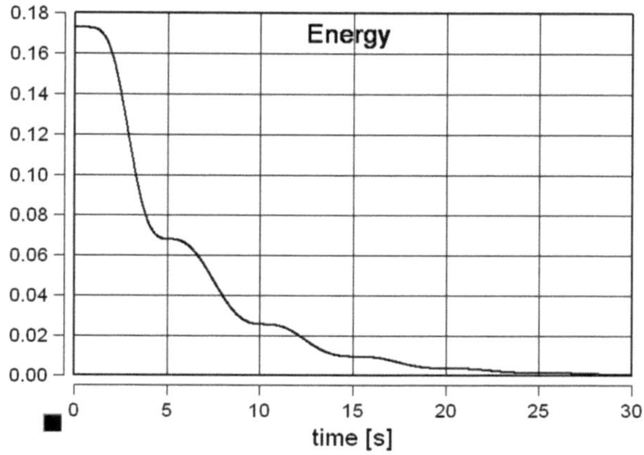

Fig. 14 Evolution of energy for $u_{C2} = 1$—simulated by SPICE

in Fig. 15. It can be seen that booth energy responses are the same. The partial state space energy time evolutions are displayed in Fig. 16.

It is possible to convert linear continuous system into discrete system. The matrix of discrete system is computed as

$$A_D = e^{A_H T_s} = L^{-1}\left\{(sI - A_H)^{-1}\right\}_{t=T_s} \tag{168}$$

where B and C are given by

$$B_D = \left(\int_{\tau=0}^{T_s} e^{A_H \tau} \mathrm{d}\tau \right) B_H,\ C_D = C_H \tag{189}$$

Numerical values of discrete system was calculated for sampling period $T_s = 0.417$ (s) and discretization method "*Zero-order hold.*" Matrices A_D, B_D, C_D

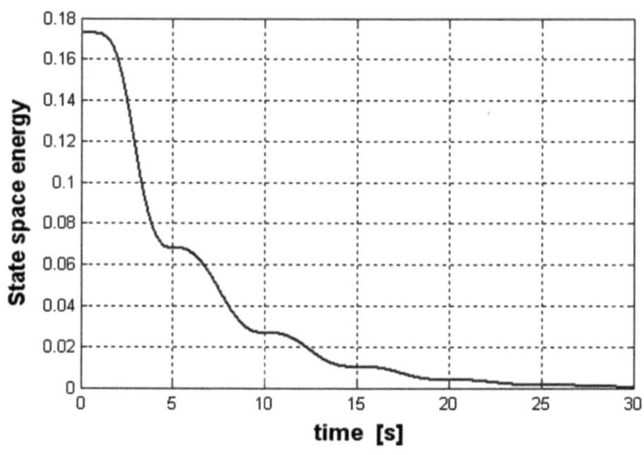

Fig. 15 State space energy approach time evolution according to Eq. (20) in Chap. Chap. 7

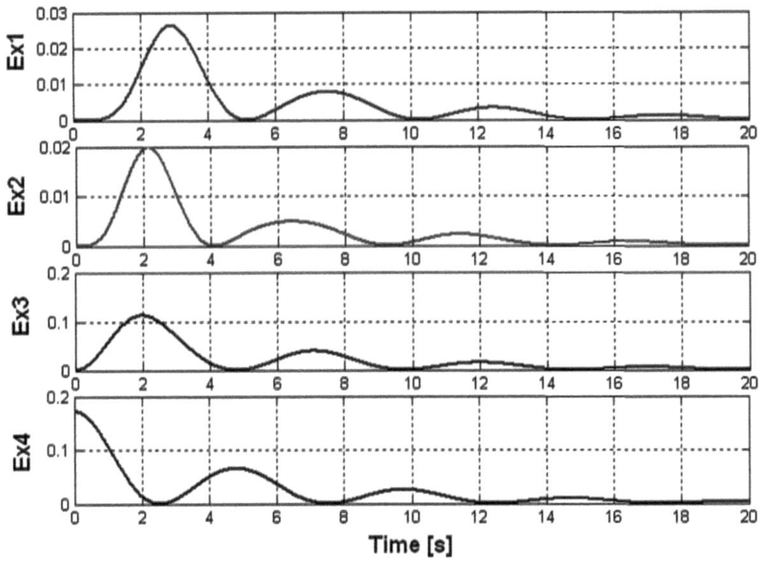

Fig. 16 The partial state space energy time evolutions

$$A_D = \begin{bmatrix} 0.519 & 0.4688 & 0.0635 & 0.0063 \\ -0.4688 & 0.8099 & 0.2257 & 0.0331 \\ 0.0635 & -0.2257 & 0.9321 & 0.2747 \\ -0.0063 & 0.0331 & -0.2747 & 0.9609 \end{bmatrix}; \quad B_D = \begin{bmatrix} 0.3198 \\ -0.1087 \\ 0.0092 \\ -0.0005 \end{bmatrix} \tag{170}$$

and transfer function:

$$F(z) = \frac{0.3198z^3 - 0.915z^2 + 0.9147z - 0.3198}{z^4 - 3.222z^3 + 4.173z^2 - 2.586z + 0.659} \tag{171}$$

The pole-zero map of discrete system is shown in Fig. 17. The time evolution of state space energy of discrete system (for state variable x_{1D}; x_{2D}; x_{3D}; x_{4D}) is presented in Fig. 18.

$$E_D(t) = \frac{1}{2}\sum_{i=1}^{4} x_{iD}^2(t) \tag{172}$$

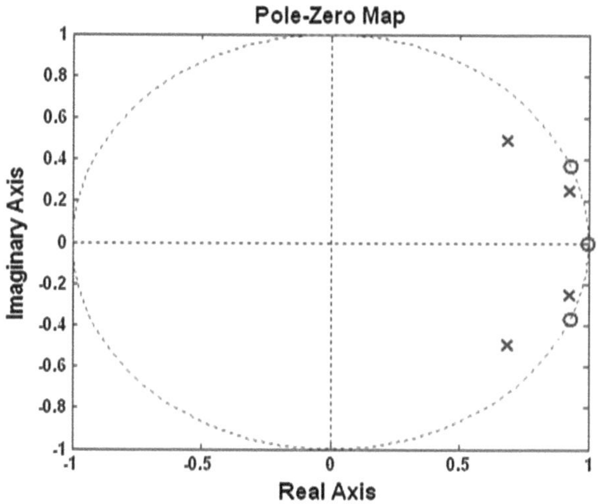

Fig. 17 The pole-zero map of discrete system. Poles: $0.6836 + 0.4967i$; $0.6836 - 0.4967i$; $0.9273 + 0.2509i$; $0.9273 - 0.2509i$. Zeros: 0.999; $0.9303 + 0.3668i$; $0.9303 - 0.3668i$

The previous results are used for *switching system simulation*. The circuit is shown in Fig. 19.

Values of resistors are $R_1 = 1\Omega$ and $R_2 = 0.5\Omega$; $C_1 = 0.4545$; $C_2 = 0.3454$; $L_1 = 1$; $L_2 = 6.368$. The states of switches are changed in $t = 6$. From $t_1 \in \langle 0;6)$ the system is charging, $t_2 \in \langle 6;50\rangle$ discharging. $\alpha_{1A} = R_1 = 1$ and $\alpha_{1B} = R2 = 0.5$ and $v = 1$. The structure of switching system (which holds state space energy approach) is shown in Fig. 20. The state space equation of the ^1S system for charging is

$$\begin{bmatrix} {}^1\dot{x}_1 \\ {}^1\dot{x}_2 \\ {}^1\dot{x}_3 \\ {}^1\dot{x}_4 \end{bmatrix}_{{}^1S} = \underbrace{\begin{bmatrix} -1 & 1.48 & 0 & 0 \\ -1.48 & 0 & 0.59 & 0 \\ 0 & -0.59 & 0 & 0.67 \\ 0 & 0 & -0.67 & 0 \end{bmatrix}}_{{}^1A} \begin{bmatrix} {}^1x_1 \\ {}^1x_2 \\ {}^1x_3 \\ {}^1x_4 \end{bmatrix} + \underbrace{\begin{bmatrix} 1 \\ 0 \\ 0 \\ 0 \end{bmatrix}}_{{}^1B} 1 \tag{173}$$

$${}^1y = \underbrace{[1 \quad 0 \quad 0 \quad 0]}_{{}^1C} \left[{}^1x_1 \quad {}^1x_2 \quad {}^1x_3 \quad {}^1x_4\right]^T$$

The state space equation of the ^2S system for discharging ($v = 0$) is (174). The ^2S system start from initial values, which are given by values of state variables, reaches by the end of charging.

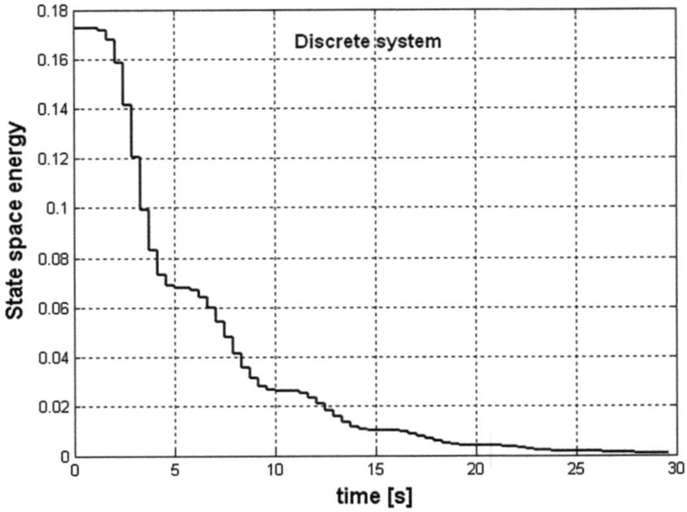

Fig. 18 State space energy time evolution of discrete system

Fig. 19 The switching system

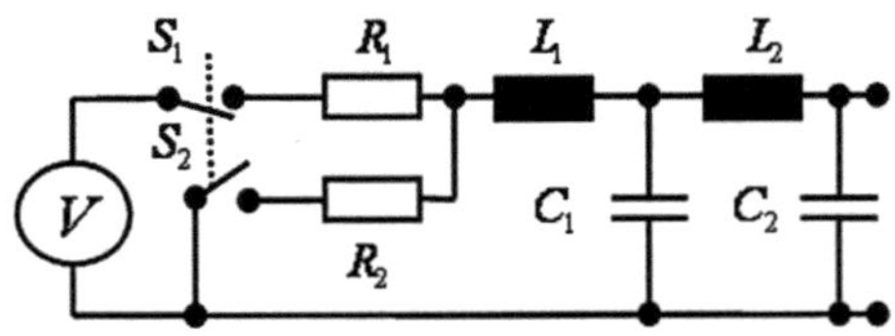

$$\begin{bmatrix} {}^2\dot{x}_1 \\ {}^2\dot{x}_2 \\ {}^2\dot{x}_3 \\ {}^2\dot{x}_4 \end{bmatrix}_{{}^2S} = \underbrace{\begin{bmatrix} -0.5 & 1.48 & 0 & 0 \\ -1.48 & 0 & 0.59 & 0 \\ 0 & -0.59 & 0 & 0.67 \\ 0 & 0 & -0.67 & 0 \end{bmatrix}}_{{}^2A} \begin{bmatrix} {}^2x_1 \\ {}^2x_2 \\ {}^2x_3 \\ {}^2x_4 \end{bmatrix} \tag{174}$$

$${}^2y = \underbrace{\begin{bmatrix} 1 & 0 & 0 & 0 \end{bmatrix}}_{{}^2C} \begin{bmatrix} {}^2x_1 & {}^2x_2 & {}^2x_3 & {}^2x_4 \end{bmatrix}^T$$

The results are checked by simulation of the system by SPICE. The dissipated powers on R_1 and R_2 (in SPICE) are given by:

$$P_{R1}(t) = R_1 i_{R1}^2(t); \quad P_{R2}(t) = R_2 i_{R2}^2(t) \tag{175}$$

The E_R is total energy dissipated on R_1 and R_2 during charging and discharging as (in SPICE):

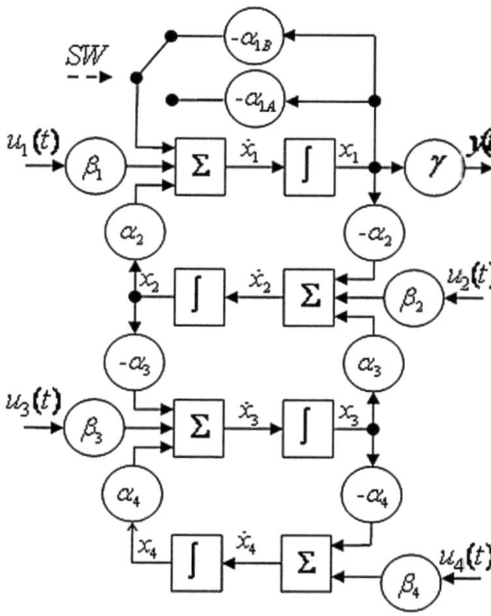

Fig. 20 The structure of switching system. The dissipation parameters are $\alpha_{1A} = R_1 = 1$ and $\alpha_{1B} = R_2 = 0.5$

$$E_R(t) = R_1 \int_0^{t_1} i_{R1}^2(t) + R_2 \int_{t_1}^{t_2} i_{R2}^2(t) \tag{176}$$

The powers dissipated on R_1 and R_2 calculated by state space energy are given by:

$$\begin{aligned} P_{R1}(t) &= \alpha_{1A} x_{1A}^2(t) = R_1 x_{1A}^2(t); \\ P_{R2}(t) &= \alpha_{1B} x_{1B}^2(t) = R_2 x_{1B}^2(t) \end{aligned} \tag{177}$$

The E_R is total energy dissipated on R_1 and R_2 during charging and discharging as:

$$\begin{aligned} E_R(t) &= \alpha_{1A} \int_0^{t_1} x_{1A}^2(t) + \alpha_{1B} \int_{t_1}^{t_2} x_{1B}^2(t) \\ &= R_1 \int_0^{t_1} x_{1A}^2(t) + R_2 \int_{t_1}^{t_2} x_{1B}^2(t) \end{aligned} \tag{178}$$

Comparison of Figs. 21 and 22 shows that the booth approach gives the same results.

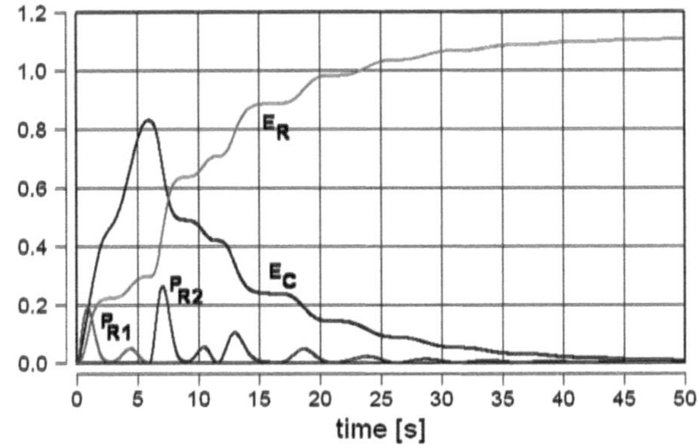

Fig. 21 Energy and power. Simulation by SPICE. P_{R1}—power dissipated on R_1 during charging. P_{R2}—power dissipated on R_2 during discharging. E_R—energy dissipated on resistors during charging and discharging. E_C—energy on C_1, C_2, L_1 and L_2 during charging and discharging

8 Switching Power Systems: Port-Hamiltonian System Approach

Port-Hamiltonian systems can be seen not only as a generalization of the classical Hamiltonian equations of motion, but also as arising naturally from a modeling network of (complex) physical systems (see [27–38]); it is, in fact, possible to describe in the port-Hamiltonian framework not only mechanical systems but also a broad class of physical systems including passive electro-mechanical systems, mechanical systems with nonholonomic constraints, and their combinations.

From the control perspective, one of the main advantages in adopting the port-Hamiltonian modeling framework is that the energy (Hamiltonian) function, which is usually a good Lyapunov function, explicitly appears in the dynamics of the system. Given a desired state of equilibrium, if the Hamiltonian of the system assumes its minimum at this configuration, then asymptotic stability can be assured by introducing a dissipative effect with the controller input. In this way, energy decreases until the minimum of energy or, equivalently, the desired equilibrium configuration is reached.

In order to introduce the class of port-Hamiltonian systems, it is good to start with a simple mechanical example to show that the equation of motion derived via the standard *Euler–Lagrange equations* can easily fit into this framework.

Calling $q = [q_1, q_2, \ldots q_n]^T$ the generalized configuration coordinates for a mechanical system with n degrees of freedom, L the Lagrangian equal to the difference $K - V$ between the kinetic energy K and the potential energy V, and $\tau = [\tau_1, \tau_2, \tau_n]^T$ the vector of generalized forces acting on the system, it is possible to describe the motion of the system via the *Euler–Lagrange equations*:

$$\frac{\mathrm{d}}{\mathrm{d}t}\left(\frac{\partial L}{\partial \dot{q}}(q, \dot{q})\right) - \frac{\partial L}{\partial q}(q, \dot{q})\tau \tag{179}$$

In standard mechanical system the kinetic energy is defined as

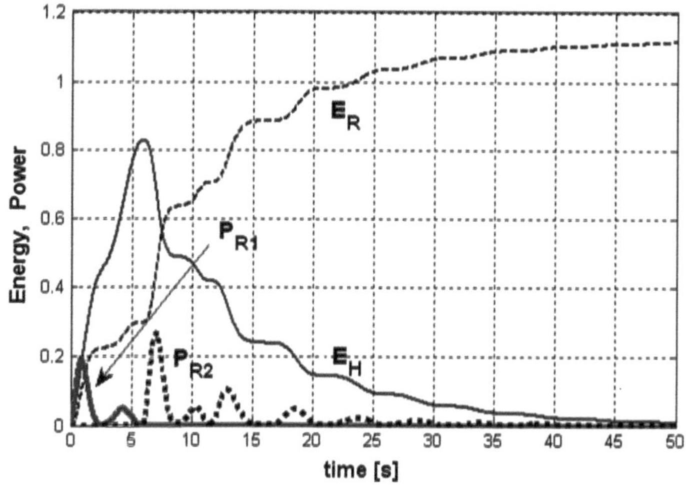

Fig. 22 Energy and power calculated by state space energy approach. P_{R1}—power dissipated on R_1 during charging. P_{R2}—power dissipated on R_2 during discharging. E_R—energy dissipated on resistors during charging and discharging. E_H—energy of state variables during charging and discharging

$$K(q,\dot{q}) = \frac{1}{2}\dot{q}^T M(q)\dot{q} \tag{180}$$

where $M(q)$ is the $n \times n$ inertia matrix, symmetric and positive definite for all q. In this case the vector of generalized momentum $p = [p_1, p_2, \ldots p_n]^T$ is defined as

$$p = \frac{\partial L}{\partial \dot{q}} = M(q)\dot{q} \tag{181}$$

Defining the state vector $(q, p)^T$ the n second order Eq. (179) transform into $2n$ first order equations

$$\begin{aligned} \dot{q} &= \frac{\partial H}{\partial p}(q,p) = M^{-1}(q)p \\ \dot{p} &= -\frac{\partial H}{\partial p}(q,p) + \tau \\ y &= \frac{\partial H}{\partial p}(q,p) = \dot{q} \end{aligned} \tag{182}$$

where

$$H(q,p) = \frac{1}{2}p^T M^{-1}(q)p + V(q) \tag{183}$$

is called Hamiltonian function of the system and represent the total energy of the system. An immediate property (conservation of energy) follows from (180):

$$\begin{aligned}\frac{d}{dt}H &= \frac{\partial^T H}{\partial q}(q,p)\dot{q} + \frac{\partial^T H}{\partial p}(q,p)\dot{p} \\ &= \frac{\partial^T H}{\partial p}(q,p)\tau = \dot{q}^T\tau = y^T u\end{aligned} \tag{184}$$

showing that the increase in internal energy of the system H is equal to the supplied work ($y^T u$ is the generalized velocity times generalized forces, i.e., power). System (180) is an example of a Hamiltonian system with collocated inputs and outputs, which more generally could be written in the form:

$$\begin{aligned}\dot{q} &= \frac{\partial H}{\partial p}(q,p) = M^{-1}(q)p \\ \dot{p} &= -\frac{\partial H}{\partial p}(q,p) + B(q)\tau \\ y &= \frac{\partial H}{\partial p}(q,p) = B^T(q)\dot{q}\end{aligned} \tag{185}$$

with, in general, $q,\ p \in \Re^n$ and $u,\ y \in \Re^m$ and $B(q)$ has proper dimension.

The class of Hamiltonian systems (184) can be further generalized to system which are described in local coordinates as

$$\left.\begin{aligned}[\dot{x}(t)] &= [J(x) - R(x)]\left[\frac{\partial H}{\partial x}(x)\right] + B(x)u(t) \\ y(t) &= B(x(t))^T\frac{\partial H}{\partial x}(x(t))\end{aligned}\right\} \tag{186}$$

where

$$J(x) + J(x)^T = 0; \quad R(x) = R(x)^T \geq 0 \tag{187}$$

Equations (186) and (187) describe a port-Hamiltonian system with dissipation. The $n \times n$ matrices $J(x)$ and $R(x)$ are called interconnection and damping matrix respectively.

Note that in this case the internal interconnection structure is given by $J(x)$ and an additional resistive structure is given by $R(x)$, which is determined by the possible presence of energy dissipation.

In this case the power balance equation reduces to

$$\begin{aligned}\frac{\mathrm{d}H}{\mathrm{d}t} &= u(t)^T y(t) - \frac{\partial^T H(x(t))}{\partial x}R(x(t))\frac{\partial H(x(t))}{\partial x} \\ &\leq u(t)^T y(t)\end{aligned} \tag{188}$$

This relation expresses a fundamental property of port-Hamiltonian systems, their passivity. Roughly speaking, the internal energy of the unforced system ($u = 0$) is non-increasing along system trajectories or, if the port variable is closed on a dissipative element, then the energy function is always a decreasing function.

If the definition of Lyapunov stability is recalled, together with the sufficient condition for the stability of an equilibrium point, then it can be deduced that the Hamiltonian is a good candidate for being a Lyapunov function.

In this part the example of port Hamiltonian-approach is used for switching system simulation.

Example 8 SEPIC DC–DC converter

Single-ended primary inductance converter (SEPIC) topology is a good choice for automotive power systems that require an output voltage between the low and high values of the input voltage. SEPIC topology fits this application because its duty ratio can be varied around 50 % to provide an output voltage that is either below or above the input voltage. Additionally, compared to flyback converters, SEPIC converters avoid the problems with leakage inductance and snubbers [39, 40].

SEPIC converters can also be used as Power Factor Preregulators (PFP). PFP have been used to improve input current waveform of off-line power supplies. There are two major approaches in implementing control circuits in PFP: the multiplier approach and the voltage-follower approach. The simplest one is the voltage-follower approach because the converter operates in discontinuous conduction mode (DCM), and only one loop control is required. SEPIC and CUK converters present a great advantage over boost and flyback topologies in DCM: an input current with low harmonic content can be obtained by correctly choosing inductors L_1, and L_2, of the converter with a fixed operation frequency.

The most basic converter is the buck converter. It is so named because it always steps down, or bucks, the input voltage. The output of the converter is given as

$$V_O = d_{\mathrm{BUCK}} \cdot V_I \tag{189}$$

where d_{BUCK} is the converter's duty cycle (on time fraction of the cycle).

Interchange the input and the output of the buck converter, and you get the second basic converter—the boost. The boost always steps up, hence its name. The output voltage is always higher than the input voltage, and is given as

$$V_O = \frac{1}{d_{\mathrm{BOOST}}} \cdot V_I \tag{190}$$

where d_{BOOST} is the converter's duty cycle.

What if you have an application where you need to both step up and step down, depending on the input and output voltages? You could use two cascaded converters, a buck and a boost. Unfortunately, this requires two separate controllers and switches. The buck–boost converter has the desired step up and step down functions:

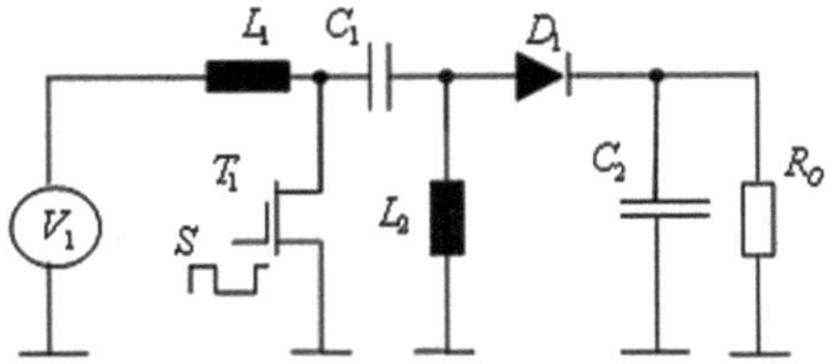

Fig. 23 Block diagram of SEPIC DC–DC converter. The SEPIC converter can both step up and step down the input voltage, while maintaining the same polarity and the same ground reference for the input and output

$$V_0 = \frac{-d_{\mathrm{BUCK}}}{d_{\mathrm{BOOST}}} \cdot V_1 \tag{191}$$

The output is inverted. A flyback converter (isolated buck–boost) requires a transformer instead of just an inductor, adding to the complexity of the development.

One converter that provides the needed input-to-output gain is the SEPIC (single-ended primary inductor converter) converter. A basic SEPIC converter with diode D_1 is shown in Fig. 23. The diode can be replaced by switch (Fig. 24). It has become popular in recent years in battery-powered systems that must step up or down depending upon the charge level of the battery. The output voltage is given as

$$V_0 = \frac{d}{1-d} \cdot V_1 \tag{192}$$

where d is the converter's duty cycle [41–43].

This suggests that if the duty cycle is greater than 0.5, a higher voltage will be regulated at the output (*boost*); if the duty cycle is less than 0.5, the regulated voltage will be lower (*buck*). Other relevant results of this analysis are that the steady-state voltage across the energy-transfer capacitor C_1 is equal to V_0 in a lossless system; the dc value of the current through the output inductor L_1 is equal to I_{OUT}; and the dc value of the current through the ground-reference inductor L_2 is $I_{\mathrm{OUT}} \times V_{\mathrm{OUT}}/V_{\mathrm{IN}}$. The energy-transfer capacitor also provides dc blocking from V_{IN} to V_{OUT}. This property can be attractive when there is a risk of a shorted output.

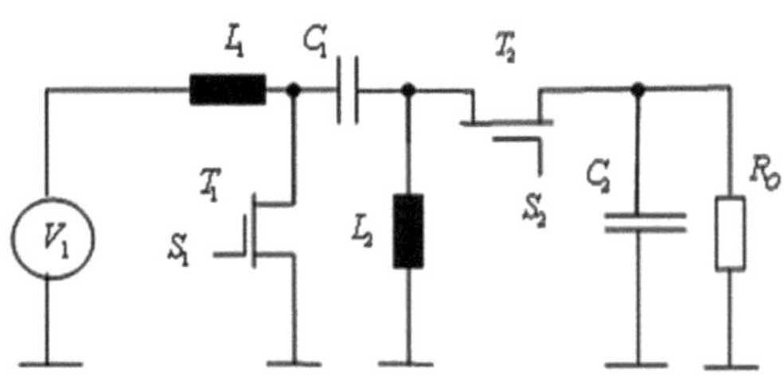

Fig. 24 Block diagram of SEPIC DC–DC converter. The diode D_1 is replaced by transistor T_2. (T_1 and T_2 is power MOSFET, $R_{\mathrm{DS}} = 10$ mΩ (max) @ 4.5 Vgs @ 25 °C *Tj*)

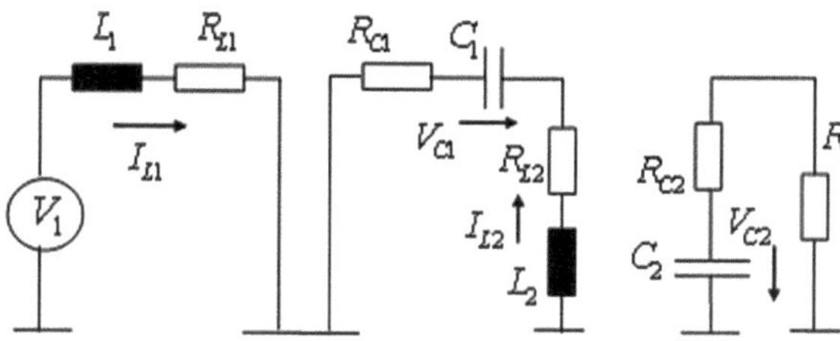

Fig. 25 When the T1 is turned on (T2 off), the input inductor is charged from the source, and the second inductor is charged from the first capacitor. No energy is supplied to the load capacitor during this time. Inductor current and capacitor voltage polarities are marked in this figure

Figure 25 shows the circuit when the power switch is turned on. The first inductor, L_1, is charged from the input voltage source V_1 during this time. The second inductor L_2 takes energy from the first capacitor C_1, and the output capacitor C_2 is left to provide the load current. The fact that both L_1 and L_2 are disconnected from the load when the switch is on leads to complex control characteristics.

When the power switch is turned off, the first inductor charges the capacitor C_1 and also provides current to the load, as shown in Fig. 26. The second inductor L_2 is also connected to the load during this time.

The output capacitor sees a pulse of current during the off time, making it inherently noisier than a buck converter. The input current is usually supplied from battery.

The SEPIC circuit diagram (according Fig. 25) for T1 on and T2 off can be described by a set of differential equations

$$
\begin{aligned}
\frac{dI_{L1}}{dt} &= \frac{1}{L_1}(-R_{L1}I_{L1} + V_1)\\
\frac{dI_{L2}}{dt} &= \frac{1}{L_2}[-(R_{C1} + R_{L2})I_{L2} + V_{C1}]\\
\frac{dV_{C1}}{dt} &= -\frac{1}{C_1}I_{L2}\\
\frac{dV_{C2}}{dt} &= -\frac{1}{C_2}\frac{1}{R_O + R_{C2}}V_{C2}
\end{aligned}
\tag{193}
$$

and circuit diagram (according Fig. 26) for T1 off and T2 on can be described by

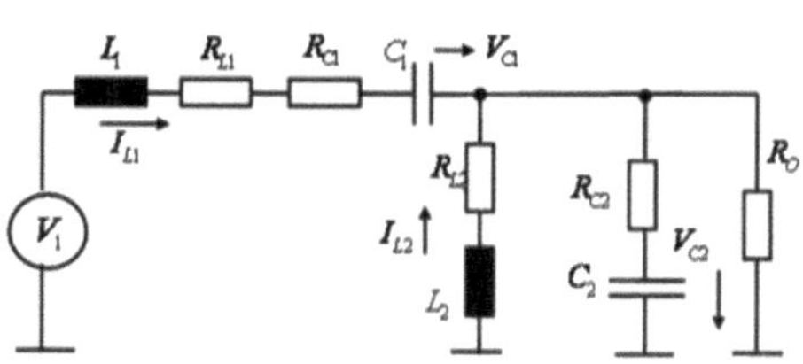

Fig. 26 When the T1 is off and T2 is on, both inductors provide current to the load capacitor

$$\begin{aligned}
\frac{dI_{L1}}{dt} &= \frac{1}{L_1}[-(R_{C1}+R_{L1}+R_{C2}R_A)I_{L1} - R_{C2}R_A I_{L2} - V_{C1} - R_A V_{C2} + V_1] \\
\frac{dI_{L2}}{dt} &= \frac{1}{L_2}[-R_{C2}R_A I_{L1} - (R_{L2}+R_{C2}R_A)I_{L2} - R_A V_{C2}] \\
\frac{dV_{C1}}{dt} &= \frac{1}{C_1}I_{L1} \\
\frac{dV_{C2}}{dt} &= \frac{1}{C_2}\left[R_A I_{L1} + R_A I_{L2} - \frac{1}{R_O+R_{C2}}V_{C2}\right]
\end{aligned} \tag{194}$$

where

$$R_A = \frac{R_O}{R_O+R_{C2}} \tag{195}$$

For $R_{C1} = R_{C2} = 0 \Rightarrow R_A = 1$ (resistors R_{C1}, R_{C2} are shifted in other resistors) Eqs. (193) and (194) are simplified (for T1 on, T2 off):

$$\begin{aligned}
\frac{dI_{L1}}{dt} &= \frac{1}{L_1}(-R_{L1}I_{L1} + V_1) \\
\frac{dI_{L2}}{dt} &= \frac{1}{L_2}[-R_{L2}I_{L2} + V_{C1}] \\
\frac{dV_{C1}}{dt} &= -\frac{1}{C_1}I_{L2} \\
\frac{dV_{C2}}{dt} &= -\frac{1}{C_2R_O}V_{C2}
\end{aligned} \tag{196}$$

and for T1 off and T2 on:

$$\begin{aligned}
\frac{dI_{L1}}{dt} &= \frac{1}{L_1}[-R_{L1}I_{L1} - V_{C1} - V_{C2} + V_1] \\
\frac{dI_{L2}}{dt} &= \frac{1}{L_2}[-R_{L2}I_{L2} - V_{C2}] \\
\frac{dV_{C1}}{dt} &= \frac{1}{C_1}I_{L1} \\
\frac{dV_{C2}}{dt} &= \frac{1}{C_2}\left[I_{L1} + I_{L2} - \frac{1}{R_O}V_{C2}\right]
\end{aligned} \tag{197}$$

For port Hamiltonian formalism we must adjust Eqs. (196) and (197) according to (186). Equation (196) (for T1 off and T2 on) is therefore written as (198)

$$J_1 - R = \underbrace{\begin{bmatrix} 0 & 0 & -\frac{1}{L_1C_1} & -\frac{1}{L_1C_2} \\ 0 & 0 & 0 & -\frac{1}{L_2C_2} \\ \frac{1}{L_1C_1} & 0 & 0 & 0 \\ \frac{1}{L_1C_2} & \frac{1}{L_2C_2} & 0 & 0 \end{bmatrix}}_{J_1} - \underbrace{\begin{bmatrix} \frac{R_{L1}}{L_1^2} & 0 & 0 & 0 \\ 0 & \frac{R_{L2}}{L_2^2} & 0 & 0 \\ 0 & 0 & 0 & 0 \\ 0 & 0 & 0 & \frac{1}{R_OC_2^2} \end{bmatrix}}_{R} \tag{198}$$

and Eq. (197) after matrix subtracting is changed to (199)

$$J_1 - R = \begin{bmatrix} -\frac{R_{L1}}{L_1^2} & 0 & -\frac{1}{L_1C_1} & -\frac{1}{L_1C_2} \\ 0 & -\frac{R_{L2}}{L_2^2} & 0 & -\frac{1}{L_2C_2} \\ \frac{1}{L_1C_1} & 0 & 0 & 0 \\ \frac{1}{L_1C_2} & \frac{1}{L_2C_2} & 0 & -\frac{1}{R_OC_2^2} \end{bmatrix} \tag{199}$$

The similar procedure is used for Eq. (196) (for T1 on and T2 off)

$$J_2 - R = \underbrace{\begin{bmatrix} 0 & 0 & 0 & 0 \\ 0 & 0 & \frac{1}{L_2C_1} & 0 \\ 0 & -\frac{1}{L_2C_1} & 0 & 0 \\ 0 & 0 & 0 & 0 \end{bmatrix}}_{J_1} - \underbrace{\begin{bmatrix} \frac{R_{L1}}{L_1^2} & 0 & 0 & 0 \\ 0 & \frac{R_{L2}}{L_2^2} & 0 & 0 \\ 0 & 0 & 0 & 0 \\ 0 & 0 & 0 & \frac{1}{R_OC_2^2} \end{bmatrix}}_{R} \tag{200}$$

and final matrix $J_2 - R$ has form (for T1 on and T2 off)

$$J_2 - R = \begin{bmatrix} -\frac{R_{L1}}{L_1^2} & 0 & 0 & 0 \\ 0 & -\frac{R_{L2}}{L_2^2} & \frac{1}{L_2C_1} & 0 \\ 0 & -\frac{1}{L_2C_1} & 0 & 0 \\ 0 & 0 & 0 & -\frac{1}{R_OC_2^2} \end{bmatrix} \tag{201}$$

The $\frac{\partial H}{\partial x}(x)$ and $B(x)$ are

$$\frac{\partial H}{\partial x}(x) = \begin{bmatrix} L_1x_1 \\ L_2x_2 \\ C_1x_3 \\ C_2x_4 \end{bmatrix} \quad B(x) = \begin{bmatrix} 1/L_1 \\ 0 \\ 0 \\ 0 \end{bmatrix} \tag{202}$$

the whole dynamical model (for T1 off and T2 on) is

$$\begin{bmatrix} \dot{x}_1 \\ \dot{x}_2 \\ \dot{x}_3 \\ \dot{x}_4 \end{bmatrix}_{S1} = \begin{bmatrix} -\frac{R_{L1}}{L_1^2} & 0 & -\frac{1}{L_1C_1} & -\frac{1}{L_1C_2} \\ 0 & -\frac{R_{L2}}{L_2^2} & 0 & -\frac{1}{L_2C_2} \\ \frac{1}{L_1C_1} & 0 & 0 & 0 \\ \frac{1}{L_1C_2} & \frac{1}{L_2C_2} & 0 & -\frac{1}{R_OC_2^2} \end{bmatrix} \begin{bmatrix} L_1x_1 \\ L_2x_2 \\ C_1x_3 \\ C_2x_4 \end{bmatrix} + \begin{bmatrix} 1/L_1 \\ 0 \\ 0 \\ 0 \end{bmatrix} V_1 \tag{203}$$

$$y = B(x)^T \frac{\partial H}{\partial x}(x) = \begin{bmatrix} 1/L_1 & 0 & 0 & 0 \end{bmatrix} \begin{bmatrix} L_1x_1 & L_2x_2 & C_1x_3 & C_2x_4 \end{bmatrix}^T$$

and dynamical model (for T1 on and T2 off) is

$$\begin{bmatrix} \dot{x}_1 \\ \dot{x}_2 \\ \dot{x}_3 \\ \dot{x}_4 \end{bmatrix}_{S2} = \begin{bmatrix} -\frac{R_{L1}}{L_1^2} & 0 & 0 & 0 \\ 0 & -\frac{R_{L2}}{L_2^2} & \frac{1}{L_2C_1} & 0 \\ 0 & -\frac{1}{L_2C_1} & 0 & 0 \\ 0 & 0 & 0 & -\frac{1}{R_OC_2^2} \end{bmatrix} \begin{bmatrix} L_1x_1 \\ L_2x_2 \\ C_1x_3 \\ C_2x_4 \end{bmatrix} + \begin{bmatrix} 1/L_1 \\ 0 \\ 0 \\ 0 \end{bmatrix} V_1 \tag{204}$$

$$y = B(x)^T \frac{\partial H}{\partial x}(x) = \begin{bmatrix} 1/L_1 & 0 & 0 & 0 \end{bmatrix} \begin{bmatrix} L_1x_1 & L_2x_2 & C_1x_3 & C_2x_4 \end{bmatrix}^T$$

The above equations of systems (S_1—for T1 off and T2 on) can be rewritten as

$$\begin{bmatrix} \dot{x}_1 \\ \dot{x}_2 \\ \dot{x}_3 \\ \dot{x}_4 \end{bmatrix}_{S1} = \begin{bmatrix} -\frac{R_{L1}}{L_1} & 0 & -\frac{1}{L_1} & -\frac{1}{L_1} \\ 0 & -\frac{R_{L2}}{L_2} & 0 & -\frac{1}{L_2} \\ \frac{1}{C_1} & 0 & 0 & 0 \\ \frac{1}{C_2} & \frac{1}{C_2} & 0 & -\frac{1}{R_OC_2} \end{bmatrix} \begin{bmatrix} x_1 \\ x_2 \\ x_3 \\ x_4 \end{bmatrix} + \begin{bmatrix} 1/L_1 \\ 0 \\ 0 \\ 0 \end{bmatrix} V \tag{205}$$

$$y = B(x)^T \frac{\partial H}{\partial x}(x) = \begin{bmatrix} 1/L_1 & 0 & 0 & 0 \end{bmatrix} \begin{bmatrix} L_1x_1 & L_2x_2 & C_1x_3 & C_2x_4 \end{bmatrix}^T$$

and system S_2—for T1 on and T2 off

$$\begin{bmatrix} \dot{x}_1 \\ \dot{x}_2 \\ \dot{x}_3 \\ \dot{x}_4 \end{bmatrix}_{S2} = \begin{bmatrix} -\frac{R_{L1}}{L_1} & 0 & 0 & 0 \\ 0 & -\frac{R_{L2}}{L_2} & \frac{1}{L_2} & 0 \\ 0 & -\frac{1}{C_1} & 0 & 0 \\ 0 & 0 & 0 & -\frac{1}{R_OC_2} \end{bmatrix} \begin{bmatrix} x_1 \\ x_2 \\ x_3 \\ x_4 \end{bmatrix} + \begin{bmatrix} 1/L_1 \\ 0 \\ 0 \\ 0 \end{bmatrix} V_1 \tag{206}$$

$$y = B(x)^T \frac{\partial H}{\partial x}(x) = \begin{bmatrix} 1/L_1 & 0 & 0 & 0 \end{bmatrix} \begin{bmatrix} L_1x_1 & L_2x_2 & C_1x_3 & C_2x_4 \end{bmatrix}^T$$

The energy stored in electrical circuit can be calculated by

$$E_S = \frac{1}{2}\left(L_1 I_{L1}^2 + L_2 I_{L2}^2 + C_1 V_{C1}^2 + C_2 V_{C2}^2\right) \tag{207}$$

And Hamiltonian energy in system is $H(x)$

$$H(x) = E_S = \frac{1}{2}\left(L_1 x_1^2 + L_2 x_2^2 + C_1 x_3^2 + C_2 x_4^2\right) \tag{208}$$

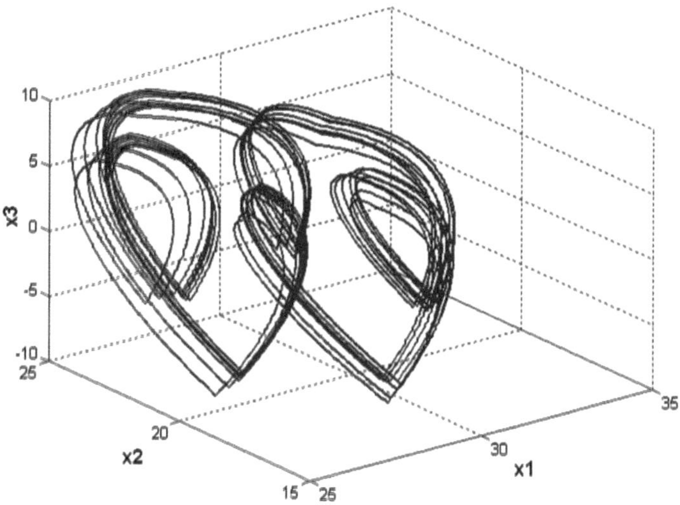

Fig. 27 Phase projection of x_1, x_2 and x_3 for SEPIC convertor

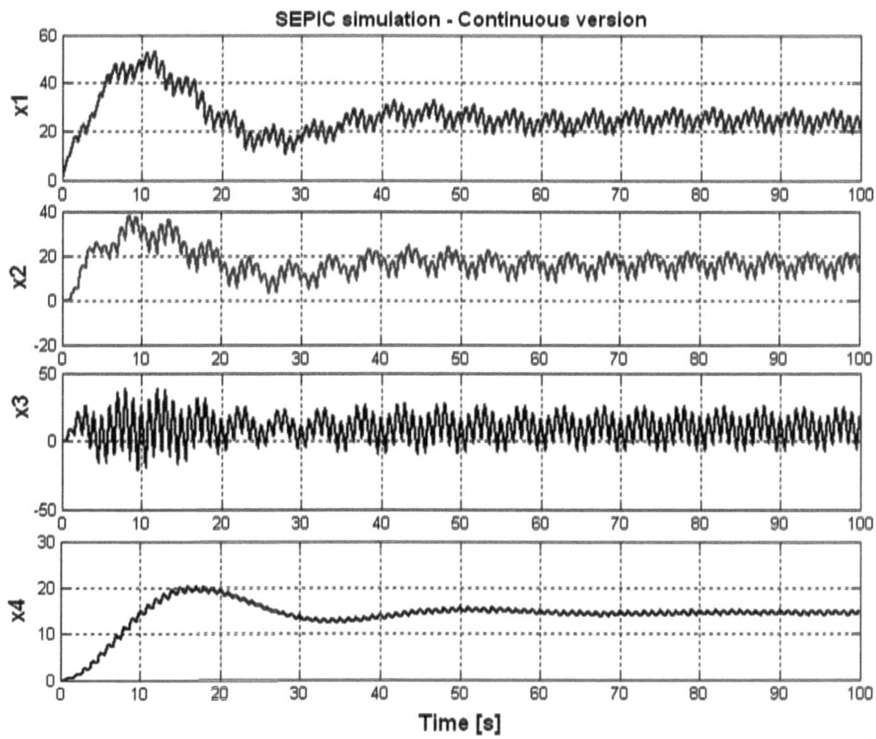

Fig. 28 The time evolution of state variables for SEPIC converter. From the *upper* to the *lower* plot: x_1, x_2, x_3, x_4

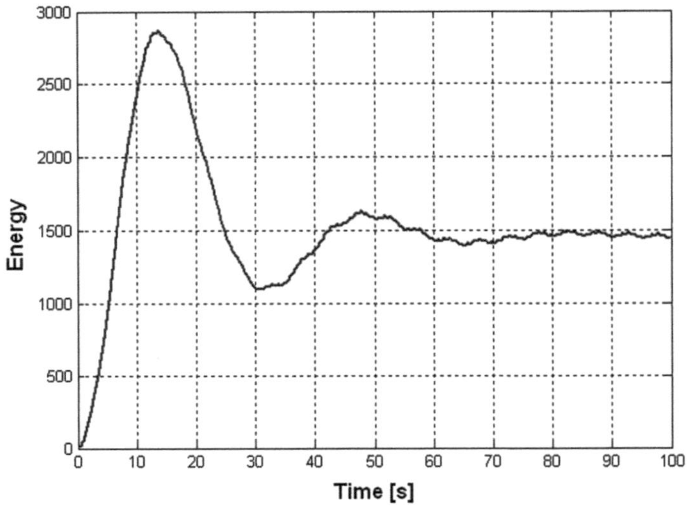

Fig. 29 The time evolution of total energy in SEPIC converter

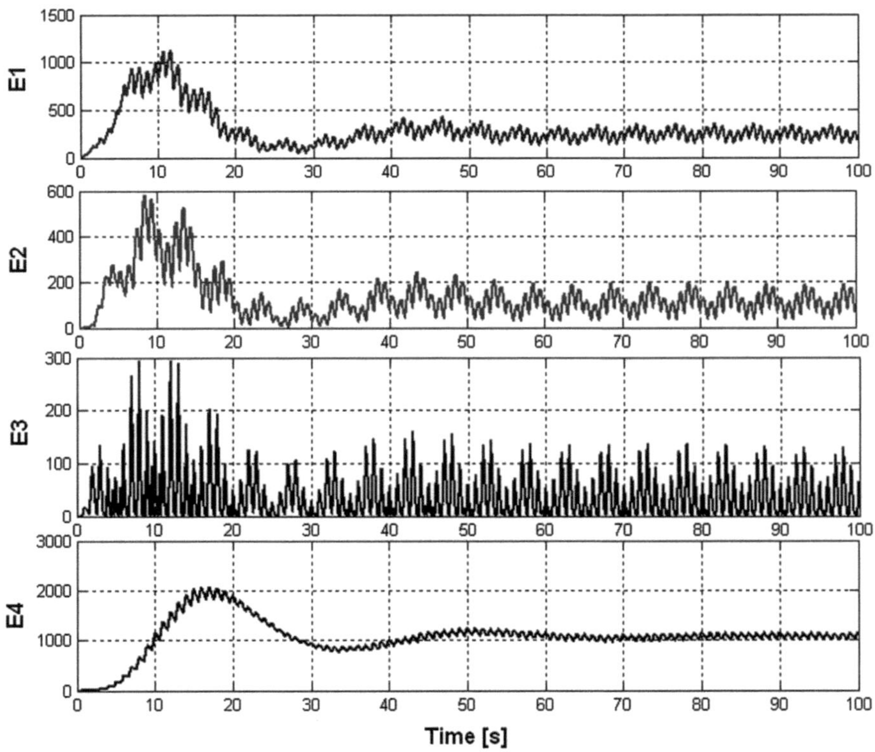

Fig. 30 Partial energy in SEPIC converter E_1, E_2, E_3, E_4. (energy in L_1, L_2, C_1, C_2)

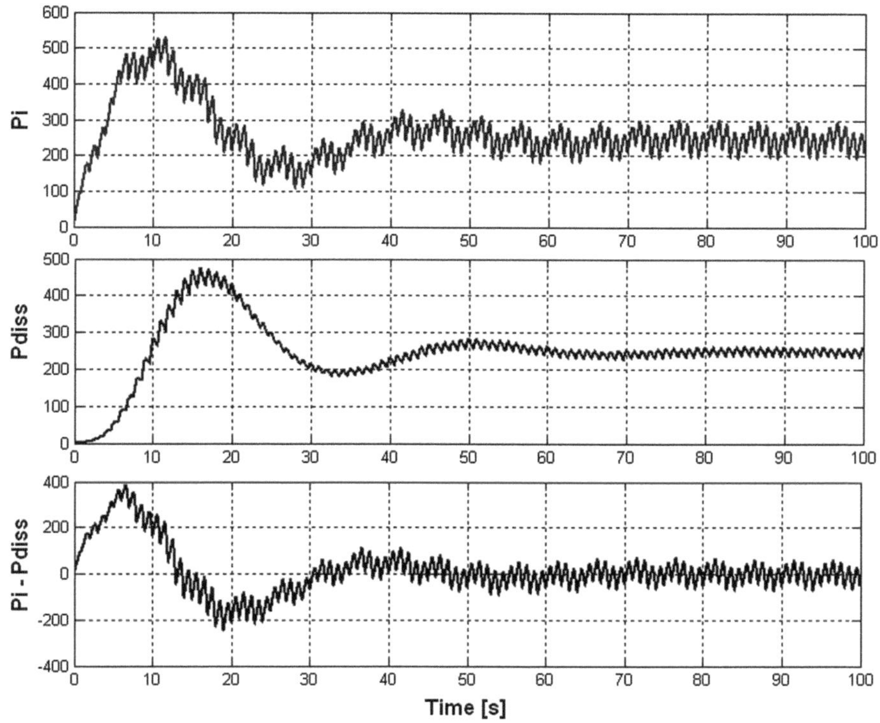

Fig. 31 The powers flow through the SEPIC converter. From the *upper* to the *lower* plot: input power, all dissipative powers and input power–dissipative powers

The SEPIC converter is simulated for parameters $V_1 = 10$; $L_1 = 0.8$ [H]; $L_2 = 0.8$ [H]; $C_1 = 0.4$ [F]; $C2 = 10$ [F]; $R_{C1} = 0$ [Ω]; $R_{C2} = 0$ [Ω]; $R_{L1} = 0.01$ [Ω]; $R_{L2} = 0.01$ [Ω]; $R_O = 0.9$ [Ω]; $d = 0.6$ (duty cycle), period $T = 1$ [s] and numerical values of system S_1 are

$$\begin{bmatrix} \dot{x}_1 \\ \dot{x}_2 \\ \dot{x}_3 \\ \dot{x}_4 \end{bmatrix}_{S1} = \begin{bmatrix} -0.0125 & 0 & -1.25 & -1.25 \\ 0 & -0.0125 & 0 & -1.25 \\ 2.5 & 0 & 0 & 0 \\ 0.1 & 0.1 & 0 & -0.1111 \end{bmatrix} \begin{bmatrix} x_1 \\ x_2 \\ x_3 \\ x_4 \end{bmatrix} + \begin{bmatrix} 1.25 \\ 0 \\ 0 \\ 0 \end{bmatrix} V_1 \tag{209}$$

$$y = [\,1.25 \quad 0 \quad 0 \quad 0\,][\,0.8x_1 \quad 0.8x_2 \quad 0.4x_3 \quad 10x_4\,]^T$$

and system S_2

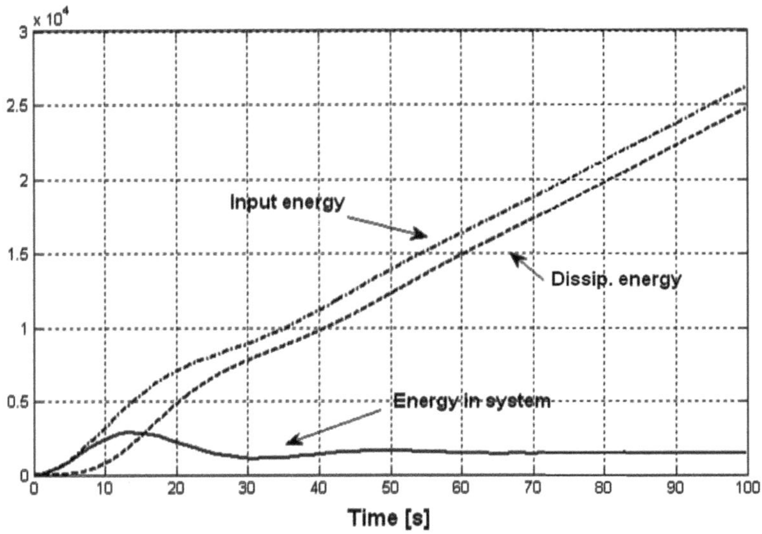

Fig. 32 The time evolution of energy flows through the SEPIC converter: Input energy, dissipated energy and energy in system

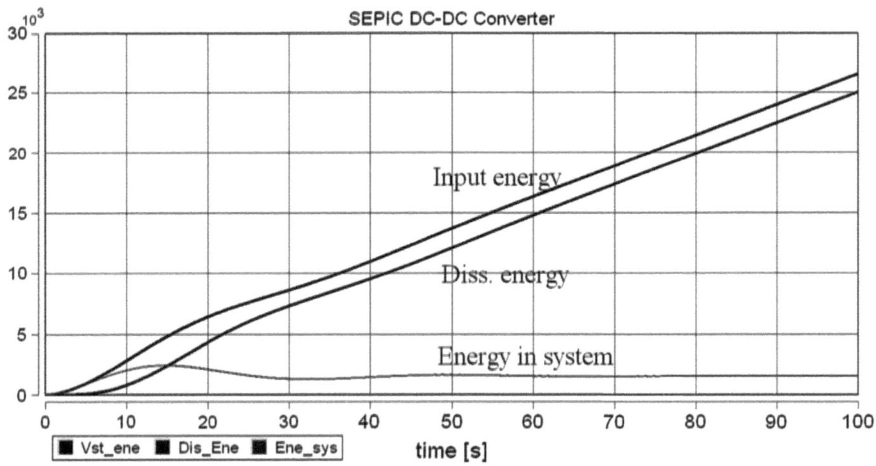

Fig. 33 The time evolution of energy flows through the SEPIC converter (calculated by SPICE): input energy, dissipated energy and energy in system

Table 1 Comparison of the state space energy and port-Hamiltonian approach: class of systems

Class of systems	State space energy	Port Hamiltonian
Linear	OK	No problems if Hamiltonian is known
Nonlinear	OK	OK
MIMO	Some problems	OK
Chaotic	Verified by variety of simulation experiment	Partly

Table 2 Comparison of the state space energy and port-Hamiltonian approach: Property

Property	State space energy	Port Hamiltonian
Basic idea	Geometric orientation	Physical orientation
State minimality	Necessary	Appropriate
Parameter minimality	Minimal description	Non-minimal description
Asymptotic stability	Straightforward (Construction of Lyapunov function, too)	By means of system passivity analysis possible
Chaotic behavior synthesis	Straightforward	Not known so far

$$\begin{bmatrix} \dot{x}_1 \\ \dot{x}_2 \\ \dot{x}_3 \\ \dot{x}_4 \end{bmatrix}_{S2} = \begin{bmatrix} -0.0125 & 0 & 0 & 0 \\ 0 & -0.0125 & 1.25 & 0 \\ 0 & -2.5 & 0 & 0 \\ 0 & 0 & 0 & -0.1111 \end{bmatrix} \begin{bmatrix} x_1 \\ x_2 \\ x_3 \\ x_4 \end{bmatrix} + \begin{bmatrix} 1.25 \\ 0 \\ 0 \\ 0 \end{bmatrix} V_1 \quad (210)$$

$$y = [\,1.25 \quad 0 \quad 0 \quad 0\,][\,0.8x_1 \quad 0.8x_2 \quad 0.4x_3 \quad 10x_4\,]^T$$

Eigenvalues of a matrix R:

$$\text{Eig\_}R = [0; 0.0111; \ 0.0156; \ 0.0156]^{\mathrm{T}} \quad (211)$$

are therefore positive semidefinite. Denominator of the transfer function of system S_1 is

$$\text{Den}_{S1} = s^4 + 0.136s^3 + 3.378s^2 + 0.389s + 0.395 \quad (212)$$

The roots of the Den_{S1} are

$$[-0.0085 + 1.8040i; -0.0085 - 1.8040i; -0.0595 + 0.3432i; -0.0595 - 0.3432i]$$

The simulation results of SEPIC converter are shown in the following figures. The state space variables are $x_1 = I_{L1}$, $x_2 = I_{L2}$, $x_3 = V_{C1}$ and $x_4 = V_{C2}$. In Fig. 27 is shown as 3D phase projection. Time evolution of x_1, x_2, x_3, x_4 is displayed in Fig. 28. Time evolution of total energy (according 208) is shown in Fig. 29 and partial energy in Fig. 30. The powers flows through the SEPIC converter according to Eq. (213) (input power, all dissipative powers and input power-dissipative powers) are shown in Fig. 31.

$$\frac{\mathrm{d}H}{\mathrm{d}t} = \underbrace{u(t)^T y(t)}_{\text{Input\_power}} - \underbrace{\frac{\partial^T H(x(t))}{\partial x} R(x(t)) \frac{\partial H(x(t))}{\partial x}}_{\text{Dissipative\_power}} \leq \underbrace{u(t)^T y(t)}_{\text{Input\_power}} \tag{213}$$

The energy balance is derived from (213) as

$$\underbrace{H[x(t)] - H[x(0)]}_{\text{Stored\_energy}} = \underbrace{\int_0^t u(s)^T y(s)\mathrm{d}s}_{\text{Supplied\_energy}} - \underbrace{\int_0^t \left(\frac{\partial H}{\partial x}(x(s))\right)^T R(x(t)) \frac{\partial H}{\partial x}(x(s))\mathrm{d}s}_{\text{Dissipated\_energy}} \tag{214}$$

In Fig. 32 the time evolution of energy that flows through the SEPIC converter is shown: Input energy, dissipated energy and energy in system (according to Eq. (214)). For confirmation of result calculated from (214), the supplied energy, dissipated energy, and energy stored in system was calculated by SPICE. All are shown in Fig. 33.

9 Conclusion

Interaction between physical systems is determined by exchange of energy and therefore, a first step toward the control interactions is to explicitly model the energetic properties of physical systems. Energy exchange plays a key role both for the description and for the control of the physical interaction. In this chapter some innovative methods for systems description by energy were presented. The state space energy-based approach seems to open a new perspective in development of sufficiently universal and more adequate abstract system representations for a broad variety of natural systems. For comparison the port-Hamiltonian approach was also introduced. Similar to the state space energy, the port-Hamiltonian description of physical systems puts into evidence all the energetic properties of the system the amount of energy stored, through the state energy variables, the energy dissipation, through the dissipative elements, the interfaces with the external world, through the power ports, and the interconnection structure along with the parts of exchange energy.

In this chapter was presented that booth approach—state space energy and port-Hamiltonian can be used to model a wide variety of physical systems by taking into account explicitly their energetic behavior and the internal interconnection structure. The main idea of the work was to introduce a new theory and present some linear and nonlinear examples. The comparisons of booth approach are simply presented in Tables 1 and 2.

Acknowledgments This research was supported by the European Regional Development Fund and Ministry of Education, Youth and Sports of the Czech Republic under project No. CZ.1.05/ 2.1.00/03.0094: Regional Innovation Centre for Electrical Engineering (RICE). M. Stork is with the Regional Innovation Centre for Electrical Engineering, University of West Bohemia, Univerzitni 22, Plzen, Czech Republic.

References

1. Mac Farlane AGJ (1970) Dynamical system models. George G. Harrap & Co. Ltd., London
2. Kalman RE (1963) Mathematical description of linear dynamical systems. SIAM J Control 1:152–192
3. Mayer D (1970) The state variable method of electrical network analysis. Acta Technica Csav 6:761–789
4. Brayton RK, Moser JK (1964) A theory of nonlinear networks-1. Q Appl Math 22:1–33
5. Brayton RK, Moser JK (1964) A theory of nonlinear networks-2. Q Appl Math 22:81–104
6. Massimo FM, Kwatny HG, Bahar LY (1980) Derivation of the Brayton-Moser equations from a topological mixed potential function. J Franklin Inst 310:259–269
7. Khalil HK (1996) Nonlinear systems. Prentice Hall, New Jersey
8. Hrusak J, Mayer D, Stork M (2004) On system structure reconstruction problem and Tellegen-like relation. In: Proceedings of 8th World Multiconf., SCI, vol VIII, Florida, USA, pp 373–378
9. Hrusak J, Mayer D, Stork M (2006) New approach to non-linear instability and chaos based on generalized Tellegen'ss principle, WMSCI 2006. Orlando, Florida, Int. Institute of Informatics and Systemics, USA, ISBN 980-6560-67-1, pp 199–206
10. Hrusak J, Mayer D, Stork M (2006) Dissipativity, minimality and physical correctness of state space representations, CITSA 2006. Orlando, Florida, USA, Int. Institute of Informatics and Systemics, ISBN 980-6560-83-3, pp 136–141
11. Hrusak J, Mayer D, Stork M (2007) Structural approach to instability and chaos in non-linear causal systems, Systems Theory and Scientific Computation. WSEAS Press, Athens, Greece, ISBN 978-960-8457-96-0. ISSN 1790-5117, pp 223–228
12. Stork M, Hrusak J, Mayer D (2006) Chaos and strange behavior detection of continuous and digital nonlinear systems based on internal system energy autocovariance, WMSCI 2006, Orlando, Florida, Int. Institute of Informatics and Systemics, USA, pp 228–233
13. Fradkov AL, Pogromsky AY (1998) Introduction to control of oscillations and chaos. World Scientific Series on Nonlinear Science, Singapore
14. Hrusak J (1969) In: Anwendung der Äquivalenz bei Stabilitätsprüfung, Tagung ü. die Regelungs-theorie, Mathematisches Forschungsinstitut, Oberwolfach, Universitaet Freiburg, West Germany
15. Hoover WG (1995) Remark on: some simple chaotic flows. Phys Rev E 51:759–760
16. Posch HA, Hoover WG, Vesely FJ (1986) Canonical dynamics of the nosé oscillator: stability, order, and chaos. Phys Rev A 33:4253–4265
17. Nosé S (1984) A unified formulation of the constant temperature molecular dynamics methods. J Chem Phys 81, Section IIB :511–519
18. Hoover WG (1988) Reversible mechanics and time's arrow. Phys Rev A 37:252–257
19. Ezra GS (2006) Reversible measure-preserving integrators for non-Hamiltonian systems. J Chem Phys 10(1063/1):2215608

20. Hrusak J, Stork M, Mayer D (2011) Generalized Tellegen's Principle and state space energy based causal systems description. In: Advances in energy research: distributed generations systems integrating renewable energy resources, Part I, Basic theory and advanced approaches, Chapter 4, NOVA Science Publ., USA, pp 95–139
21. Jeltsema D, Scherpen JMA (2007) A power-based description of standard mechanical systems. Syst Control Lett 56:349–356
22. Mayer D, Hrusak J (2003) On correctness and asymptotic stability in causal system theory. In: Proceedings of the 7th World Multiconf. Systemics, Cybernetics and Informatics, vol XIII, Orlando, USA, pp 355–360
23. Hrusak J, Stork M, Mayer D (2005) Dissipation normal form, conservativity, instability and chaotic behavior of continuous-time strictly causal systems. WSEAS Trans Syst 4:915–920
24. Stork M, Hrusak J, Mayer D (2005) Continuous and digital nonlinear systems, chaos and strange behavior detection, simulations and experiments. WSEAS Trans Circuits Syst 4:395–405
25. Hrusak J, Stork M, Mayer D (2008) Dissipation Normal forms and further applications of Lyapunov-Tellegen's principle. In: 12th WSEAS international conference on systems, WSEAS Press, Heraclion, Greece
26. Stork M, Hrusak J, Mayer D (2010) Nonlinearly coupled oscillators and state space energy approach. In: 14th WSEAS international conference on SYSTEMS Corfu Island, Greece
27. Tabuada P, Pappas GJ (2003) Abstractions of Hamiltonian control systems. Autom IFAC 39:2025–2033
28. Jeltsema D, Ortega R, Scherpen JMA (2003) On passivity and power-balance inequalities of nonlinear RLC circuits. IEEE Trans. Circuits Syst Fund Theory Appl 50:1174–1178
29. Maschke B, van der Schft AJ (1994) A Hamiltonian approach to stabilization of nonholonomic mechanical systems. In: Proceedings of the 33rd IEEE conference on decision and control, pp 2950–2954
30. Maschke B, van der Schft AJ (1997) Interconnected mechanical systems. In: Part I: Geometry of interconnection and implicit Hamiltonian systems. Imperial College Press, London
31. Maschke B and van der Schft AJ (1997) Interconnected mechanical systems. In: Part II: the dynamics of spatial mechanical networks. Imperial College Press, London
32. Maschke B, van der Schft AJ, Breedveld PC (1992) An intrinsic Hamiltonian formulation of network dynamics: non-standard poisson structures and gyrators. J Franklin 329:923–966
33. Maschke BM, van der Schaft AJ, Breedveld PC (1995) An intrinsic hamiltonian formulation of the dynamics of lc-circuits. IEEE Trans Circ Syst 42:73–82
34. Maschke B, van der Schft AJ (1992) Port-controlled hamiltonian system: modelling origins and system theoretic approach. In: Proceedings of 2nd IFAC NOLCOS, Bordeaux, pp 282–288
35. Ortega R, Loria A, Nicklasson PJ, Sira-Ramirez H (1998) Passivity-based control of Euler–Lagrange systems. Springer, London
36. Ortega R, van der Schaft AJ, Mareels I, Maschke B (2001) Putting energy back in control. IEEE Control Syst Mag 21:18–33
37. Ortega R, van der Schaft AJ, Maschke B, Escobar G (1999) Energy-shaping of port-controlled hamiltonian systems by interconnection. In: Proceedings of IEEE conference on decision and Control, Phoenix, AZ, USA, pp 1646–1651
38. Van der Schaft AJ (2000) L2-gain and passivity techniques in nonlinear control. Springer, London
39. Čuk S, Middlebrook RD (1983) Coupled-inductor and other extensions of a new optimum topology switching DC–DC converter. Advances in switched-mode power conversion, vols I and II. Tesla Co, Irvine
40. Chung HS, Tse KK, Hui SY, Mok CM, Ho MT (2003) A novel maximum power point tracking technique for solar panels using a SEPIC or Cuk converter. IEEE Trans Power Electron 18:717–724
41. Maged N, Nashed F (2004) Design of a digital PWM controller for a soft switching SEPIC converter. J Power Electron 4:152–160

42. Adar D, Rahav G, Ben-Yaakov S (1997) A unified behavioral average model of SEPIC converters with coupled inductors. IEEE PESC'97, 1:441–446
43. Vorperian V (1990) Simplified analysis of PWM converters using model of PWM switch, Parts I(CCM) and II (DCM). IEEE Trans Aerosp Electron Syst 26:497–505

Power Systems Stability Analysis Based on Classical Techniques in Work

Naser Mahdavi Tabatabaei, Aysen Demiroren, Naser Taheri, Ahmad Hashemi and Narges Sadat Boushehri

Abstract This chapter presents a linearized Phillips–Heffron model of a parallel AC/DC power system in order to studying power system stability. In addition, a supplementary controller for a modeling back-to-back voltage source converter (BtB VSC) HVDC to damp low-frequency oscillations in a weakly connected system is proposed. Also, input controllability measurement for BtB VSC HVDC is investigated using relative gain array (RGA), singular value decomposition (SVD) and damping function (DF) and a supplementary controller is designed based on phase compensating method. In addition, a supplementary controller for a novel modeling VSC HVDC to damp low-frequency oscillations in a weakly connected system is proposed. The potential of the VSC HVDC supplementary controllers to enhance the dynamic stability is evaluated by measuring the electromechanical controllability through SVD analysis. The presented control scheme not only performs damping oscillations but also the voltage and power flow control can be achieved. Simulation results obtained by MATLAB verify the effectiveness of the VSC HVDC and its control strategy for enhancing dynamical stability. Moreover, a linearized model of a power system installed with a UPFC has been

N. Mahdavi Tabatabaei (✉) · N. S. Boushehri
Electrical Engineering Department, Seraj Higher Education Institute, Tabriz, Iran
e-mail: n.m.tabatabaei@gmail.com

N. S. Boushehri
e-mail: nargesboush@yahoo.com

N. Mahdavi Tabatabaei · A. Demiroren
Electrical Engineering Department, Istanbul Technical University, Maslak, Istanbul, Turkey
e-mail: demiroren@itu.edu.tr

N. Taheri
Electrical Engineering Department, Islamic Azad University, Quchan Branch, Quchan, Iran
e-mail: n.taheri.1362@gmail.com

A. Hashemi
Electrical Engineering Department, Sama Technical and Vocational College, Islamic Azad University, Kermanshah Branch, Kermanshah, Iran
e-mail: ahmad.hashemi.v@gmail.com

N. Bizon et al. (eds.), *Analysis, Control and Optimal Operations in Hybrid Power Systems*, Green Energy and Technology, DOI: 10.1007/978-1-4471-5538-6_4,

presented. UPFC has four control loops that, by adding an extra signal to one of them, increases dynamic stability and load angle oscillations are damped. To increase stability, a novel online adaptive controllers have been used analytically to identify power system parameters. Suitable operation of adaptive controllers to decrease rotor speed oscillations against input mechanical torque disturbances is confirmed by the simulation results.

Keywords Power system dynamic stability · Power system stabilizer (PSS) · Phillips–Heffron model · Low-frequency oscillations · Supplementary controller · Voltage source converter (VSC) · Single-machine infinite bus (SMIB) · Back-to-back VSC (BtB VSC) · Unified power flow controller (UPFC) · Adaptive controller

1 Introduction

HVDC transmission was first used commercially in 1954 when a DC submarine cable linked Gotland Island with mainland Sweden. It was an under cable, 96 km long with ratings of 100 kV and 20 MW. That HVDC system and the ones that followed for almost two decades used mercury arc valves for converting from AC to DC and vice versa starting with the Eel River back-to-back converter application in 1972, mercury arc valves were replaced by thyristor in commercial HVDC applications. Thyristors are silicon-based power semiconductors and were initially known as silicon controlled rectifiers (SCRs). Similar to diodes they allow only unidirectional current flow. However, a small current injected at the gate of a forward biased thyristor permits control of the start of current conduction. Once the thyristor is conducting, it will cease conduction only if the voltage across the thyristor reverses and the current it is conducting droops to zero. The HVDC controls send coordinated signals to the gates of groups of thyristors so that by switching the DC current through various paths at various times as described later, the AC–DC or DC–AC conversion process is realized. Approximately, 50 HVDC transmission projects have been built around the world since 1954, at present, under construction or being planned. The deregulation and restructuring of the electric utility industry currently in progress may encourage more applications of HVDC transmission, perhaps for nontraditional purposes.

As power demand grows rapidly and expansion in transmission and generation is restricted with the limited availability of resources and the strict environmental constraints, power systems are today much more loaded than before. This causes the power systems to be operated near their stability limits. Power system stabilizers (PSSs) aid in maintaining power system stability and improving dynamic performance by providing a supplementary signal to the excitation system [1].

However, PSSs may adversely affect voltage profile, may result in leading power factor, and may not be able to suppress oscillations resulting from severe disturbances, especially those three-phase faults which may occur at the generator terminals [1]. The availability of flexible AC transmission system (FACTS) controllers, such as static VAR compensators (SVC), thyristor control series compensators (TCSC), static synchronous compensators (STATCOM), and unified power flow controller (UPFC), can be applied to damping oscillations by adding a supplementary signal for main control loops [2–4].

Extremely fast control action associated with FACTS-device operations, they have been very promising candidates for utilization in power system damping enhancement. It has been observed that utilizing a feedback supplementary control, in addition to the FACTS-device primary control, can considerably improve system damping and can also improve system voltage profile, which is advantageous over PSSs [1].

Recently VSC HVDC systems have greatly increased. VSC HVDC transmission links are used in power systems for various purposes, for example, interconnecting two neighboring systems of using different frequencies, improving system transient stability, etc. Many existing VSC HVDC control methods provide various forms of modulation for damping power system oscillation and improving dynamic performance [5–7]. Power system stabilizers (PSSs) aid in maintaining power system stability and improving dynamic performance by providing a supplementary signal to the excitation system [4]. However, PSSs may adversely affect voltage profile, may result in leading power factor, and may not be able to suppress oscillations resulting from severe disturbances [8].

A unified power flow controller (UPFC) is one the FACTS devices which can control power system parameters such as terminal voltage, line impedance, and phase angle [9, 10]. Recently, researchers have presented dynamic UPFC models in order to design a suitable controller for power flow, voltage and damping controls [11–14]. Wang has presented a modified linearized Heffron–Phillips model of a power system installed with a UPFC [9, 15]. Wang has not presented a systematic approach to design the damping controllers. Furthermore, no effort seems to have been made to identify the most suitable UPFC control parameters, in order to arrive at a robust damping controller and has not used the deviation of active and reactive powers, ΔP_e and ΔQ_e as the input control signals. Abido has used the PSO control to design a controller and this manner not only is an offline procedure, but also depends strongly on the selection of the primary conditions of control systems [16, 17].

2 Analyzing of Power System Stability in BtB VSC SMIB Network

Recently, HVDC systems have greatly increased. They interconnect large power systems offering numerous technical and economic benefits. This interest results from functional characteristics and performance that include for example nonsynchronous interconnection, control of power flow, and modulation to increase stability limits [18]. It is well-known that the transient stability of the AC systems in a composite AC–DC system can be improved by taking advantage of the fast controllability of HVDC converters [19–24]. There are, therefore, good reasons for constructing HVDC links in close proximity to HVAC lines. Most available control theories assume that a control structure is given at the outset. The first step for an engineer is to determine the variable that should be controlled, the variable that should be measured, and the input that should be manipulated.

Power system which is equipped with parallel AC–DC system is a multivariable system and it is important to choose best input to apply supplementary controller signal to damp oscillations. It was proposed to using singular value decomposition (SVD) to measure controllability of BtB VSC HVDC inputs in [25, 26]. These papers show that, phase angel of rectifier has most controllability rather to others input.

In this section, an approach is presented to model parallel AC/DC power system namely Phillips–Heffron model based d-q algorithm in order to studying system dynamical stability. In addition, a block diagram representation is formed to analyze the system stability characteristics. By this modeling approach, it is possible to analyze the small-signal stability of the system and low-frequency oscillation phenomena which is caused by external disturbances such as variation of input torque and fault occurring.

Also selection of a suitable signal for applying to a power system is done based on relative gain array (RGA), singular value decomposition (SVD) and Damping function (DF). Simulation results carried by MATLAB, show the proposed strategy has fast dynamic response.

2.1 Configuration of Power System

Figure 1 shows a single-machine infinite bus (SMIB) network equipped with a BtB VSC HVDC. The four input control signals to the HVDC are M_r, PH_r, M_i, PH_i where M_r, M_i are the amplitude modulation ratio and PH_r, PH_i are phase angle of the control signals of each VSC, respectively.

By applying Park's transformation and neglecting the resistance and transients of the coupling transformers, the BtB VSC HVDC can be modeled:

$$\begin{bmatrix} V_{\mathrm{Ld}} \\ V_{\mathrm{Lq}} \end{bmatrix} = \begin{bmatrix} 0 & -X_s \\ X_s & 0 \end{bmatrix} \begin{bmatrix} I_{\mathrm{lod}} \\ I_{\mathrm{loq}} \end{bmatrix} + \begin{bmatrix} \frac{M_r V_{\mathrm{dc}} \cos(PH_r)}{2} \\ \frac{M_r V_{\mathrm{dc}} \sin(PH_r)}{2} \end{bmatrix} \tag{1}$$

$$\begin{bmatrix} V_{\mathrm{bd}} \\ V_{\mathrm{bq}} \end{bmatrix} = \begin{bmatrix} 0 & -X_{\mathrm{sp}} \\ X_{\mathrm{sp}} & 0 \end{bmatrix} \begin{bmatrix} I_{\mathrm{bopd}} \\ I_{\mathrm{bopq}} \end{bmatrix} + \begin{bmatrix} \frac{M_i V_{\mathrm{dc}} \cos(PH_i)}{2} \\ \frac{M_i V_{\mathrm{dc}} \sin(PH_i)}{2} \end{bmatrix} \tag{2}$$

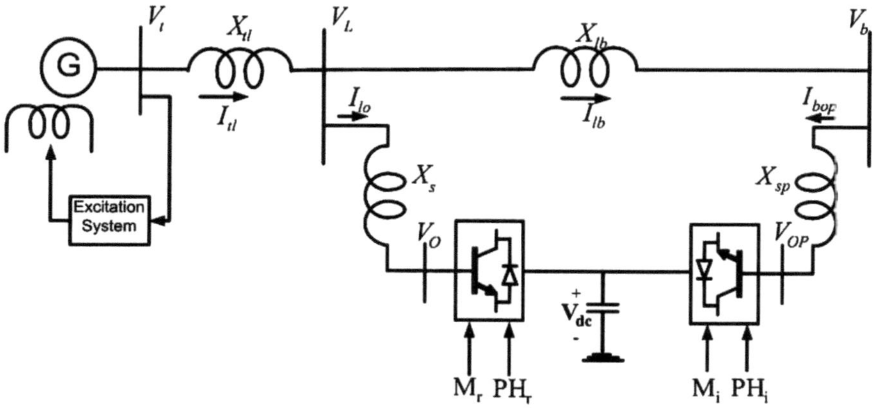

Fig. 1 A SMIB network equipped with a BtB VSC HVDC

$$C_{\mathrm{dc}}\dot{V}_{\mathrm{dc}} = \left[\frac{M_r}{2}(I_{\mathrm{lod}}\cos(PH_r) + I_{\mathrm{loq}}\sin(\mathrm{PH_r})) + \frac{M_i}{2}(I_{\mathrm{bopd}}\cos(PH_i) + I_{\mathrm{bopq}}\sin(PH_i))\right] \tag{3}$$

where $V_L, V_b, I_{\mathrm{lo}}$ and I_{bo} are the middle bus voltage, infinite bus voltage, flowed current to rectifier, and inverter, respectively. C_{dc} and V_{dc} are the DC link capacitance and voltage, respectively.

The nonlinear model of the SMIB system of Fig. 1 is:

$$\dot{\delta} = \omega_b(\omega - 1) \tag{4}$$

$$\dot{\omega} = \frac{(P_m - P_e - D\omega)}{M} \tag{5}$$

$$\dot{E}_q{}' = \frac{(E_{fd} - (x_d - x_d')I_{td} - E_q')}{T_{\mathrm{do}}'} \tag{6}$$

$$\dot{E}_{\mathrm{fd}} = \frac{(K_A(V_{\mathrm{ref}} - V_t) - E_{\mathrm{fd}})}{T_A} \tag{7}$$

where

$$P_e = V_{\mathrm{td}}I_{\mathrm{td}} + V_{\mathrm{tq}}I_{\mathrm{tq}}$$

$$V_t = \sqrt{V_{\mathrm{td}}^2 + V_{\mathrm{tq}}^2}$$

$$V_{\mathrm{td}} = x_q I_{\mathrm{tq}}$$

$$V_{tq} = E'_q - x'_d I_{td}$$

$$I_{td} = I_{lod} + I_{bd}$$

$$I_{tq} = I_{loq} + I_{bq}$$

and P_m and P_e are the input and output power, respectively; M and D are the inertia constant and damping coefficient, respectively; ω_b the synchronous speed; δ and ω the rotor angle and speed, respectively; E'_q, E_{fd} and V_t the generator internal, field, and terminal voltages, respectively; T'_{do} the open circuit field time constant; x_d, x'_d and X_q the d-axis, d-axis transient reactance, and q-axis reactance, respectively; K_A and T_A the exciter gain and time constant, respectively; V_{ref} the reference voltage. Effects of PSS can be considered by adding u_{pss} to the Eq. (7) which this is not a goal in this section. Also from Fig. 1 we have:

$$\bar{V}_t = jX_{tl}\bar{I}_{tl} + \bar{V}_l \tag{8}$$

$$\bar{V}_t = jX_{tl}\bar{I}_{tl} + jX_{lb}\bar{I}_{lb} + \bar{V}_b \tag{9}$$

$$\bar{I}_{lb} = \bar{I}_{tl} - \frac{\bar{V}_t - jX_{tl}\bar{I}_{tl} - \bar{V}_o}{jX_s} \tag{10}$$

where $\bar{I}_{tl}$, $\bar{V}_o$, $\bar{I}_{lb}$ and $\bar{V}_b$ are the armature current, rectifier voltage, infinite bus current, and voltage, respectively. From Eqs. (8)–(10), we can have:

$$I_{tlq} = \frac{\frac{X_{lb}}{X_s}\frac{M_r}{2} V_{dc}\cos(PH_r) + V_b \sin(\delta)}{ZX_q + A} \tag{11}$$

$$I_{tld} = \frac{ZE'_q - \frac{X_{lb}}{X_s}\frac{M_r}{2} V_{dc}\sin(PH_r) + V_b \cos(\delta)}{Zx'_d + A} \tag{12}$$

and for inverter side:

$$I_{bopd} = \frac{V_b \cos(\delta) - \frac{M_i}{2} V_{dc}\sin(PH_i)}{X_{sp}} \tag{13}$$

$$I_{bopq} = -\frac{V_b \sin(\delta) - \frac{M_i}{2} V_{dc}\cos(PH_i)}{X_{sp}} \tag{14}$$

By linearising (1)–(7), (11)–(14):

$$\Delta\dot{\delta} = \omega_b \Delta\omega \tag{15}$$

$$\Delta\dot{\omega} = \frac{(\Delta P_m - \Delta P_e - D\Delta\omega)}{M} \tag{16}$$

$$\Delta\dot{E}_q{}' = \frac{(\Delta E_{\mathrm{fd}} - (x_d - x_d')\Delta I_{\mathrm{td}} - \Delta E_q')}{T_{do}'} \tag{17}$$

$$\Delta\dot{E}_{\mathrm{fd}} = \frac{(K_A\Delta V_t - \Delta E_{\mathrm{fd}})}{T_A} \tag{18}$$

where:

$$\Delta\dot{V}_t = K_5\Delta\delta + K_6\Delta E_q' + K_{V_{\mathrm{dc}}}\Delta V_{\mathrm{dc}} + K_{VM_r}\Delta M_r + K_{VPH_r}\Delta PH_r \tag{19}$$

$$\Delta P_e = K_1\Delta\delta + K_2\Delta E_q' + K_{\mathrm{pdc}}\Delta V_{\mathrm{dc}} + K_{pM_r}\Delta M_r + K_{pPH_r}\Delta PH_r \tag{20}$$

$$\Delta E_q = K_3\Delta E_q' + K_4\Delta\delta + K_{qPH_r}\Delta PH_r + K_{qM_r}\Delta M_r + K_{\mathrm{qdc}}\Delta V_{\mathrm{dc}} \tag{21}$$

$$\Delta\dot{V}_{\mathrm{dc}} = q_1\Delta\delta + q_2\Delta E_q' + q_3\Delta V_{\mathrm{dc}} + q_4\Delta M_r + q_5\Delta PH_r + q_6\Delta M_i + q_7\Delta PH_i \tag{22}$$

Substitute (19)–(22) in (15)–(18) we can obtain the state variable of the power system installed with the BtB VSC HVDC to be:

$$\begin{bmatrix}\Delta\dot{\delta}\\ \Delta\dot{\omega}\\ \Delta\dot{E}_q'\\ \Delta\dot{E}_{\mathrm{fd}}\\ \Delta\dot{V}_{\mathrm{dc}}\end{bmatrix} = \begin{bmatrix} 0 & \omega_b & 0 & 0 & 0\\ -\frac{K_1}{M} & -\frac{D}{M} & -\frac{K_2}{M} & 0 & -\frac{K_{\mathrm{pdc}}}{M}\\ -\frac{K_4}{T_{\mathrm{do}}'} & 0 & -\frac{K_3}{T_{\mathrm{do}}'} & \frac{1}{T_{\mathrm{do}}'} & -\frac{K_{\mathrm{qdc}}}{M}\\ -\frac{K_A T_A}{T_A} & 0 & -\frac{K_A K_6}{T_A} & -\frac{1}{T_A} & -\frac{K_A K_{V_{\mathrm{dc}}}}{T_A}\\ q_1 & 0 & q_2 & 0 & q_3\end{bmatrix} \cdot \begin{bmatrix}\Delta\delta\\ \Delta\omega\\ \Delta E_q'\\ \Delta E_{\mathrm{fd}}\\ \Delta V_{\mathrm{dc}}\end{bmatrix} + \begin{bmatrix} 0 & 0 & 0 & 0\\ -\frac{K_{\mathrm{pM_r}}}{M} & -\frac{K_{pPH_r}}{M} & 0 & 0\\ -\frac{K_{qM_r}}{T_{\mathrm{do}}'} & -\frac{K_{qPH_r}}{T_{\mathrm{do}}'} & 0 & 0\\ -\frac{K_A K_{VM_r}}{T_A} & -\frac{K_A K_{VPH_r}}{T_A} & 0 & 0\\ q_4 & q_5 & q_6 & q_7\end{bmatrix} \cdot \begin{bmatrix}\Delta M_r\\ \Delta PH_r\\ \Delta M_i\\ \Delta PH_i\end{bmatrix} \tag{23}$$

where $\Delta M_i, \Delta M_r, \Delta PH_i$ and ΔPH_r are the linearization of the input control signals of the BtB VSC HVDC. The linearised dynamic model of (23) can be shown by Fig. 2. In this figure $K_{\mathrm{pu}}, K_{\mathrm{qu}}, K_{\mathrm{vu}}, K_q$ and ΔU are defined below:

$$K_{\mathrm{pu}} = [K_{pM_r}, K_{pPH_r}, 0, 0]^T$$

$$K_{\mathrm{qu}} = [K_{qM_r}, K_{qPH_r}, 0, 0]^T$$

$$K_{\mathrm{vu}} = [K_{VM_r}, K_{VPH_r}, 0, 0]^T$$

$$K_q = [q_4, q_5, q_6, q_7]^T$$

$$\Delta U = [\Delta M_r, \Delta PH_r, \Delta M_i, \Delta PH_i]^T$$

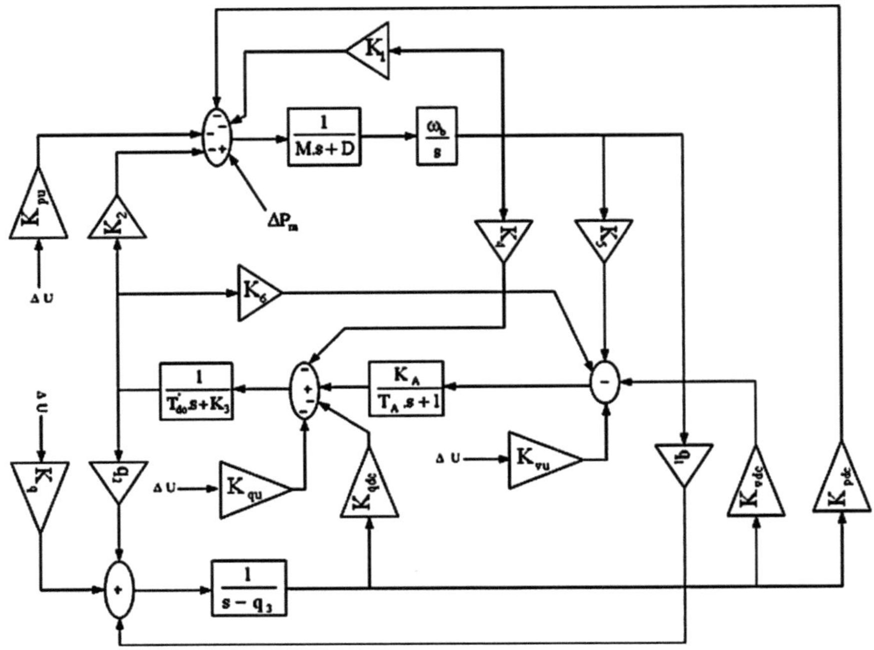

Fig. 2 Phillips–Heffron model of power system installed with HVDC

It can be seen that the configuration of the Phillips–Heffron model is exactly the same as that installed with SVC, TCSC, TCPS, UPFC and STATCOM.

Also from (23) it can be seen that there are four choice of input control signals of the BtB VSC HVDC to superimpose on the damping function of the BtB VSC HVDC ΔM_i, ΔM_r, ΔPH_i and ΔPH_r. Therefore, in designing the damping controller of the BtB VSC HVDC, besides setting its parameters, the selection of the input control signal of the BtB VSC HVDC to superimpose on the damping function of the BtB VSC HVDC is also important.

2.2 Input Controllability Measurement

2.2.1 Singular Value Decomposition (SVD)

To measure the controllability of the EM mode by a given input (control signal), the singular value decomposition (SVD) is employed [25, 26]. Mathematically, if G is an $m \times n$ complex matrix, then there exist unitary matrices U and V with dimensions of $m \times m$ and $n \times n$, respectively, such that:

$$G = U\Sigma V^H$$

where $\Sigma = \begin{bmatrix} \Sigma_1 & 0 \\ 0 & 0 \end{bmatrix}$, $\Sigma_1 = \mathrm{diag}(\sigma_1, \ldots, \sigma_r)$ with $\sigma_1 \geq \cdots \geq \sigma_r \geq 0$ where $r = \min\{m, n\}$ and $\sigma_1, \ldots, \sigma_r$ are the singular values of G.

The minimum singular value σ_r represents the distance of the matrix G from all the matrices with a rank of $r - 1$. This property can be used to quantify modal controllability. The matrix H can be written as $H = [h_1 h_2 h_3 h_4]$ where h_i is a column vector corresponding to the ith input. The minimum singular value, $\sigma_{\min}$ of the matrix $[\lambda I - A, h_i]$ indicates the capability of the ith input to control the mode associated with the eigenvalue λ. Actually, the higher $\sigma_{\min}$, the higher the controllability of this mode by the input considered. As such, the controllability of the EM mode can be examined with all inputs in order to identify the most effective one to control the mode.

2.2.2 Relative Gain Array (RGA)

The relative gain array (RGA) was introduced in Ref. [27]. It provides the designer with a quick assessment of interaction among the control loops of a multivariable system. The RGA of an $l \times m$ matrix G is defined as:

$$\mathrm{RGA}(G) = \Lambda(G) = G \times (G')^T$$

where G' is pseudoinverse, $(.)^T$ means transpose and $\times$ denotes the element of multiplication. The RGA has a number of interesting control properties [27]. RGA can exhibit the interaction of different input/output pair on the total system. This property can be used in the measurement of input controllability:

The ith row sum of the RGA is equal to the square of the ith output projection, and the jth column sum of the RGA is equal to the square of the jth input projection. For more information readers can refer to Refs. [27, 28].

2.2.3 Damping Function (DF)

The linearised model of the power system installed with the BtB VDC HVDC can be expressed by Fig. 3 [29–31], where H(s) is the transfer function of the HVDC damping controller. From Fig. 3 we can obtain the electric torque provided by the HVDC damping controller to the electromechanical oscillation loop of the generator to be:

$$\Delta T_{\mathrm{HVDC}} = \frac{K_c(\lambda_0) K_0(\lambda_0) H(\lambda_0)}{1 - K_{\mathrm{IL}}(\lambda_0) H(\lambda_0)} \Delta\omega \tag{24}$$

An ideal HVDC damping controller should contribute a pure positive damping torque to the electromechanical oscillation loop with $\Delta T_{\mathrm{HVDC}} = D_{\mathrm{HVDC}} \Delta\omega$ that is:

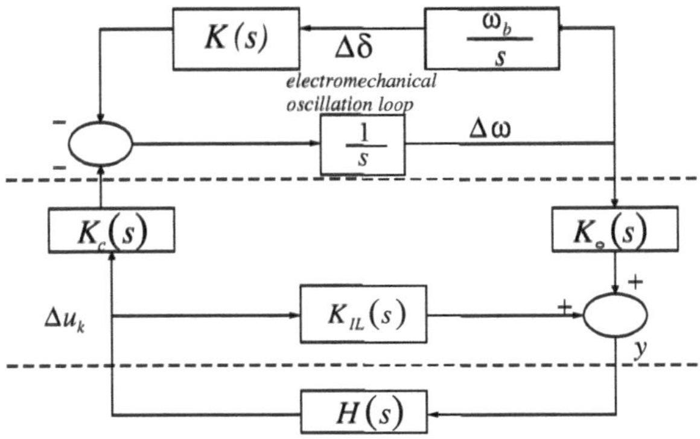

Fig. 3 Closed-loop system installed with UPFC dumping controller

$$D_{\mathrm{HVDC}} = \frac{K_c(\lambda_0)K_0(\lambda_0)H(\lambda_0)}{1 - K_{\mathrm{IL}}(\lambda_0)H(\lambda_0)} \tag{25}$$

which results in:

$$D_{\mathrm{HVDC}} = [K_c(\lambda_0)K_0(\lambda_0) + D_{\mathrm{IPFC}}K_{\mathrm{IL}}(\lambda_0)]H(\lambda_0) = F(\lambda_0)H(\lambda_0) \tag{26}$$

$F(\lambda_0)$ which is named as the forward path of the BtB VSC HVDC damping controller, has a decisive influence on the effectiveness of the HVDC damping controller. If we assume the set of the operating conditions of the power system is $\Omega(\mu), F(\lambda_0)$ can be denoted as the function of system operating condition μ and input control signal of the HVDCu_k. The criterion of the selection can be Ref. [26]:

$$\mu_{\mathrm{selected}} = \min_{\mu} F(\lambda_0, \mu, u_k), \quad \mu \in \Omega(\mu) \tag{27}$$

$$\begin{gathered} u_{\mathrm{selected}} = \max_{u_k} F(\lambda_0, \mu_{\mathrm{selected}}, u_k) \\ u_k \in \{M_r, M_i, \varphi_r, \varphi_i\} \end{gathered} \tag{28}$$

$$\begin{gathered} u_{\mathrm{selected}} = \min_{u_k}\{\max_{\mu} F(\lambda_0, \mu, u_k) - \min_{\mu} F(\lambda_0, \mu, u_k)\} \\ u_k \in \{M_r, M_i, \varphi_r, \varphi_i\},\ \mu \in \Omega(\mu) \end{gathered} \tag{29}$$

(a) Equation (27) requires that the operating condition, where the HVDC damping control is least effective, is selected for the design of the controller.
(b) For the efficient operation of the HVDC damping function. The required damping should be provided at minimum control cost.
(c) A good design of damping controller requires that it provides a steady damping over all the range of power system operating conditions.

Furthermore, from Eq. (26) we can see that the phase compensation method can be used to set the parameters of the HVDC damping controller.

2.3 Neural Network Controller

This paper an adaptive neural controller (NN) is used as shown in Fig. 4. This adaptive neural controller is consisted from two separate neural networks as identifier [29] and controller described in following.

Structure of neural identifier is shown in Fig. 4. This network has four neuron at hidden and one at output layer. f is activation function that is hyperbolic tangent in this paper. It is trained using error back propagation method that described in detail in following [26]. Cost function is defined as:

$$E_{\text{id}} = \frac{1}{2}(\Delta\omega - \Delta\hat{\omega})^2 = \frac{1}{2}e_{\text{id}}^2 \tag{30}$$

where $\Delta\omega$ and $\Delta\hat{\omega}$ are power system (i.e., rotor speed deviation) and neural identifier

output, respectively.

$$\frac{\partial E_{\text{id}}}{\partial w_{\text{oh}}^{\text{id}}} = \frac{\partial E_{\text{id}}}{\partial e_{\text{id}}} \frac{\partial e_{\text{id}}}{\partial(\Delta\hat{\omega})} \frac{\partial(\Delta\hat{\omega})}{\partial v} \frac{\partial v}{\partial w_{\text{oh}}^{\text{id}}} \tag{31}$$

where $w_{\text{oh}}^{\text{id}}$ are weights between output and hidden layer. Using Eq. (28), the sensitive coefficient of output neuron is calculated and output weights are updated according Eq. (26).

$$w_{\text{oh,New}}^{\text{id}} = w_{\text{oh,Old}}^{\text{id}} - \eta \frac{\partial E_{\text{id}}}{w_{\text{oh}}^{\text{id}}} \tag{32}$$

Structure of neural controller is shown in Fig. 5. This is a feed forward network including four neuron at hidden and one neuron at output layer. Back propagation method used to train this network as described in following [26]. Cost function to training this network is:

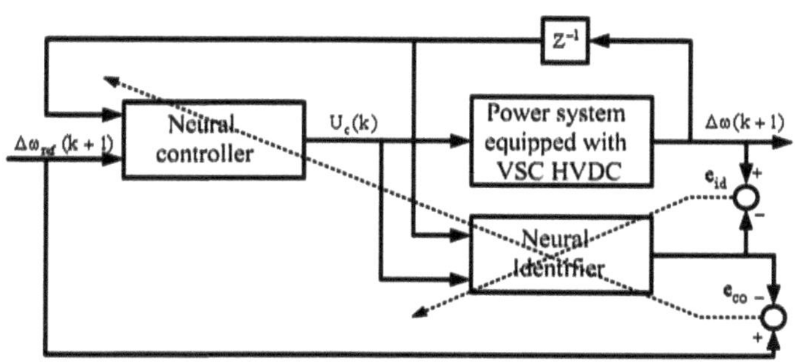

Fig. 4 Structure of the online neural controller

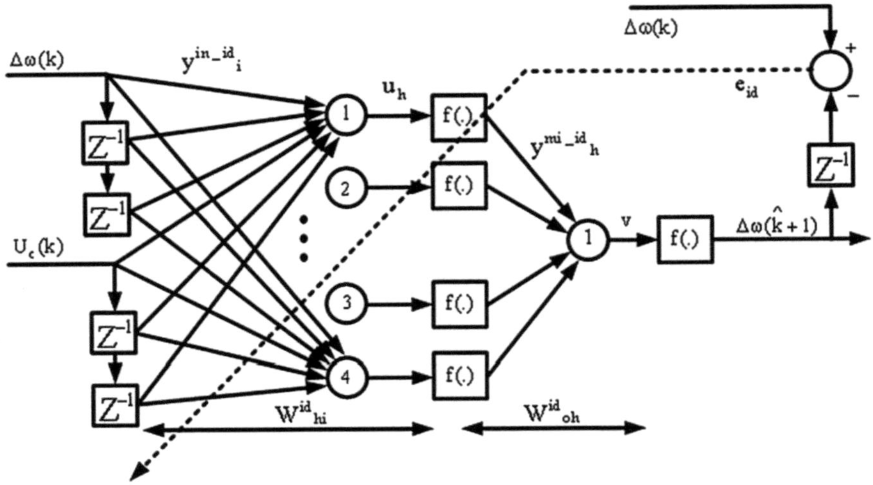

Fig. 5 Structure of the online neural identifier

$$E_{\rm co} = \frac{1}{2}(0 - \Delta\hat{\omega})^2 = \frac{1}{2}\Delta\hat{\omega}^2 = \frac{1}{2}e_{\rm co}^2 \tag{33}$$

$$\frac{\partial E_{\rm co}}{\partial w_{\rm oh}^{\rm co}} = \frac{\partial E_{\rm co}}{\partial e_{\rm co}} \frac{\partial e_{\rm co}}{\partial(\Delta\hat{\omega})} \frac{\partial(\Delta\hat{\omega})}{\partial v} \frac{\partial v}{\partial w_{\rm oh}^{\rm co}} \tag{34}$$

where V, $w_{\rm oh}^{\rm co}$ are the neural identifier output and the weights between output and hidden layer of neural controller. Using Eqs. (27)–(31), it is possible to calculate the sensitive coefficient in output neuron of neural controller and correct the middle and output weights of neural controller (Fig. 6).

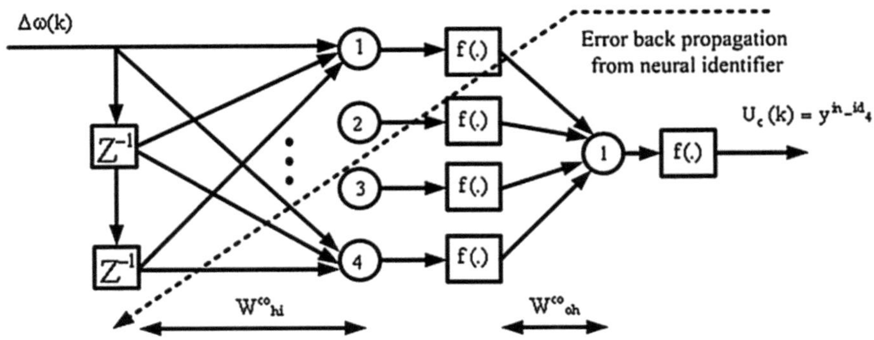

Fig. 6 Structure of the online neural controller

3 Simulation Results

Power system information is given in Appendix 1. Constant coefficients in Eq. (23) are calculated according information's which given in [26, 28]. For given information, characteristic equation is:

$$\Delta(s) = s^5 + 67.13s^4 + 684.7s^3 + 799.9s^2 + 9089s + 28180$$

with poles: $-54.8593, 1.2442 \pm 3.8027j, -2.6505, -12.1114$.

According to the above, there are two poles with positive real part and power system is unstable (for $\Delta P_e = 0.05$). Lead/lag compensator are designed using the phase compensation method. Parameters of these controllers are given in Table 1.

To assess the effectiveness of the damping controller two different conditions are considered according Table 2. Rotor speed deviation and electrical power for suddenly change in mechanical power $(\Delta P_m = 0.05)$.

For SVD analysis, P_e range is from 0.01 to 1.5 pu and $Q_e = [-0.3, 0, 0.3]$. At each loading condition, the system model is linearized, the EM mode is identified, and the SVD-based controllability measure is implemented. For comparison purposes, the minimum singular value for all inputs at $Q_e = -0.3, 0.0$ and 0.3 pu are shown in Fig. 7, respectively. From these figures, the following can be noticed:

- EM mode controllability via φ_r always is higher than that of any other input.
- The capabilities of $\varphi_r, u_{PSS}, M_r, \varphi_i$ to control the EM mode is higher than of M_i.
- All control signals have low EM controllability in low load condition except φ_r.

The RGA results are shown in Fig. 8. It can be result that between VSC HVDC inputs, phase angel of rectifier has good controllability in low frequencies (0–5 Hz) rather to other inputs. This input is selected to apply supplementary NN controller.

Figures 9 and 10 show the damping of the oscillation mode over the $\Omega(\mu)$. It can be seen that, at the heavy load operating condition, the oscillation mode is of poorest damping. Therefore, this operating condition is selected to design the damping controller.

Table 1 Parameters of designed controllers

	M_r	PH_r
T_1	0.2	0.14
T_2	0.31	0.41
K_{dc}	–53.01	–72.1

Table 2 Load condition

Load condition	P_e(pu)	Q_e(pu)
μ_1	0.8	0.015
μ_2	1.1	0.3

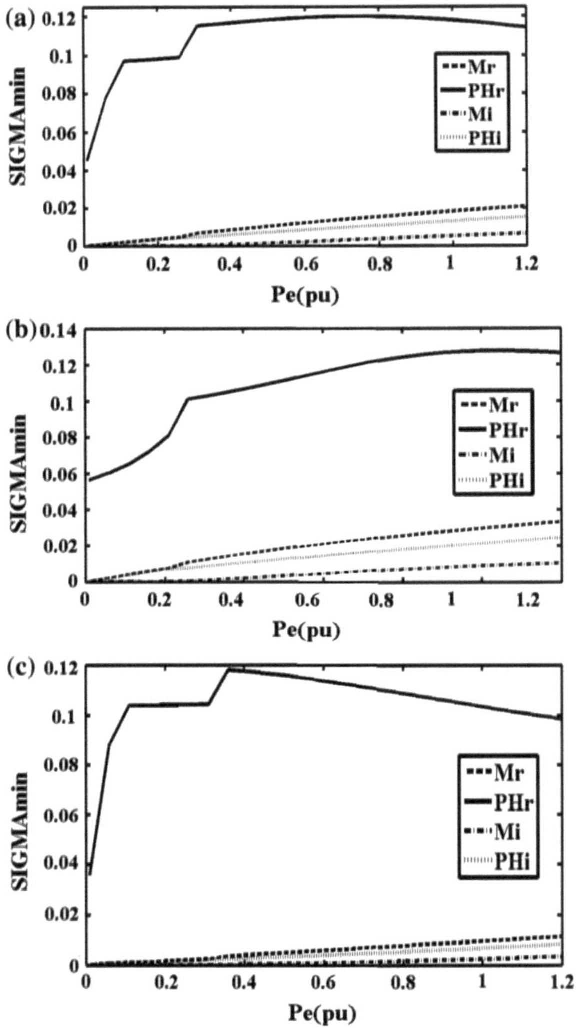

Fig. 7 SVD results for five inputs. **a** $Q_e = 0$, **b** $Q_e = 0.3$, **c** $Q_e = -0.3$

- The results of calculation of the forward path $F(\lambda)$ over $\Omega(\mu)$ as shown by Fig. 9. With $u_k = M_i$ we have $F(\lambda) \approx 0$ over $\Omega(\lambda)$. Therefore, the oscillation mode controllability is not controllable if the input control signal is chosen to be M_i. In the following M_i will not be included in the discussion.
- According to the criteria of Eq. (24), it can be seen that the operating condition to be selected for the design of the HVDC damping controller is $P_e = 0.1$ pu.
- Figure 10 indicates $u_k = \varphi_r$ is most effective input control signal. So the criteria of Eq. (24) lead to the selection of the input control signal for the UPFC damping controller as $u_k = \varphi_r$.

Fig. 8 RGA results

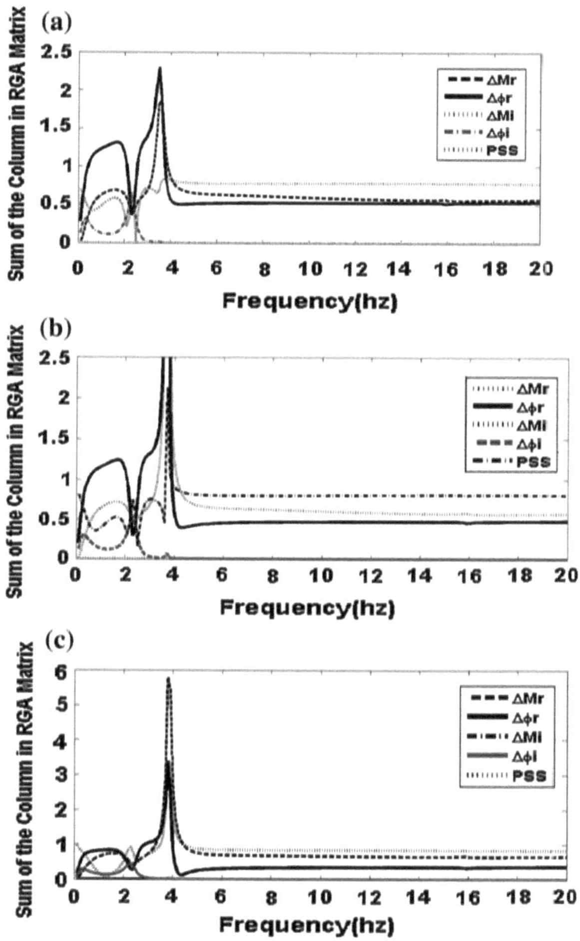

- The results of applying the criteria of Eq. (29) are:

(a) With $u_k = \varphi_r$: $\frac{\max\limits_{\mu} F(\lambda_0,\mu,u_k) - \min\limits_{\mu} F(\lambda_0,\mu,u_k)}{\min\limits_{\mu} F(\lambda_0,\mu,u_k)} = 0.22$

(b) With $u_k = M_r$: $\frac{\max\limits_{\mu} F(\lambda_0,\mu,u_k) - \min\limits_{\mu} F(\lambda_0,\mu,u_k)}{\min\limits_{\mu} F(\lambda_0,\mu,u_k)} = 3.166$

(c) With $u_k = \varphi_i$: $\frac{\max\limits_{\mu} F(\lambda_0,\mu,u_k) - \min\limits_{\mu} F(\lambda_0,\mu,u_k)}{\min\limits_{\mu} F(\lambda_0,\mu,u_k)} = 1.5$

Therefore, with $u_k = \varphi_r$, the HVDC damping controller provides the smoothest damping to the oscillation mode. The final result of selection is:

$$\mu_{\text{selected}} : P_e = 0.1 \text{ and } u_k = \varphi_r$$

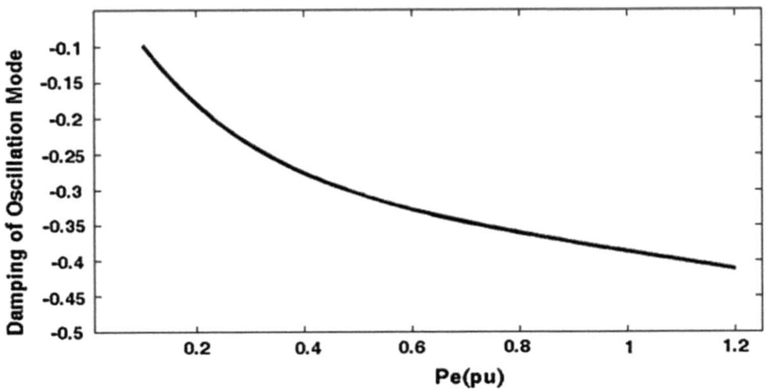

Fig. 9 Damping of oscillation mode over the operating condition

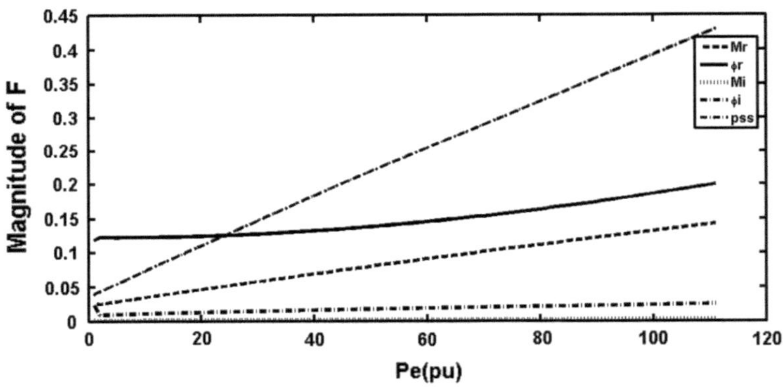

Fig. 10 Magnitude of $F(\lambda)$ over $\Omega(\mu)$

According to Fig. 9 in μ_{selected}, magnitude of F is very low. With attention to Eq. (24), this reason decrease D_{HVDC}. So, we select $\mu : P_e = 0.9\,\text{pu}$ for damping controller designing.

Designed controllers are applied to every input in Fig. 2 individually. It is observed that if controller is applied to ΔPH_r, rotor speed oscillation is damped better than others input (because of RGA, SVD and DF results). This supplementary controller is caused the system stability and damping oscillations as shown in Fig. 11. The Fig. 11 also shows that supplementary neural controller has good operation rather the lead/lag compensator.

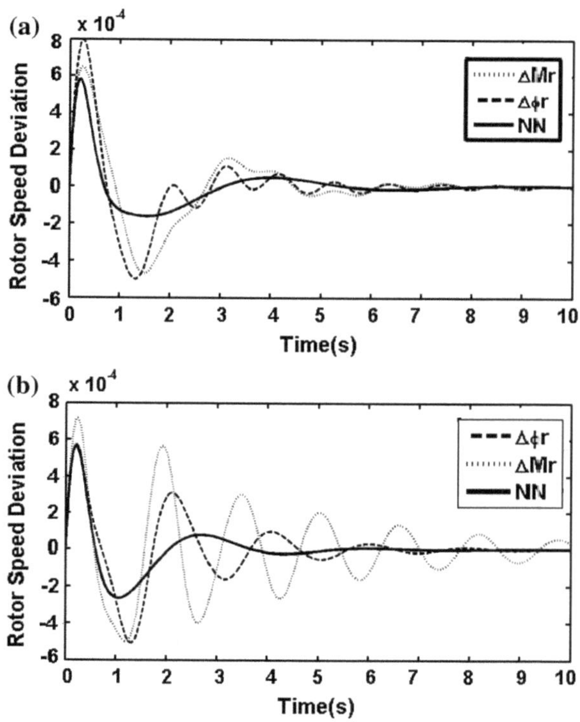

Fig. 11 Rotor speed deviation. **a** μ_1 condition, **b** μ_2 condition

4 Enhancing of Power System Stability in VSC SMIB Network

VSC HVDC systems may improve the transient and dynamic performances of interconnected AC/DC system due to its fast electronic control of power flow and also the transient stability of AC systems in an AC–DC system can be improved by taking advantage of the fast controllability of VSC HVDC converters [20–24].

In this section, a novel approach is presented to model parallel AC/DC power system namely Phillips–Heffron model based d-q algorithm in order to studying system dynamical stability. In addition, a block diagram representation is formed to analyze the system stability characteristics. By this modeling approach, it is possible to analyze the small-signal stability of the system and low-frequency oscillation phenomena which is caused by external disturbances such as variation of input torque and fault occurring.

In order to enhance dynamical stability of power system, a supplementary signal which is the same as that applied for FACTS devices, is superimposed on the main input control signals in this paper. To measure the controllability of VSC HVDC supplementary controller by a given input (control signal), the singular value decomposition (SVD) is employed.

4.1 Configuration of Power System

Figure 12 shows a SMIB network equipped with a detailed VSC HVDC. As it can be seen the infinite bus is supplied by HVAC parallel connected with a VSC HVDC power transmission system. The VSC HVDC consists of two coupling transformer, two three-phase IGBT based voltage source converters (VSCs). These two converters are connected either back-to-back or joined by a DC cable, depending on the application. The AC side of each converter is connected to the line through a coupling transformer. The first voltage source converter behaves as a rectifier. It regulates the DC link voltage and maintains the magnitude of the voltage at the connected terminal. The second voltage source converter acts as a controlled voltage source, which controls power flow in VSC HVDC feeder. The four input control signals to the VSC HVDC are M_r, φ_r, M_i, φ_i where M_r, M_i are the amplitude modulation ratio and φ_r, φ_i are phase angle of the control signals of each VSC, respectively.

4.1.1 Power System Nonlinear Model

By applying Park's transformation and neglecting the resistance and transients of the coupling transformers, the detailed VSC HVDC can be modeled similar to Eqs. (1)–(3):

$$\begin{bmatrix} V_{\mathrm{ld}} \\ V_{\mathrm{lq}} \end{bmatrix} = \begin{bmatrix} 0 & x_r \\ -x_r & 0 \end{bmatrix} \begin{bmatrix} I_{\mathrm{rd}} \\ I_{\mathrm{lrq}} \end{bmatrix} + \begin{bmatrix} \dfrac{M_r V_{\mathrm{dcr}} \cos(\varphi_r)}{2} \\ \dfrac{M_r V_{\mathrm{dcr}} \sin(\varphi_r)}{2} \end{bmatrix} \tag{35}$$

$$\begin{bmatrix} V_{\mathrm{bd}} \\ V_{\mathrm{bq}} \end{bmatrix} = \begin{bmatrix} 0 & x_i \\ -x_i & 0 \end{bmatrix} \begin{bmatrix} I_{\mathrm{id}} \\ I_{\mathrm{iq}} \end{bmatrix} + \begin{bmatrix} \dfrac{M_i V_{\mathrm{dci}} \cos(\varphi_i)}{2} \\ \dfrac{M_i V_{\mathrm{dci}} \sin(\varphi_i)}{2} \end{bmatrix} \tag{36}$$

$$C_{\mathrm{dc}} \dot{V}_{\mathrm{dc}} = -(I_1 + I_2) \tag{37}$$

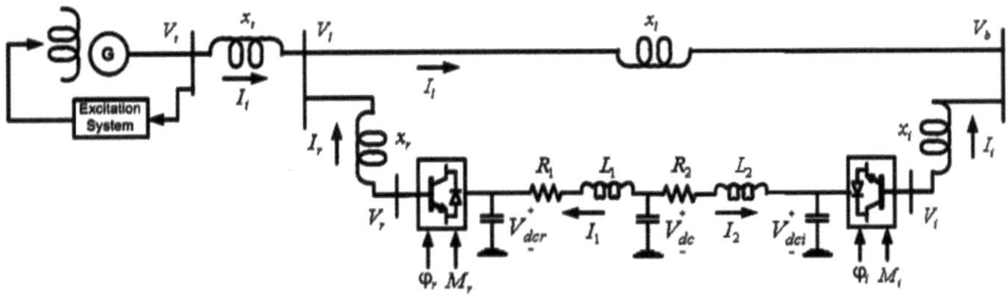

Fig. 12 A SMIB network equipped with a detailed VSC HVDC

$$L_1 \frac{dI_1}{dt} = V_{dc} - V_{dcr} - R_1 I_1 \tag{38}$$

$$L_2 \frac{dI_2}{dt} = V_{dc} - V_{dci} - R_2 I_2 \tag{39}$$

where V_l, V_b, I_r and I_i are the middle bus voltage, infinite bus voltage, flowed current to rectifier, and inverter, respectively. C And V_{dc} are the DC link capacitance and voltage, respectively. C_r, C_i, V_{dcr} and V_{dci} are the DC capacitances and voltages of rectifier and inverter, respectively. The nonlinear model of the SMIB system of Fig. 12 similar to Eqs. (4)–(7) is:

$$\dot{\delta} = \omega_b(\omega - 1) \tag{40}$$

$$\dot{\omega} = \frac{(P_m - P_e - D\omega)}{M} \tag{41}$$

$$\dot{E}'_q = \frac{(E_{fd} - (x_d - x'_d)I_t - E'_q)}{T'_{do}} \tag{42}$$

$$\dot{E}_{fd} = \frac{(K_A(V_{ref} - V_t + u_{PSS}) - E_{fd})}{T_A} \tag{43}$$

where

$$P_e = V_{td}I_{td} + V_{tq}I_{tq}$$

$$V_t = \sqrt{V_{td}^2 + V_{tq}^2}$$

$$V_{td} = x_q I_{tq}$$

$$V_{tq} = E'_q - x'_d I_{td}$$

$$I_{td} = I_{ld} - I_{rd}$$

$$I_{tq} = I_{lq} - I_{rq}$$

where P_m and P_e are the input and output power, respectively; M and D the inertia constant and damping coefficient, respectively; ω_b the synchronous speed; δ and ω the rotor angle and speed, respectively; E'_q, E_{fd} and V_t the generator internal, field, and terminal voltages, respectively; T'_{do} the open circuit field time constant; x_d, x'_d and x_q the d-axis, d-axis transient reactance, and q-axis reactance, respectively; K_A and T_A the exciter gain and time constant, respectively; V_{ref} the reference voltage. Effects of PSS can be considered by adding u_{pss} to the Eq. (43). Also, from Fig. 12, similar to Eqs. (8)–(10), we have:

$$\bar{V}_t = jx_t\bar{I}_t + \bar{V}_l \tag{44}$$

$$\bar{V}_t = jx_t\bar{I}_t + jx_l\bar{I}_l + \bar{V}_b \tag{45}$$

$$\bar{I}_l = \bar{I}_t - \frac{\bar{V}_t - jx_t\bar{I}_t - \bar{V}_r}{jx_r} \tag{46}$$

where $\bar{I}_t$, $\bar{V}_r$, $\bar{I}_l$ and $\bar{V}_b$ are the armature current, rectifier voltage, infinite bus current and voltage, respectively. From Eqs. (44)–(46) we can have:

$$I_{\text{tq}} = \frac{\frac{x_l}{x_r}\frac{M_r}{2}V_{\text{dcr}}\cos(\varphi_r) + V_b\sin(\delta)}{Zx_q + A} \tag{47}$$

$$I_{\text{td}} = \frac{ZE'_q - \frac{x_l}{x_r}\frac{M_r}{2}V_{\text{dcr}}\sin(\varphi_r) - V_b\cos(\delta)}{Zx'_d + A} \tag{48}$$

and for inverter side:

$$I_{\text{id}} = \frac{-V_b\cos(\delta) + \frac{M_i}{2}V_{\text{dci}}\sin(\varphi_i)}{x_i} \tag{49}$$

$$I_{\text{iq}} = \frac{V_b\sin(\delta) - \frac{M_i}{2}V_{\text{dci}}\cos(\varphi_i)}{x_i} \tag{50}$$

By linearising Eqs. (35)–(41) and (47)–(50), similar to Eqs. (15)–(18):

$$\Delta\dot{\delta} = \omega_b\Delta\omega \tag{51}$$

$$\Delta\dot{\omega} = \frac{(\Delta P_m - \Delta P_e - D\Delta\omega)}{M} \tag{52}$$

$$\Delta\dot{E}'_q = \frac{(\Delta E_{\text{fd}} - (x_d - x'_d)\Delta I_{\text{td}} - \Delta E'_q)}{T'_{\text{do}}} \tag{53}$$

$$\Delta\dot{E}_{\text{fd}} = \frac{(K_A(\Delta V_t + \Delta u_{\text{PSS}}) - \Delta E_{\text{fd}})}{T_A} \tag{54}$$

where

$$\Delta\dot{V}_t = K_5\Delta\delta + K_6\Delta E'_q + K_{\text{vdcr}}\Delta V_{\text{dcr}} + K_{vM_r}\Delta M_r + K_{v\varphi r}\Delta\varphi_r \tag{55}$$

$$\Delta P_e = K_1\Delta\delta + K_2\Delta E'_q + K_{\text{pdcr}}\Delta V_{\text{dc}} + K_{pM_r}\Delta M_r + K_{p\varphi r}\Delta\varphi_r \tag{56}$$

$$\Delta E_q = K_3\Delta E'_q + K_4\Delta\delta + K_{q\varphi r}\Delta\varphi_r + K_{qM_r}\Delta M_r + K_{\text{qdcr}}\Delta V_{\text{dcr}} \tag{57}$$

$$\Delta\dot{V}_{\text{dcr}} = \frac{C_{31}}{C_r}\Delta\delta + \frac{C_{32}}{C_r}\Delta E'_q + \frac{C_{33}}{C_r}\Delta V_{\text{dcr}} + \frac{1}{C_r}\Delta I_1 + \frac{C_{34}}{C_r}\Delta M_r + \frac{C_{35}}{C_r}\Delta\varphi_r \tag{58}$$

Substitute Eqs. (55)–(57) in (51)–(54), we can obtain the state variable of the power system installed with the detailed VSC HVDC to be (state-space model):

$$\dot{X} = \mathrm{AX} + \mathrm{BU} \tag{59}$$

where $X = [\Delta\delta, \Delta\omega, \Delta E'_q, \Delta E_{\mathrm{fd}}, \Delta V_{\mathrm{dcr}}, \Delta_{\mathrm{I1}}, \Delta V_{\mathrm{dc}}, \Delta I_2, \Delta V_{\mathrm{dci}}]^T$ and $U = [\Delta M_r, \Delta\varphi_r, \Delta M_i, \Delta\varphi_i, u_{\mathrm{PSS}}]^T$ where $\Delta M_i, \Delta M_r, \Delta\varphi_i, \Delta\varphi_r$ and u_{PSS} are the linearization of the input control signals of the VSC HVDC and PSS output, respectively. The linearised dynamic model of Eq. (59) can be shown by Fig. 13. In this figure K_{pu}, K_{qu}, K_{vu}, K_r and K_i are defined below:

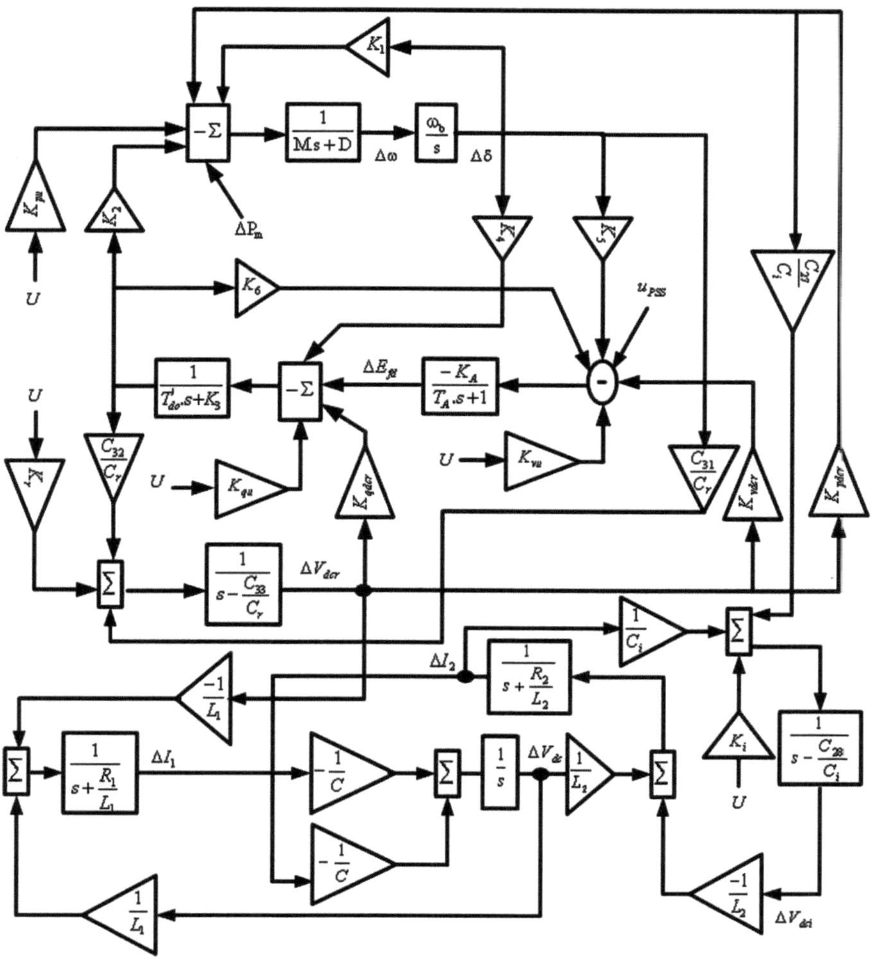

Fig. 13 VSC HVDC block diagram based Eq. (59)

$$K_{pu} = [K_{pM_r}, K_{p\varphi r}, 0, 0, 0]^T,\ K_{qu} = [K_{qM_r}, K_{q\varphi r}, 0, 0, 0]^T,\ K_{vu} = [K_{vM_r}, K_{v\varphi r}, 0, 0, 0]^T$$

$$K_r = [\frac{C_{34}}{C_r}, \frac{C_{35}}{C_r}, 0, 0, 0]^T,\ K_i = [0, 0, \frac{C_{29}}{C_i}, \frac{C_{30}}{C_i}, 0]$$

$$A = \begin{bmatrix} 0 & \omega b & 0 & 0 & 0 & 0 & 0 & 0 & 0 \\ -\frac{K_1}{M} & -\frac{D}{M} & -\frac{K_2}{M} & 0 & -\frac{K_{pdcr}}{M} & 0 & 0 & 0 & 0 \\ -\frac{K_4}{T'_{do}} & 0 & -\frac{K_3}{T'_{do}} & \frac{1}{T'_{do}} & -\frac{K_{qdcr}}{T'_{do}} & 0 & 0 & 0 & 0 \\ -\frac{K_A K_5}{T_A} & 0 & -\frac{K_A K_6}{T_A} & -\frac{1}{T_A} & -\frac{K_A K_{vdcr}}{T_A} & 0 & 0 & 0 & 0 \\ \frac{C_{31}}{C_r} & 0 & \frac{C_{32}}{C_r} & 0 & \frac{C_{33}}{C_r} & \frac{1}{C_r} & 0 & 0 & 0 \\ 0 & 0 & 0 & 0 & -\frac{1}{L_1} & -\frac{R_1}{L_1} & \frac{1}{L_1} & 0 & 0 \\ 0 & 0 & 0 & 0 & 0 & -\frac{1}{C} & 0 & -\frac{1}{C} & 0 \\ 0 & 0 & 0 & 0 & 0 & 0 & \frac{1}{L_2} & -\frac{R_2}{L_2} & -\frac{1}{L_2} \\ \frac{C_{27}}{C_i} & 0 & 0 & 0 & 0 & 0 & 0 & -\frac{1}{C_i} & \frac{C_{28}}{C_i} \end{bmatrix}$$

$$B = \begin{bmatrix} 0 & 0 & 0 & 0 & 0 \\ -\frac{K_{pM_r}}{M} & -\frac{K_{p\varphi r}}{M} & 0 & 0 & 0 \\ -\frac{K_{qM_r}}{T'_{do}} & -\frac{K_{q\varphi r}}{T'_{do}} & 0 & 0 & 0 \\ -\frac{K_{vM_r} K_A}{T_A} & -\frac{K_{v\varphi r} K_A}{T_A} & 0 & 0 & \frac{K_A}{T_A} \\ \frac{C_{34}}{C_r} & \frac{C_{35}}{C_r} & 0 & 0 & 0 \\ 0 & 0 & 0 & 0 & 0 \\ 0 & 0 & 0 & 0 & 0 \\ 0 & 0 & 0 & 0 & 0 \\ 0 & 0 & \frac{C_{29}}{C_i} & \frac{C_{30}}{C_i} & 0 \end{bmatrix}$$

It can be seen that the configuration of the Phillips–Heffron model is exactly the same as that installed with SVC, TCSC, TCPS, UPFC, and STATCOM.

Also from Eq. (59) it can be seen that there five choice of input control signals of the VSC HVDC to superimpose on the damping function of the VSC HVDC $\Delta M_i, \Delta M_r, \Delta\varphi_i, \Delta\varphi_r$ and u_{PSS}.therefore, in designing the damping controller of the VSC HVDC, besides setting its parameters, the selection of the input control signal of the VSC HVDC to superimpose on the DF of the VSC HVDC is also important.

4.1.2 Controllability Measure

To measure the controllability of the EM mode by a given input (control signal), the singular value decomposition (SVD) is employed [8]. Mathematically, if G is an $m \times n$ complex matrix, then there exist unitary matrices $U$and V with dimensions of $m \times m$ and$n \times n$, respectively, such that:

$$G = U\Sigma V^H \quad \text{where} \quad \Sigma = \begin{bmatrix} \Sigma_1 & 0 \\ 0 & 0 \end{bmatrix}, \quad \Sigma_1 = \text{diag}(\sigma_1, \ldots, \sigma_r) \quad \text{with} \quad \sigma_1 \geq \cdots$$

$\geq \sigma_r \geq 0$ where $r = \min\{m, n\}$ and$\sigma_1, \ldots, \sigma_r$ are the singular values of G.

The minimum singular value σ_r represents the distance of the matrix G from all the matrices with a rank of $r - 1$. This property can be used to quantify modal controllability [32, 33]. The matrix Hcan be written as $H = [h_1 h_2 h_3 h_4]$ where h_i is a column vector corresponding to the ith input.

The minimum singular value, $\sigma_{\min}$ of the matrix $[\lambda I - A, h_i]$ indicates the capability of the ith input to control the mode associated with the eigenvalue λ. Actually, the higher $\sigma_{\min}$, the higher the controllability of this mode by the input considered. As such, the controllability of the EM mode can be examined with all inputs in order to identify the most effective one to control the mode.

4.1.3 Design of Damping Controllers

The damping controllers are designed to produce an electrical torque in phase with the speed deviation. The four control parameters of the VSC HVDC ($\Delta M_i, \Delta M_r, \Delta\varphi_i$ and $\Delta\varphi_r$) can be modulated in order to produce the damping torque. The speed deviation $\Delta\omega$ is considered as input to the damping controllers.

The structure of VSC HVDC based damping controller is shown in Fig. 3. It consists of gain, signal washout ,and phase compensator blocks. The parameters of the damping controller are obtained using the phase compensation technique 9. The detailed step-by-step procedure for computing the parameters of the damping controllers using phase compensation technique is given below:

(1) Computation of oscillation natural frequency $\omega_n = \sqrt{\frac{K_1 \omega_0}{M}}$ from mechanical loop
(2) Computation of $\angle GEPA$ (Phase lag between Δu,ΔP_e) at $s = j\omega_n$. Let it be γ.
(3) Design of phase lead/lag compensator G_C:

The phase lead/lag compensator G_C is designed to provide the required degree of phase compensation. For 100 % phase compensation,

$$\angle G_C(j\omega_n) + \angle \mathrm{GEPA}(j\omega_n) = 0 \tag{60}$$

Assuming one lead-lag network, $T_1 = aT_2$, the transfer function of the phase compensator becomes,

$$G_C(s) = \frac{1 + saT_2}{1 + sT_2} \tag{61}$$

Since the phase angle compensated by the lead-lag network is equal to $-\gamma$, the parameters a and T_2 are computed as,

$$a = \frac{1 + \sin(\gamma)}{1 - \sin(\gamma)}, \quad T_2 = \frac{1}{\omega_n \sqrt{a}} \tag{62}$$

(4) Computation of optimum gain K_{dc}.

The required gain setting K_{dc} for the desired value of damping ratio $\zeta = 0.5$ is obtained as,

$$K_{dc} = \frac{2\zeta\omega_n M}{|G_C(s)||\mathrm{GEPA}(s)|} \tag{63}$$

where $|G_C(s)|$ and $|\mathrm{GEPA}(s)|$ are evaluated at $s = j\omega_n$.

The signal washout is the high pass filter that prevents steady changes in the speed from modifying the VSC HVDC input parameter. The value of the washout time constant T_w should be high enough to allow signals associated with oscillations in rotor speed to pass unchanged. From the viewpoint of the washout function, the value of T_w is not critical and may be in the range of 1–20 s.

4.2 Simulation Results

As Power system information is given in Appendix 2. Constant coefficients of the linearised dynamic model of Eq. (59) are calculated according information who given in Appendix 3. For given information, poles of the VSC HVDC system are:

$-17.0984 + 6.6503i, -17.0984 - 6.6503i,$
$-0.1670 + 15.8881i, -0.1670 - 15.8881i, -0.0182 + 3.6842i, -0.0182$
$-3.6842i, 0.7675 + 2.3900i, 0.7675 - 2.3900i, -1.4115.$

According above, there are two poles with positive real part and power system is unstable. SVD is employed to measure the controllability of the electromechanical mode (EM) mode from each of the five inputs: M_r, φ_r, M_i, φ_i, u_{PSS}. The minimum singular value $\sigma_{\min}$ is estimated over a wide range of operating conditions. For SVD analysis, P_e ranges from 0.01 to 1.5 Pu and $Q_e = [-0.3, 0, 0.3]$.

At each loading condition, the system model is linearized, the EM mode is identified, and the SVD-based controllability measure is implemented. For comparison purposes, the minimum singular value for all inputs at $Q_e = -0.3$, 0.0 and 0.3Pu is shown in Fig. 14, respectively. From these figures, the following can be noticed:

- EM mode controllability via is φ_r always higher than that of any other input.
- The capabilities of φ_r, u_{PSS}, M_r, φ_i to control the EM mode is higher than of M_i.
- All control signals have low EM controllability in low load condition except φ_r.

To assess the effectiveness of the proposed stabilizers two different conditions are considered according Table 3. Rotor speed deviation and load angle for suddenly change in input power $(\Delta P_m = 0.1)$ at $t = 3$ s is shown in Fig. 15. Supplementary controller is applied φ_r input in Fig. 12 according to SVD results. Simulation results are shown in Figs. 15 and 16.

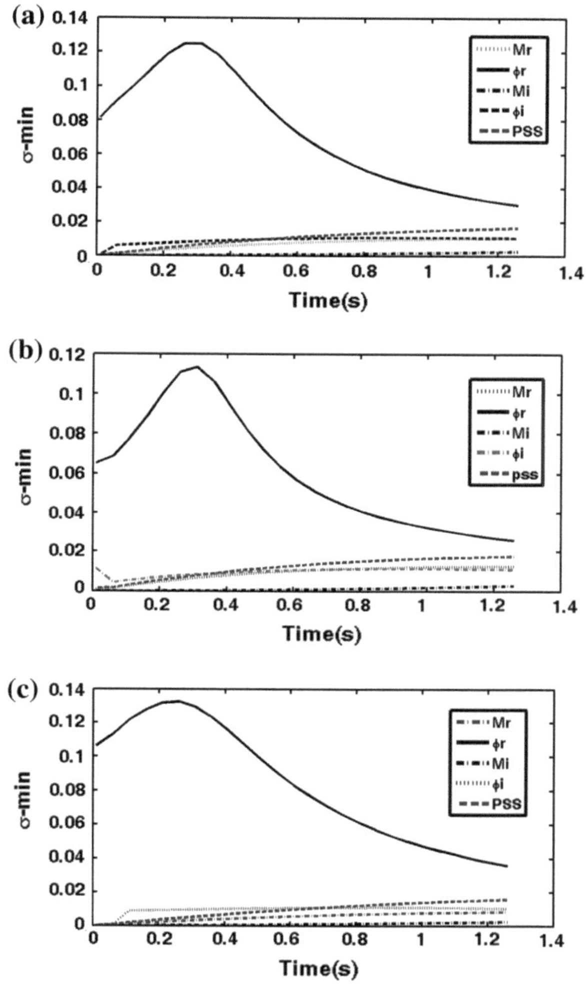

Fig. 14 SVD results for five inputs. **a** $Q_e = 0$, **b** $Q_e = 0.3$, **c** $Q_e = -0.3$

Table 3 Load condition

Loading	P_e(pu)	Q_e(pu)
Nominal	0.9	0.1
Heavy	1.1	0.4

5 Improvement of Power System Stability in UPFC SMIB Network

An adaptive controller is able to control a nonlinear system with fast changing dynamics, like the power system better, since the dynamics of a power system are continually identified by a model. Advantages of online adaptive controllers over

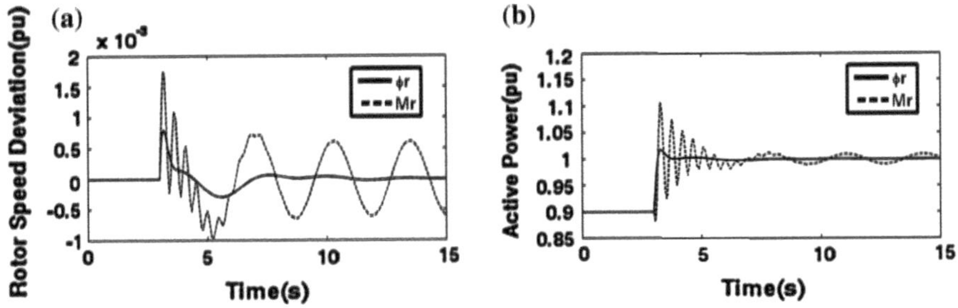

Fig. 15 Nominal results. **a** Rotor speed deviation, **b** Active power

Fig. 16 Heavy load results. **a** Rotor speed deviation for φ_r. **b** Active power for φ_r, **c** Rotor speed deviation for M_r

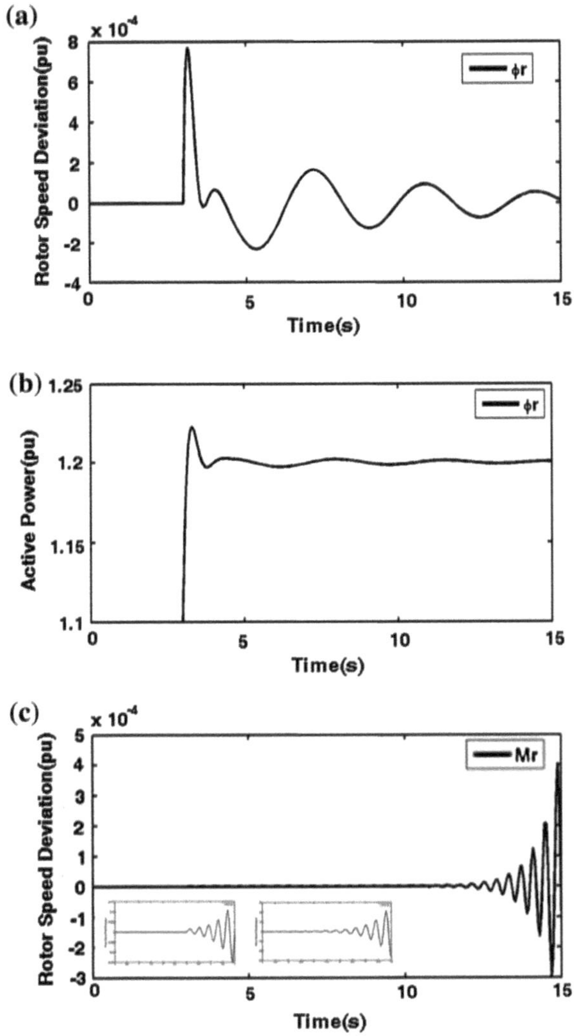

conventional controllers are that they are able to adapt to changes in system operating conditions automatically, unlike conventional controllers whose performance is degraded by such changes and require retuning in order to provide the desired performance [11]. In [34], an adaptive based controller for STATCOM has been provided and has been used as a VAR compensator in [35–39].

5.1 The Power System Case Study

Figure 17 shows a single-machine-infinite bus (SMIB) system installed with UPFC. The static excitation system model type IEEE-ST1A has been considered. The UPFC considered here is assumed to be based on pulse width modulation (PWM) converters. The UPFC is a combination of a static synchronous compensator (STATCOM) and a static synchronous series compensator (SSSC) which is coupled via a common dc link.

5.2 State-Space Equations of Power System

If the general pulse width modulation (PWM) is adopted for GTO-based VSCs, the three-phase dynamic differential equations of the UPFC, similar to Eqs. (51)–(54), are [17]:

$$\Delta\dot{\delta} = \omega_b \Delta\omega \tag{64}$$

$$\Delta\dot{\omega} = \frac{(\Delta P_m - \Delta P_e - D\Delta\omega)}{M} \tag{65}$$

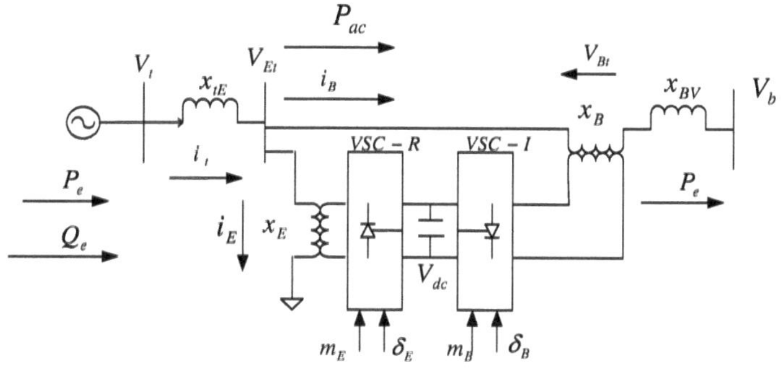

Fig. 17 UPFC installed in a SMIB system

$$\Delta\dot{E}'_q = \frac{(\Delta E_{\text{fd}} - (x_d - x'_d)\Delta I_{\text{td}} - \Delta E'_q)}{T'_{\text{do}}} \tag{66}$$

$$\Delta\dot{E}_{\text{fd}} = \frac{(K_A(\Delta V_{\text{ref}} - \Delta V_t + \Delta u_{\text{PSS}}) - \Delta E_{\text{fd}})}{T_A} \tag{67}$$

$$\Delta\dot{V}_{\text{dc}} = K_7\Delta\delta + K_8\Delta E'_q - K_9\Delta V_{\text{dc}} + K_{\text{ce}}\Delta m_E + K_{c\delta e}\Delta\delta_E + K_{\text{cb}}\Delta m_B + K_{c\delta b}\Delta\delta_B \tag{68}$$

The equations below can be obtained with a line arising from Eq. (68).

$$\Delta P_e = K_1\Delta\delta + K_2\Delta E'_q + K_{\text{qd}}\Delta V_{\text{dc}} + K_{\text{qe}}\Delta m_E + K_{q\delta e}\Delta\delta_E + K_{\text{qb}}\Delta m_B + K_{q\delta b}\Delta\delta_B \tag{69}$$

$$\Delta E'_q = K_4\Delta\delta + K_3\Delta E'_q + K_{\text{qd}}\Delta V_{\text{dc}} + K_{\text{qe}}\Delta m_E + K_{q\delta e}\Delta\delta_E + K_{\text{qb}}\Delta m_B + K_{q\delta b}\Delta\delta_B \tag{70}$$

$$\Delta V_t = K_5\Delta\delta + K_6\Delta E'_q + K_{\text{vd}}\Delta V_{\text{dc}} + K_{\text{ve}}\Delta m_E + K_{v\delta e}\Delta\delta_E + K_{\text{vb}}\Delta m_B + K_{v\delta b}\Delta\delta_B \tag{71}$$

$$\Delta V_{\text{dc}} = K_7\Delta\delta + K_8\Delta E'_q - K_9\Delta V_{\text{dc}} + K_{\text{ce}}\Delta m_E + K_{c\delta e}\Delta\delta_E + K_{\text{cb}}\Delta m_B + K_{c\delta b}\Delta\delta_B \tag{72}$$

The state-space equations of the system can be calculated by combination of Eqs. (69)–(72) with Eq. (68):

$$\dot{X} = \text{AX} + \text{BU} \tag{73}$$

and

$$X = [\Delta\delta, \Delta\omega, \Delta E'_q, \Delta E_{\text{fd}}, \Delta V_{\text{dc}}]^T$$

$$U = [\Delta u_{\text{pss}},\ \Delta m_E,\ \Delta\delta_E,\ \Delta m_B,\ \Delta\delta_B]^T$$

where Δm_E, Δm_B, $\Delta\delta_E$ and $\Delta\delta_B$ are a linearization of the input control signal of the UPFC and the equations related to the *K* parameters have been presented in Appendix 6. The linearized dynamic model of Eqs. (69)–(72) can be seen in Fig. 18, where there is only one input control signal for *U*. The Fig. 18 includes the UPFC relating the pertinent variables of electric torque, speed, angle, terminal voltage, field voltage, flux linkages, UPFC control parameters, and dc link voltage.

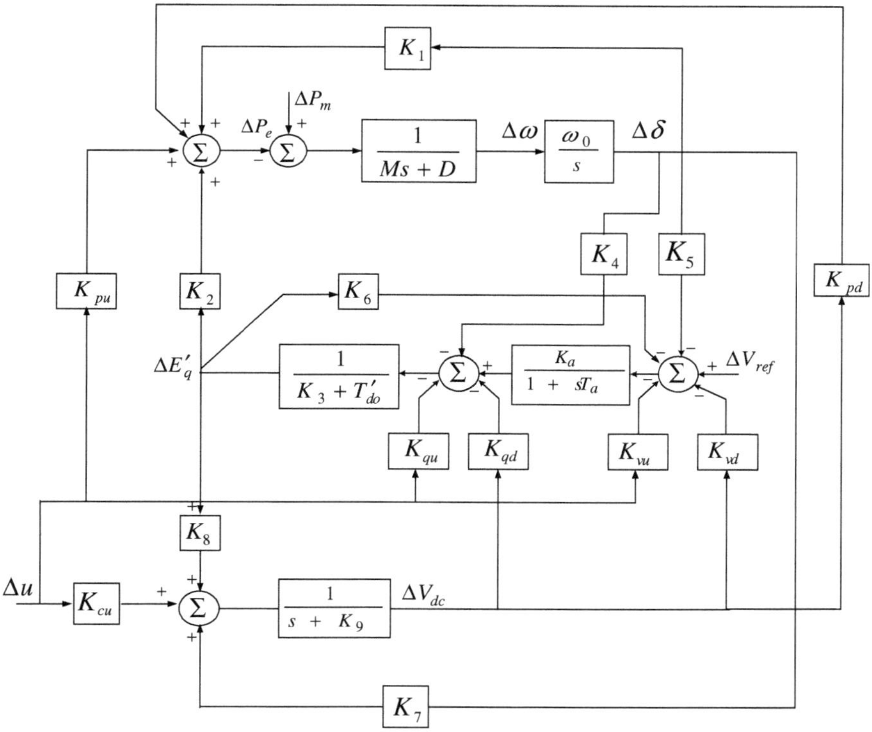

Fig. 18 Modified Heffron–Phillips model of SMIB system with UPFC

$$A = \begin{bmatrix} 0 & \omega_b & 0 & 0 & 0 \\ -\frac{K_1}{M} & -\frac{D}{M} & -\frac{K_2}{M} & 0 & -\frac{K_{\mathrm{pd}}}{M} \\ -\frac{K_4}{M} & 0 & -\frac{K_3}{T'_{\mathrm{do}}} & \frac{1}{T'_{\mathrm{do}}} & -\frac{K_{\mathrm{qd}}}{T'_{\mathrm{do}}} \\ -\frac{K_A K_5}{T_A} & 0 & -\frac{K_A K_6}{T_A} & -\frac{1}{T_A} & -\frac{K_A K_{\mathrm{pd}}}{T_A} \\ K_7 & 0 & K_8 & 0 & -K_9 \end{bmatrix}$$

$$B = \begin{bmatrix} 0 & 0 & 0 & 0 & 0 \\ 0 & -\frac{K_{\mathrm{pe}}}{M} & -\frac{K_{p\delta e}}{M} & -\frac{K_{\mathrm{pb}}}{M} & -\frac{K_{p\delta b}}{M} \\ 0 & -\frac{K_{\mathrm{qe}}}{T'_{\mathrm{do}}} & -\frac{K_{q\delta e}}{T'_{\mathrm{do}}} & -\frac{K_{\mathrm{qb}}}{T'_{\mathrm{do}}} & -\frac{K_{q\delta b}}{T'_{\mathrm{do}}} \\ \frac{K_A}{T_A} & -\frac{K_A K_{\mathrm{ve}}}{T_A} & -\frac{K_A K_{v\delta e}}{T_A} & -\frac{K_A K_{\mathrm{vb}}}{T_A} & -\frac{K_A K_{v\delta b}}{T_A} \\ 0 & K_{\mathrm{ce}} & K_{c\delta e} & K_{\mathrm{cb}} & K_{c\delta b} \end{bmatrix}$$

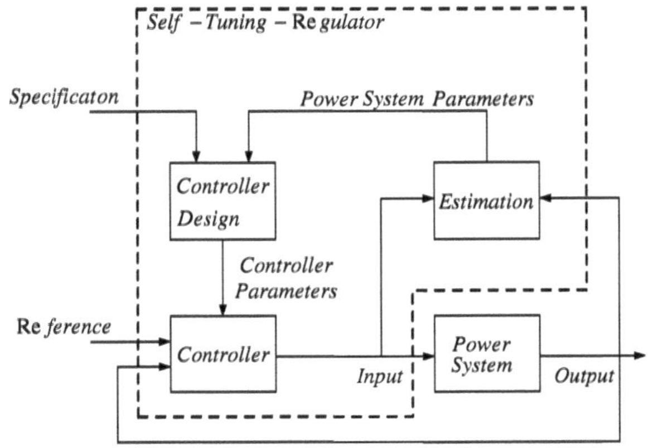

Fig. 19 Block diagram of self-tuning regulator

5.3 Adaptive Controller

Figure 19 shows a block diagram of a process with a self-tuning regulator (STR). The parameters of the power system transfer function are estimated by estimation block with samples taken from input $\Delta\delta_E$ and output $\Delta\omega$ with a specified sampling time [35, 36]. It has been shown as the discrete mod of the state equation of the power system (73) as follows:

$$H(q) = \frac{\Delta\omega(q)}{\Delta\delta_E} = \frac{B(q)}{A(q)} = \frac{b_0q^4 + b_1q^3 + b_2q^2 + b_3q + b_4}{q^5 + a_1q^4 + a_2q^3 + a_3q^2 + a_4q + a_5} \tag{74}$$

The block labeled controller design contains the computation's Diophantine equation required to perform a design of a controller with a specified method and few design parameters that can be chosen externally. The recursive least-square method (RLS) will be used for parameter estimation and the design method is a deterministic pole placement (MDPP). A general linear controller can be described:

$$Ru(t) = Tu_c(t) - Sy(t) \tag{75}$$

where *R*, *S*, and *T* are polynomials. A block diagram of the closed-loop system is shown in Fig. 20. General equations of *R*, *S*, and *T* are polynomials and have been calculated by MDPP as follows:

$$\begin{aligned} R(q) &= q^4 + r_1q^3 + r_2q^2 + r_3q + r_4 \\ S(q) &= s_0q^4 + s_1q^3 + s_2q^2 + s_3q + s_4 \\ T(q) &= t_0q^4 + t_1q^3 + t_2q^2 + t_3q + t_4 \end{aligned} \tag{76}$$

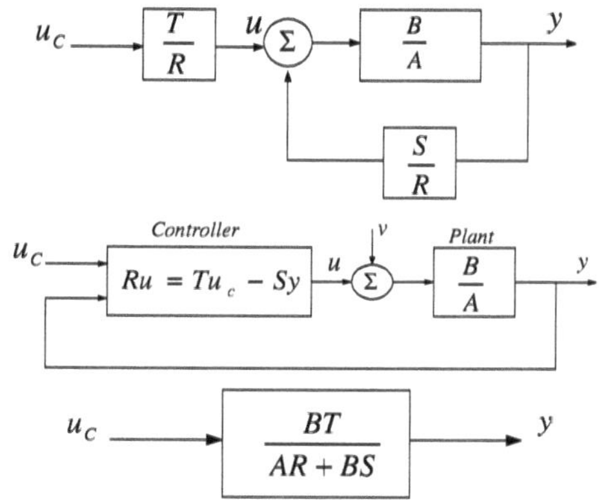

Fig. 20 A general linear controller with two degrees of freedom

The closed-loop characteristic polynomial is thus:

$$\mathrm{AR} + \mathrm{BS} = A_C \tag{77}$$

The key idea of the design method is to specify the desired closed-loop characteristic polynomial A_C. The polynomial R and S can then be solved from Eq. (77). In the design procedure we consider polynomial $A_C$to be a design parameter that is chosen to give desired properties to the closed-loop system. Equation (78), which plays a fundamental role in algebra, is called the Diophantine equation. The equation always has solutions if polynomials A and B do not have common factors. The solution may be poorly conditioned if the polynomials have factors that are closed. The solution can be obtained by introducing polynomials with unknown coefficients and solving the linear equations obtained.

In fact, in an offline state, the adaptive controller parameters are as according to Fig. 21.

$\frac{B_m}{A_m}$ is the desired transfer function of power system. B_m and A_m polynomials must be chosen in the way that the adaptive controller can omit the perturbation in input mechanical torque with suitable speed. The desired transfer function used in this paper according to Eq. (74) offers as below:

$$\frac{y(q)}{u(q)} = \frac{B_m(q)}{A_m(q)} = \frac{q^4}{q^5 + a_1 q^4 + a_2 q^3 + a_3 q^2 + a_4 q + a_5} \tag{78}$$

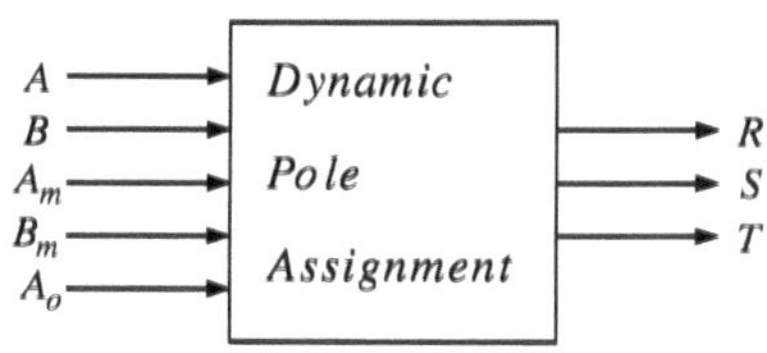

Fig. 21 Block diagram of offline adaptive controller inputs and outputs

According to Fig. 20, designing an adaptive offline controller (MDPP technique) consists of the three following steps:

(1) Selection polynomials of A_m, B_m and A_o as below:

$$^{\circ}A_m = {^{\circ}A} = n \tag{79}$$

$$^{\circ}B_m = {^{\circ}B} = m \tag{80}$$

$$^{\circ}A_o = {^{\circ}A} - {^{\circ}B^+} - 1 \tag{81}$$

$$B_m = B^+B^- \tag{82}$$

That B^+ and B^- are strongly and poorly damped roots polynomials.

(2) The Diophantine equation is formed as below and will be solved for finding R' and S polynomials:

$$AR' + B^-S = A_oA_m \tag{83}$$

(3) Calculating R and T control as below:

$$R = R'B^+ \tag{84}$$

$$T = A_oB'_m \tag{85}$$

But the online control design consists of the three following steps:

1. Selection polynomials of A_m,B_m and A_o.
2. Calculation of $\hat{\theta}$ matrix with RLS as in the equations:

$$y(q) = \frac{B}{A}u(q) \tag{86}$$

$$A(q)y(q) = B(q)u(q) \tag{87}$$

$$\begin{aligned} &y(q) + a_1y(q-1) + a_2y(q-2) + \cdots + a_ny(q-n) \\ &= b_1u(q+m-n-1) + \cdots + b_mu(q-m) \end{aligned} \tag{88}$$

$$\begin{aligned} y(q) = &\left[-y(q-1) \quad \cdots \quad -y(q-n) \quad u(q+m-n-1) \quad \cdots \quad u(q-m) \right] \\ &\cdot \left[a_1 \quad \cdots \quad a_n \quad b_1 \quad \cdots \quad b_m \right]^T \end{aligned} \tag{89}$$

$$y(q) = \varphi^T(q-1)\theta \tag{90}$$

$$K(q) = P(q)\varphi(q) = P(q-1)\varphi(q)[I + \varphi^T(q)P(q-1)\varphi(q)]^{-1} \tag{91}$$

$$\begin{aligned} P(q) &= P(q-1) - P(q-1)\varphi(q)[I + \varphi^T(q)P(q-1)\varphi(q)]^{-1}\varphi^T(q)P(q-1) \\ &= [I - K(q)\varphi^T(q)]P(q-1) \end{aligned} \tag{92}$$

$$\hat{\theta}(q) = \hat{\theta}(q-1) + K(q)[y(q) - \varphi^T(q)\hat{\theta}(q-1)] \tag{93}$$

3. Calculation of R, $S$and T polynomials with MDPP.

5.4 Simulation Results

The linearized model of the case study system in Fig. 17 with parameters is shown in Appendix 4 and have been simulated with MATLAB/SIMULINK. In order to examine the robustness of the damping controllers to a step load perturbation, it has been applied a step duration in mechanical power (i.e., $\Delta P_m = 0.01$ pu) to the system seen in Fig. 18. The Fig. 22 is related to an estimation of the control reference system in the online adaptive controller at nominal condition calculated by

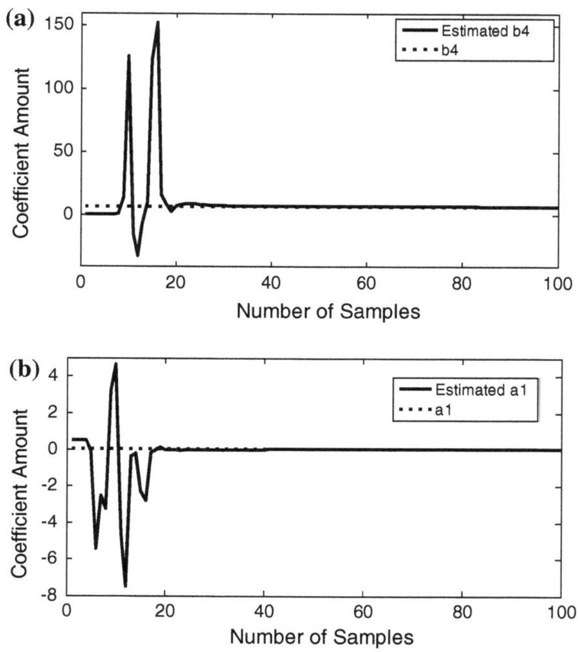

Fig. 22 Adaptive controller polynomial coefficients of RLS estimated plant at nominal operating condition. **a** b_4, **b** a_1

RLS technique. Some of the coefficients of the transfer function of the power system in Eq. (76) and their estimation by RLS technique have been shown in Fig. 22. It can be seen that the estimation of transfer function coefficients have been converged to the polynomials of the reference power system model at less than 20 iterations.

After estimation of the transfer function of the reference control model, in order to calculate the online adaptive controller polynomial coefficients, the Diophantine equation must be solved. In the following, it has been shown some of the parameters R and S in Fig. 23 at a nominal condition. It can be considered that the coefficients have been converged at less than 20 iterations samples have been taken from the input and output of the transfer function of the case study with sampling time $T_s = 0.01$ s for adaptive control designing. The desired transfer function of Eq. (79) has been presented in Appendix 5. The Fig. 24 shows the dynamic responses of $\Delta\omega$ with adaptive controller at nominal operating loads due to$\Delta P_m = 0.01$ pu perturbation., it can be seen that the dynamic response of the system equipped with an adaptive controller (Fig. 24) has adequate quality because short settling time as 0.1 s. Also, the dynamic response of the system equipped with the adaptive controller (Fig. 24) has an agreeable small peek amplitude amount.

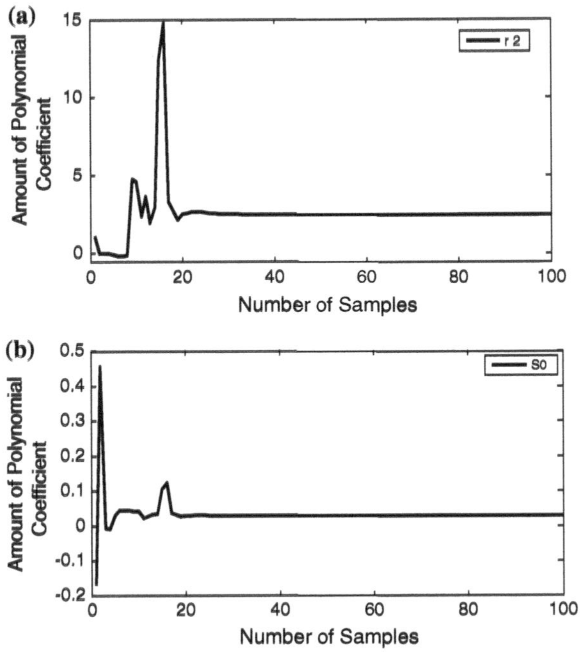

Fig. 23 Adaptive controller parameters calculated with Diophantine equation at nominal operating condition. **a** r_2, **b** s_0

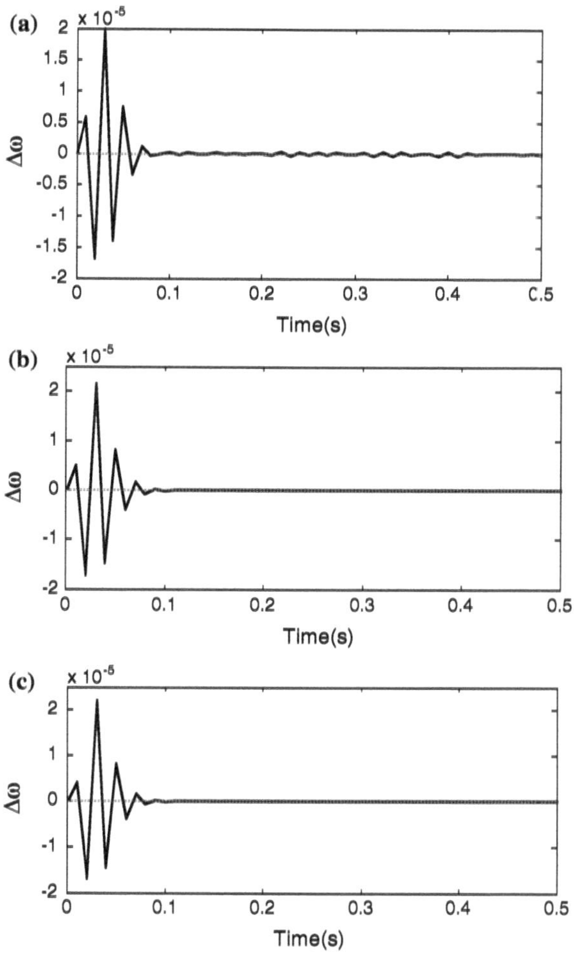

Fig. 24 Dynamic responses of $\Delta\omega$ with adaptive controller at different operating conditions due to $\Delta P_m = 0.01$ pu. **a** Light, **b** Nominal, **c** Heavy

6 Conclusions

In this chapter, a novel dynamic model based VSC HVDC is considered and a supplementary controller, which is the same as that applied for HVDC, is designed for improve power system stability and oscillation damping. SVD, RGA, DF have been employed to evaluate the EM mode controllability to the four VSC HVDC input which illustrated that the EM mode has best controllability via the fire angle of rectifier. Simulation results carried by MATLAB, show the proposed strategy has fast dynamic response.

In addition, a novel dynamic model based VSC HVDC is considered and a supplementary controller, which is the same as that applied for UPFC, is designed for improve power system stability and oscillation damping. SVD has been employed to evaluate the EM mode controllability to the four VSC HVDC input. SVD illustrated that the EM mode has best controllability via the fire angle of rectifier. Simulation results carried by MATLAB, show the proposed strategy has fast dynamic response.

Moreover, a UPFC has been used for dynamic stability improvement and state-space equations has been applied for the design of damping controllers. Simulation results operated by MATLAB/SIMULINK show that using an input speed deviation signal decreases speed oscillations effectively. According to the simulation results, the designed adaptive controller for the system has the perfect effect in oscillation damping and dynamic stability improvement.

Appendices

Appendix 1. Parameters of Test System (pu) for BtB VSC HVDC Network

Machine and Exciter: $\begin{matrix} x_d = 1,\ x_q = 0.6,\ x'_d = 0.3,\ D = 0,\ M = 8, \\ T'_{\text{do}} = 5.044,\ f = 60,\ v_{\text{ref}} = 1,\ K_A = 120,\ T_A = 0.015 \end{matrix}$

Transmission line and transformer reactance: $X_{\text{tl}} = 0.15,\ X_{\text{lb}} = 0.6,\ X_{\text{sp}} = X_s = 0.15$

BtB VSC HVDC: $V_{\text{dc}} = 3,\ C_{\text{dc}} = 1.$

Neural controller: two multilayer feed forward neural network with activation function: $a\ \tanh(bx)$.

Hidden and output layer for identifier includes 4 and 1 neuron, respectively with $a = b = 1.2,\ Eta = 0.4.$

Hidden and output layer for controller includes 3 and 1 neuron, respectively with $a = 17, b = 0.65,\ Eta = 0.22.$

Appendix 2. Parameters of Test System (pu) for VSC HVDC Network

Machine and Exciter: $\begin{matrix} x_d = 1,\ x_q = 0.6,\ x'_d = 0.3,\ D = 0,\ M = 8, \\ T'_{\text{do}} = 5.044,\ f = 60,\ v_{\text{ref}} = 1,\ K_A = 120,\ T_A = 0.015 \end{matrix}$

Transmission line and transformer reactance: $x_t = 0.1,\ x_l = 1,\ x_r = x_i = 0.15$

VSC HVDC: $V_{\text{dcr}} = 2, V_{\text{dci}} = 1.95,\ C_r = C_i = 2,\ L_1 = L_2 = 0.09,\ C = 0.09$

Appendix 3. Constant Coefficients of the Linearised Dynamic Model of Eq. (59)

$$Z = 1 + \frac{x_l}{x_r},\ A = x_t + x_l + \frac{x_t}{x_r},\ [A] = A + Zx'_d,\ [B] = A + Zx_q$$

$$C_1 = \frac{V_b \cos(\delta)}{[B]},\ C_2 = -\frac{x_l M_r V_{\mathrm{dc}} \sin(\varphi_r)}{2x_r[B]},\ C_3 = \frac{x_l V_{\mathrm{dc}} \cos(\varphi_r)}{2x_r[B]},\ C_4 = \frac{x_l M_r \cos(\varphi_r)}{2x_r[B]}$$

$$C_5 = \frac{Z}{A},\ C_6 = \frac{V_b \sin(\delta)}{[A]},\ C_7 = -\frac{x_l M_r V_{\mathrm{dc}} \cos(\varphi_r)}{2x_r[A]},\ C8 = -\frac{x_l V_{\mathrm{dc}} \sin(\varphi_r)}{2x_r[A]},$$
$$C_9 = -\frac{x_l M_r V_{\mathrm{dc}} \cos(\varphi_r)}{2X_r[A]}$$

$$C_a = (x_q - x'_d)I_t,\ C_b = E'_q + (x_q - x'_d),\ K_1 = C_b C_1 + C_a C_6,\ K_2 = I_t(1 + (x_q - x'_d)C_5)$$

$$K_{\mathrm{pdcr}} = C_b C_4 + C_a C_9,\ K_{pM_r} = C_b C_3 + C_a C_8,\ K_{p\varphi r} = C_b C_2 + C_a C_7$$

$$K_3 = 1 + JC_5,\ K_4 = JC_6,\ K_{q\varphi r} = JC_7,\ K_{qM_r} = JC_8,\ K_{\mathrm{qdcr}} = JC_9$$

$$K_5 = L(V_{\mathrm{td}} x_q C_1 - V_{\mathrm{tq}} x'_d C_6),\ K_6 = LV_{\mathrm{tq}}(1 - x'_d C_5)$$

$$K_{V_{\mathrm{dc}}r} = L(V_{\mathrm{td}} x_q C_4 - V_{\mathrm{tq}} x'_d C_9), K_{VM_r} = L(V_{\mathrm{td}} x_q C_3 - V_{\mathrm{tq}} x'_d C_8),$$
$$K_{V\varphi r} = L(V_{\mathrm{td}} x_q C_2 - V_{\mathrm{tq}} x'_d C_7)$$

$$x_d - x'_d = J,\ L = \frac{1}{V_t},\ E = \frac{x'_d + x_t}{x_r},\ F = \frac{x_q + x_t}{x_r}$$

$$C_{10} = EC_5 - \frac{1}{x_r},\ C_{11} = EC_6,\ C_{12} = EC_7 - \frac{M_r}{2x_r} V_{\mathrm{dcr}} \sin(\varphi_r),$$
$$C_{13} = \frac{1}{2x_r} M_r \cos(\varphi_r) + EC_8$$

$$C_{14} = \frac{1}{2x_r} \cos(\varphi_r) + EC_9,\ C_{15} = FC_1,\ C_{16} = \frac{1}{2x_r} V_{\mathrm{dcr}} \sin(\varphi_r) + FC_2,$$
$$C_{17} = -\frac{1}{2x_r} M_r \cos(\varphi_r) + FC_4$$

$$C_{18} = FC_3 - \frac{1}{2x_r} V_{\mathrm{dcr}} \cos(\varphi_r),\ C_{19} = \frac{1}{x_i} V_{\mathrm{bd}},\ C_{20} = \frac{1}{2x_i} M_i \sin(\varphi_i),$$
$$C_{21} = \frac{1}{2x_i} V_{\mathrm{dcr}} \sin(\varphi_i)$$

$$C_{22} = \frac{1}{2x_i} M_i V_{\mathrm{dci}} \cos(\varphi_i),\ C_{23} = \frac{1}{x_i} V_{\mathrm{bq}},\ C_{24} = -\frac{1}{2x_i} M_i \cos(\varphi_i),$$
$$C_{25} = -\frac{1}{2x_i} V_{\mathrm{dci}} \cos(\varphi_i),\ C_{26} = \frac{1}{2x_i} V_{\mathrm{dci}} \sin(\varphi_i)$$

$f_1 = [0.5\cos(\varphi_i)I_{\mathrm{id}} + 0.5\sin(\varphi_i)I_{\mathrm{iq}}]$, $f_2 = -[-0.5\sin(\varphi_i)I_{\mathrm{id}} + 0.5\cos(\varphi_i)I_{\mathrm{iq}}]$

$f_3 = -0.5M_i\cos(\varphi_i)$, $f_4 = -0.5M_i\sin(\varphi_i)$, $f_5 = -[0.5\cos(\varphi_r)I_{\mathrm{rd}} + 0.5\sin(\varphi_i)I_{\mathrm{rq}}]$

$f_6 = -[-0.5\sin(\varphi_i)I_{\mathrm{rd}} + 0.5\cos(\varphi_i)I_{\mathrm{rq}}]$, $f_7 = -0.5M_r\cos(\varphi_r)$, $f_4 = -0.5M_r\sin(\varphi_r)$

$C_{27} = f_3C_{19} + f_4C_{23}$, $C_{28} = f_3C_{20} + f_4C_{24}$, $C_{29} = f_1 + f_3C_{21} + f_4C_{25}$,
$C_{30} = f_2 + f_3C_{22} + f_4C_{26}$

$C_{31} = f_7C_{11} + f_8C_{15}$, $C_{32} = f_7C_{10}$, $C_{33} = f_7C_{14} + f_8C_{17}$, $C_{34} = f_5 + f_7C_{13} + f_8C_{18}$,
$C_{35} = f_6 + f_7C_{12} + f_8C_{16}$

Appendix 4. The Test System Parameters for UPFC HVDC Network

Generator: $M = 2H = 8.0$ MJ/MVA, $D = 0.0$, $T'_{\mathrm{do}} = 5.044$ s, $x_d = 1.0$ pu, $x_q = 0.6$ pu, $x'_d = 0.3$ pu

Excitation System: $K_a = 100$, $T_a = 0.01$ s
Transformer: $X_{\mathrm{tE}} = 0.1$ pu, $X_E = X_B = 0.1$ pu, $X_E = X_B = 0.1$ pu
Transmission Line: $X_{\mathrm{BV}} = 0.3$ pu, $X_e = X_{\mathrm{BV}} + X_B + X_{tE} = 0.5$ pu
Operating Condition: $V_t = 1.0$ pu, $P_e = 0.8$ pu, $V_b = 1.0$ pu, $f = 60$ Hz
DC Link: $V_{\mathrm{dc}} = 2$ pu, $C_{\mathrm{dc}} = 1$ pu

Appendix 5. Adaptive Controller Parameters for UPFC HVDC Network

$$A_m = (q - 0.01)(q - 0.03)(q - 0.02)(q - 0.1)(q + 0.1)$$
$$B_m = q^4$$
$$A_o = 1$$
$$^\circ B_m = {}^\circ B = m = 4$$
$$^\circ A_m = {}^\circ A = n = 5$$

Appendix 6. K Parameters for UPFC HVDC Network

$$K_1 = \frac{(V_{\mathrm{td}} - I_{\mathrm{tq}}x'_d)(x_{dE} - x_{dt})V_b\sin\delta}{x_{d\Sigma}} + \frac{(x_qI_{\mathrm{td}} + V_{\mathrm{tq}})(x_{\mathrm{qt}} - x_{\mathrm{qE}})V_b\cos\delta}{x_{q\Sigma}}$$

$$K_2 = \frac{-(x_{\mathrm{BB}} + x_E)V_{\mathrm{td}}}{x_{d\Sigma}x_d} + \frac{(x_{\mathrm{BB}} + x_E)x'_d I_{\mathrm{tq}}}{x_{d\Sigma}}$$

$$K_3 = 1 + \frac{(x'_d - x_d)(x_{\mathrm{BB}} + x_E)}{x_{d\Sigma}}$$

$$K_4 = \frac{(x'_d - x_d)(x_{dE} - x_{dt})V_b \sin\delta}{x_{d\Sigma}}$$

$$K_5 = \frac{V_{\mathrm{td}}x_q(x_{\mathrm{qt}} - x_{qE})V_b\cos\delta}{V_t x_{q\Sigma}} - \frac{V_{\mathrm{tq}}x'_d(x_{dtE} - x_{dt})V_b\sin\delta}{V_t x_{d\Sigma}}$$

$$K_6 = \frac{V_{\mathrm{tq}}(x_{d\Sigma} + x'_d(x_{\mathrm{BB}} + x_E))}{V_t x_{d\Sigma}}$$

$$\begin{aligned}K_7 =& 0.25C_{\mathrm{dc}}(V_b\sin\delta(m_E\cos\delta_E x_{dE} - m_B\cos\delta_B x_{dt}))\\ &- \frac{m_B\cos\delta_B x_{dt}}{x_{d\Sigma}}V_b\cos\delta(m_B\sin\delta_B x_{qt} - m_E\sin\delta_E x_{\mathrm{qe}}\end{aligned}$$

$$K_8 = -0.25\frac{m_B\cos\delta_B x_{\mathrm{E}} + m_E\cos\delta_E x_{\mathrm{BB}}}{x_{d\Sigma}}$$

$$\begin{aligned}K_9 = 0.25C_{\mathrm{dc}}(&\frac{m_B\sin\delta_B(m_B\cos\delta_B x_{dt} - m_E\cos\delta_E x_{dE})}{2x_{d\Sigma}}\\ &+ \frac{m_E\sin\delta_E(m_E\cos\delta_E x_{Bd} - m_B\cos\delta_B x_{dt})}{2x_{d\Sigma}}\\ &\frac{m_B\cos\delta_B(m_B\sin\delta_B x_{qt} - m_E\sin\delta_E x_{qE})}{2x_{q\Sigma}}\\ &+ \frac{m_E\cos\delta_E(-m_B\sin\delta_B x_{qE} - m_E\sin\delta_E x_{Bq})}{2x_{q\Sigma}}\end{aligned}$$

$$\begin{aligned}K_{\mathrm{pe}} =& \frac{(V_{\mathrm{td}} - I_{\mathrm{tq}}x'_d)(x_{Bd} - x_{dE})V_{\mathrm{dc}}\sin\delta_E}{2x_{d\Sigma}}\\ &+ \frac{(x_q I_{\mathrm{td}} + V_{\mathrm{tq}})(x_{Bq} - x_{qE})V_{\mathrm{dc}}\cos\delta_E}{2x_{q\Sigma}}\end{aligned}$$

$$\begin{aligned}K_{p\delta E} =& \frac{(V_{\mathrm{td}} - I_{\mathrm{tq}}x'_d)(x_{Bd} - x_{dE})V_{\mathrm{dc}}m_E\cos\delta_E}{2x_{d\Sigma}}\\ &+ \frac{(x_q I_{\mathrm{td}} + V_{\mathrm{tq}})(-x_{Bq} + x_{qE})V_{\mathrm{dc}}m_E\sin\delta_E}{2x_{q\Sigma}}\end{aligned}$$

$$K_{\mathrm{pb}} = \frac{\left(V_{\mathrm{td}} - I_{\mathrm{tq}}x'_d\right)(x_{dt} - x_{dE})x_{\mathrm{dc}}\sin\delta_B}{2x_{d\Sigma}} + \frac{(x_q I_{\mathrm{td}} + V_{\mathrm{tq}})(x_{\mathrm{qt}} - x_{qE})V_{\mathrm{dc}}\cos\delta_B}{2x_{q\Sigma}}$$

$$K_{p\delta B} = \frac{\left(V_{\mathrm{td}} - I_{\mathrm{tq}}x'_d\right)(x_{dE} + x_{\mathrm{dt}})V_{\mathrm{dc}}m_B\cos\delta_B}{2x_{d\Sigma}} + \frac{(x_q I_{\mathrm{td}} + V_{\mathrm{tq}})(-x_{\mathrm{qt}} + x_{\mathrm{qE}})V_{\mathrm{dc}}m_B\sin\delta_B}{2x_{q\Sigma}}$$

$$K_{\mathrm{pd}} = (V_{\mathrm{td}} - I_{\mathrm{tq}}x'_d)(\frac{(x_{dt} - x_{dE})m_B\sin\delta_B}{2x_{d\Sigma}} + \frac{(x_{Bd} - x_{dE})m_E\sin\delta_E}{2x_{d\Sigma}} + (x_q I_{\mathrm{td}} + V_{\mathrm{td}})(\frac{(x_{\mathrm{qt}} - x_{qE})m_B\cos\delta_B}{2x_{q\Sigma}} + \frac{(x_{Bq} - x_{qE})m_E\cos\delta_E}{2x_{q\Sigma}}$$

$$K_{\mathrm{qe}} = -\frac{\left(x'_d - x_d\right)(x_{Bd} - x_{dE})V_{\mathrm{dc}}\sin\delta_E}{2x_{d\Sigma}}$$

$$K_{q\delta e} = -\frac{\left(x'_d - x_d\right)(x_{Bd} - x_{dE})m_E V_{\mathrm{dc}}\cos\delta_E}{2x_{d\Sigma}}$$

$$K_{\mathrm{qb}} = -\frac{\left(x'_d - x_d\right)(x_{dt} - x_{dE})V_{\mathrm{dc}}\sin\delta_B}{2x_{d\Sigma}}$$

$$K_{q\delta B} = -\frac{\left(x'_d - x_d\right)(x_{dE} - x_{dt})m_B V_{\mathrm{dc}}\cos\delta_B}{2x_{\Sigma}}$$

$$K_{\mathrm{qe}} = -\left(x'_d - x_d\right)(\frac{(x_{Bd} - x_{dE})m_E\sin\delta_E}{2x_{d\Sigma}} + \frac{(x_{dt} - x_{dE})m_B\sin\delta_B}{2x_{d\Sigma}})$$

$$K_{\mathrm{ve}} = \frac{V_{\mathrm{td}}(x_{Bq} - x_{qE})V_{\mathrm{dc}}\cos\delta_E}{2V_t x_{q\Sigma}} - \frac{V_{tq}(x_{Bd} - x_{dE})V_{\mathrm{dc}}\sin\delta_E}{2V_t x_{d\Sigma}}$$

$$K_{v\delta E} = \frac{V_{\mathrm{td}}x_q(x_{qE} - x_{Bq})m_E V_{\mathrm{dc}}\sin\delta_E}{2V_t x_{q\Sigma}} - \frac{V_{\mathrm{tq}}x'_d(x_{Bd} - x_{dE})m_E V_{\mathrm{dc}}\cos\delta_E}{2V_t x_{q\Sigma}}$$

$$K_{\mathrm{vb}} = \frac{V_{td}x_q(x_{\mathrm{qt}} - x_{qE})V_{\mathrm{dc}}\cos\delta_E}{2V_t x_{q\Sigma}} - \frac{V_{\mathrm{tq}}x'_d(x_{dt} - x_{dE})V_{\mathrm{dc}}\sin\delta_E}{2V_t x_{d\Sigma}}$$

$$K_{v\delta b} = \frac{V_{td}x_q(x_{qE} - x_{qt})m_B V_{dc} \sin \delta_E}{2V_t x_{q\Sigma}} + \frac{V_{tq} m_B x_d'(x_{dE} + x_{dt}) V_{dc} \cos \delta_E}{2V_t x_{d\Sigma}}$$

$$K_{vd} = \frac{V_{td}x_q(x_{Bq} - x_{qE})m_E \cos \delta_E}{2V_t x_{q\Sigma}} + \frac{(x_{qt} - x_{qE})m_B \cos \delta_B}{2x_{q\Sigma}}$$
$$\frac{V_{tq} m_E x_d'(x_{Bd} - x_{dE}) \sin \delta_E}{2V_t x_{q\Sigma}} + \frac{m_B(x_{dt} - x_{qE}) \sin \delta_E}{2x_{d\Sigma}}$$

$$K_{ce} = 0.25C_{dc}\frac{V_{dc} \sin \delta_E(m_E \cos \delta_E x_{Bd} - m_B \cos \delta_B x_{dE})}{2x_{d\Sigma}} + \frac{V_{dc} \cos \delta_E(m_E \sin \delta_E x_{Bq} - m_B \sin \delta_B x_{qE})}{2x_{q\Sigma}}$$

$$K_{c\delta e} = \frac{0.25m_E}{C_{dc}}(\cos \delta_E I_{Eq} - \sin \delta_E I_{Ed}) + \frac{0.25}{C_{dc}}(m_E V_{dc} \cos \delta_E \frac{(m_E \cos \delta_E x_{Bd} - m_B \cos \delta_B x_{dE})}{2x_{d\Sigma}} + m_E V_{dc} \sin \delta_E \frac{(m_B \sin \delta_B x_{qE} + m_E \sin \delta_E x_{Bq})}{2x_{q\Sigma}})$$

$$K_{cb} = 0.25C_{dc}\frac{V_{dc} \sin \delta_B(-m_E \cos \delta_E x_{dE} + m_B \cos \delta_B x_{dt})}{2x_{d\Sigma}} + \frac{V_{dc} \cos \delta_B(m_B \sin \delta_E x_{qt} - m_E \sin \delta_E x_{qE})}{2x_{q\Sigma}}$$

$$K_{c\delta B} = \frac{0.25m_B}{C_{dc}}(\cos \delta_B I_{Bq} - \sin \delta_B I_{Bd}) + \frac{0.25}{C_{dc}}(m_B V_{dc} \cos \delta_B \frac{(m_E \cos \delta_E x_{dE} + m_B \cos \delta_B x_{dt})}{2x_{d\Sigma}} + m_B V_{dc} \sin \delta_B \frac{(-m_B \sin \delta_E x_{qt} + m_E \sin \delta_E x_{qE})}{2x_{q\Sigma}})$$

References

1. Chaturvedi DK, Malik OP, Kalra PK (2004) Performance of a generalized neuron based PSS in a multimachine power system, IEEE Trans 0885-8969/04
2. Wang HF, Swift FJ (1997) A unified model for the analysis of FACTS devices in damping power system oscillations, part I: single-machine infinite-bus power systems. IEEE Trans Power Delivery (12)2:941–946
3. Yang N, Liu Q, McCalley JD (1998) TCSC controller design for damping interarea oscillations. IEEE Trans Power Syst (13)4:1304–1310
4. Uzunovic E, Canizares CA, Reeve J (1999) EMTP studies of UPFC power oscillation damping. In: Proceedings of NAPS'99, San Luis Obispo, California, pp 155–163

5. Hsu YY, Wang L (1988) Damping of a parallel AC–DC power system using PID power system stabilizers and rectifier current regulators. IEEE Trans Energy Convers 3(3):540–547
6. Badran SM, Choudhry MA (1993) Design of modulation controllers for AC/DC power systems. IEEE Trans Power Syst 8(4):1490–1496
7. Hao C (1993) Design of modulation controller to damp power oscillations of parallel AC line in the Tianshenqiao to Guangdong HVDC transmission. IEEE Trans, 244–247
8. Al-Awami AT, Abdel-Magid YL, Abido MA (2007) A particle-swarm-based approach of power system stability enhancement with unified power flow controller. Electr Power Energy Syst 29:251–259
9. Wang HF (1999) Damping function of unified power flow controller. IEE Proc Gen Trans Distrib (146)1:81–87
10. Qin C, Du WJ, Wang HF, Xu Q, Ju P (2005) Controllable parameter region and variable-parameter design of decoupling unified power flow controller. Transmission and distribution conference and exhibition, Asia and Pacific, IEEE/PES, Dalian, China
11. Kalyani RP, Venayagamoorthy GK (2004) Two separately continually online trained neurocontrollers for a unified power flow controller. IEEE international conference on intelligent sensing and information processing, Cat. No. 04EX783, pp 243–248
12. Keri AJF, Lombard X, Edris AA (1999) Unified power flow controller: modeling and analysis, IEEE Trans Power Delivery (14)2:648–654
13. Rouco L (2001) Coordinated design of multiple controllers for damping power system oscillation. Elec Power Energy Syst 21:517–530
14. Pal BC (2002) Robust damping of interarea oscillations with unified power flow controller. IEE Proc Gen Trans Distrib (149)6:733–738
15. Wang HF (1999) Application of modeling UPFC into multi-machine power systems. IEE Proc Gen Trans Disturb (146)3:306–312
16. Abido MA, Al-Awami AT, Abdel-Magid YL (2006) Simultaneous design of damping controllers and internal controllers of a unified power flow controller. IEEE power engineering society general meeting, Montreal
17. Abido MA (2001) Particle swarm optimization for multimachine power system stabilizer design. IEEE Power Eng Soc Summer Meet 3:1346–1351
18. Villablanca XM, Valle JD, Rojas J, Abarca J, Rojas W (2000) A modified back-to-back HVDC system for 36-pulse operation. IEEE Trans (15)2:641–645
19. Padiyar KR (1990) HVDC power transmission systems, technology and system interactions. Wiley, India
20. Hammad A, Taylor C (1991) HVDC controllers for system dynamic performance. IEEE Trans Power Syst (6)2:743–752
21. Gjerdc JO, Flolo R, Gjengeddal T (1996) Use of HVDC and FACTS components for enhancement of power system stability. 8th mediterranean electro technical conference, MELECON vol 96, issue 2, pp 802–808, May 1996
22. Huang GM, Krishnaswamy V (2002) HVDC controls for power system stability. IEEE Power Eng Soc Summer Meeting 1:597–602
23. Baker M, Abbott K, Gemmell B (2002) Frequency and system damping assistance from HVDC and FACTS controller. IEEE Power Eng Soc Summer Meeting 2:770–773
24. Corsi S et al (2002) Emergency stability controls through HVDC links. IEEE Power Eng Soc Summer Meeting 2:774–779
25. Banaei MR, Taheri N (2009) HVDC based damping controllers for power system stability. Telecommunications energy conference, INTELEC09, Incheon, Korea
26. Taheri N, Banaei MR (2010) A supplementary neural controller for novel modeling of VSC HVDC to enhance dynamic stability in a power system. PEDSTC conference, 17–18 Feb
27. Skogestad S, Postletwaite I (2005) Multivariable feedback control. Wiley, New York
28. Bristol EH (1966) On a new measure of interactions for multivariable process control. IEEE Trans Autom Control 11:133–134
29. Larsen EV, Sanchez-Gasca JJ, Chow JH (1995) Concept for design of FACTS controllers to damp power swings. IEEE Trans PWRS 2(10):948–956

30. Shamsollahi P, Malik OP (1999) Application of neural adaptive power system stabilizer in a multi-machine power system. IEEE Trans Energy Convers (14)3:731–736
31. Tabatabaei NM, Taheri N, Nezhadali Baghan A (2011) Controllability of BTB VSC HVDC inputs based on power system stability. International conference on engineering and mathematics (ENMA 2001), Bilbao, Spain, 23–24 Jun 2011, pp 85–95
32. Hamdan AMA (1999) An investigation of the significance of singular value decomposition in power system dynamics. Electr Power Energy Syst 21:417–424
33. Abdel-Magid YL, Abido MA (2004) Robust coordinated design of excitation and TCSC-based stabilizers using genetic algorithm. Electr Power Syst Res 69:129–141
34. Nazarpour D, Hosseini SH, Gharehpetian GB (2005) An adaptive STATCOM based stabilizer for damping generator oscillations, ELECO Conference, Bursa, Turkey, 7–11 Dec 2005, pp 60-64
35. Cheng CH, Hsu YY (1992) Damping of generator oscillations using an adaptive static VAR compensator. IEEE Trans Power Syst (7)2:718
36. Tabatabaei NM, Hashemi A, Taheri N, Sadikoglu FM (2011) A novel on line adaptive based stabilizer for dynamic stability improvement with UPFC. 7th international conference on technical and physical problems of power engineering (ICTPE-2011), Near East University, Lefkosa, Northern Cyprus, 7–9 Jul 2011, vol 57, Code 03EPS31, pp 280–285
37. Tabatabaei NM, Taheri N, Nezhadali Baghan A, Khudiev AT (2011) A novel modeling and controlling of HVDC transmission system for enhancing dynamic stability in a power system. 7th international conference on technical and physical problems of power engineering (ICTPE-2011), Near East University, Lefkosa, Northern Cyprus, 7–9 Jul, vol 33, Code 03EPS07, pp 158–164
38. Yam CM, Haque MH (2005) A SVD based controller of UPFC for power flow control. 15th power system computation conference, Aug 2005, Session 12, Paper 2, pp 1–7
39. Nabavi-Niaki A, Iravani MR (1996) Steady-state and dynamic models of unified power flow controller (UPFC) for power system studies. IEEE Trans Power Syst (11)4:1937–1943

Optimal Design of UPFC Based Damping Controller Using PSO and QPSO

Hossein Shayeghi and Amin Safari

Abstract The Unified Power Flow Controller (UPFC) is regarded as one of the most versatile devices in the FACTS device family which has the ability to control the power flow in the transmission line, improve the transient stability, mitigate system oscillation, and provide voltage support. In this book chapter, the problem of UPFC based damping controller is formulated as an optimization problem which is solved using classic and Quantum-behaved Particle Swarm Optimization technique (QPSO). Two different objective functions are proposed in this work for the UPFC based damping controller design problem. The first objective function is the eigenvalues based comprising the damping factor, and the damping ratio of the lightly damped electromechanical modes, while the second is the time domain-based multi-objective function. The performance of the proposed controllers under different disturbances and loading conditions is investigated for a single machine infinite bus and multi-machine power systems. The results of the proposed controllers are demonstrated through eigenvalue analysis and nonlinear time domain simulation.

1 Introduction

As power demand grows rapidly and expansion in transmission and generation is restricted with the limited availability of resources and the strict environmental constraints, power systems are today much more loaded than before. This causes

---

H. Shayeghi (✉)
Technical Engineering Department, University of Mohaghegh Ardabili, Ardabil, Iran
e-mail: hshayeghi@gmail.com

A. Safari
Electrical Engineering Department, Azarbaijan Shahid Madani University, Tabriz, Iran
e-mail: asafari1650@yahoo.comr

N. Bizon et al. (eds.), *Analysis, Control and Optimal Operations in Hybrid Power Systems*, Green Energy and Technology, DOI: 10.1007/978-1-4471-5538-6_5,

the power systems to be operated near their stability limits. In addition, interconnection between remotely located power systems gives rise to low frequency oscillations in the range of 0.2–2.0 Hz. If not well damped, these oscillations may keep growing in magnitude until loss of synchronism results [1, 2]. In order to damp these power system oscillations and increase system oscillations stability, the installation of Power System Stabilizer (PSS) is both economical and effective. PSSs have been used for many years to add damping to electromechanical oscillations. However, PSSs suffer a drawback of being liable to cause great variations in the voltage profile and they may even result in leading power factor operation and losing system stability under severe disturbances, especially those three-phase faults which may occur at the generator terminals [3].

In recent years, the fast progress in the field of power electronics had opened new opportunities for the application of the Flexible Alternative Current Transmission Systems (FACTS) devices as one of the most effective ways to improve power system operation controllability and power transfer limits. FACTS-devices provide a better adaptation to vary operational conditions and improve the usage of existing installations. The basic applications of FACTS-devices are: power flow control, increase of transmission capability, voltage control, reactive power compensation, stability improvement, power quality improvement, power conditioning, flicker mitigation, interconnection of renewable, and distributed generation and storages [4]. Through the modulation of bus voltage, phase shift between buses, and transmission line reactance, FACTS-devices can cause a substantial increase in power transfer limits during steady-state. It has been observed that utilizing a feedback supplementary control, in addition to the FACTS-device primary control, can considerably improve system damping and can also improve system voltage profile, which is advantageous over PSSs.

The UPFC is regarded as one of the most versatile devices in the FACTS device family which has the ability to control the power flow in the transmission line, improve the transient stability, mitigate system oscillation, and provide voltage support. The UPFC is a combination of a static compensator and static series compensation. It acts as a shunt compensating and a phase shifting device simultaneously. The UPFC consists of a shunt and a series transformer, which are connected via two voltage source converters with a common DC-capacitor. The DC-circuit allows the active power exchange between shunt and series transformer to control the phase shift of the series voltage. This setup, as shown in Fig. 1, provides the full controllability for voltage and power flow [5].

Several trials have been reported in the literature of dynamic models of UPFC in order to design suitable controllers for power flow, voltage, and damping controls [6]. Nabavi-Niaki and Iravani [7] developed a steady-state model, a small-signal linearized dynamic model, and a state-space large-signal model of a UPFC. Wang [8–10] presents the establishment of the linearized Phillips-Heffron model of a power system installed with a UPFC. Rouco [11] developed a novel unified Phillips-Heffron model for a power system equipped with a SVC, a TCSC and a TCPS. Damping torque coefficient analysis has been performed based on the proposed model to study the effect of FACTS controllers damping for different

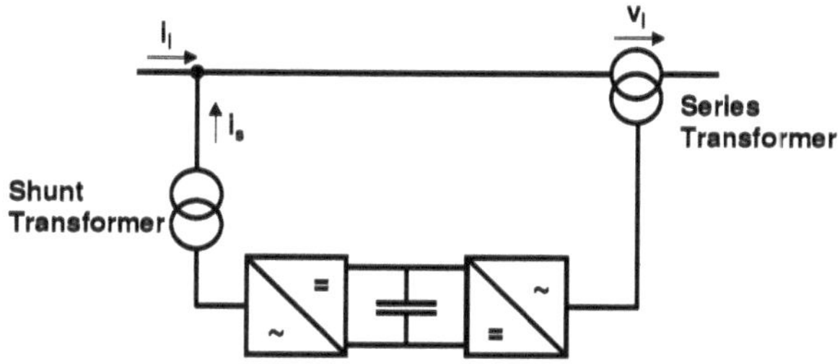

Fig. 1 Principal configuration of an UPFC

loading conditions. Some authors suggested neural networks [12] and fuzzy control methodologies [13] to cope with system loading condition to enhance the system damping performance using the UPFC. However, the parameter's adjustments of these controllers need some trial and error. Despite the potential of the modern control techniques with different structures, power system utilities still prefer the conventional lead-lag damping controller structure. The reasons behind that might be the ease of online tuning and the lack of assurance of the stability related to some adaptive or variable structure techniques. A conventional damping control design approach considers a single operating condition of the system [14]. The controllers obtained from these approaches are simple but tend to lack robustness since, at times; they fail to produce adequate damping at other operating conditions.

Recently, global optimization techniques like Genetic Algorithms (GA) and rule-based bacteria foraging [15–17] have been applied for the UPFC based damping controller parameter optimization. These evolutionary algorithms are heuristic population-based search procedures that incorporate random variation and selection operators. Although these approaches seem to be good methods for the solution of the controller parameter optimization problem, however, when the system has a highly epistatic objective function (i.e., where parameters being optimized are highly correlated), and the number of parameters to be optimized is large, then they have degraded efficiency to obtain the global optimum solution and also simulation process takes a lot of computing time. In order to overcome these drawbacks, classic and quantum-behaved particle swarm optimization (CPSO and QPSO) techniques are proposed in this research. The PSO technique is used for optimal tuning of controller parameter to improve optimization synthesis and the speed of algorithms convergence. The PSO is a novel population-based metaheuristic, which utilizes the swarm intelligence generated by the cooperation and competition between the particles in a swarm and has emerged as a useful tool for engineering optimization. It has also been found to be robust in solving problems featuring non-linearity, non-differentiability, and high dimensionality [18–21]. Unlike the other heuristic techniques, it has a flexible and well-balanced mechanism to enhance the global and local exploration abilities. Also, it suffices to specify the objective function and to place finite bounds on the optimized parameters. A CPSO and QPSO for the design of the damping controller parameters at different operating conditions are proposed in [22]. Our basic motivation behind applying these methods in power system model with UPFC is that these can achieve the desired performance under different operating conditions.

2 PSO and QPSO Technique

2.1 Particle Swarm Optimization

Particle swarm optimization is one of the latest evolutionary optimization techniques developed by Eberhart and Kennedy [19]. The PSO is a population-based method and is described by its developers as an optimization paradigm, which models the social behavior of the birds flocking or fish schooling for food. Therefore, PSO works with a population of potential solutions rather than with a single individual [20]. The advantages of the PSO algorithm are its relative simplicity and stable convergence characteristic with good computational efficiency. The PSO has also been found to be robust in solving problem featuring non-linearity, non-differentiability, multiple optima, and high-dimensionality [20–22]. In the PSO technique a number of simple entities, the particles, are placed in the search space of some problem or function, and each evaluates the objective function at its current location. Each particle then determines its movement through the search space by combining some aspect of the history of its own current and best locations by those of one or more members of the swarm with some random perturbations. The next iteration takes place after all particles have been moved. Eventually, the swarm as a whole, like a flock of birds collectively foraging for food, is likely to move close to an optimum of the fitness function. In the PSO technique, the trajectory of each individual in the search space is adjusted by dynamically altering the velocity of each particle, according to its own flying experience and the flying experience of the other particles in the search space. The position and velocity vectors of the *i*th particle in the *D*-dimensional search space can be represented as $x_i = (x_{i1}, x_{i2}, \ldots, x_{iD})$ and $v_i = (v_{i1}, v_{i2}, \ldots, v_{iD})$, respectively. According to a user-defined fitness function, let us assume that the best position of each particle, which corresponds to the best fitness value (pbest) obtained by that particle at time, be $P_i = (P_{i1}, P_{i2}, \ldots, P_{iD})$, and the global version of the PSO keeps track of the overall best value (gbest), and its location, obtained thus far by any particle in the population. Then, the new velocities and the positions of the particles for the next fitness evaluation are calculated using the following two equations:

$$v_{id} = w \times v_{id} + c_1 \times r_1.(P_{id} - x_{id}) + c_2 \times r_2.(P_{gd} - x_{id}) \tag{1}$$

$$x_{id} = x_{id} + cv_{id} \tag{2}$$

where, P_{id} and P_{gd} are *pbest* and *gbest*. The positive constants c_1 and c_2 are the cognitive and social components that are the acceleration constants responsible for varying the particle velocity toward *pbest* and *gbest*, respectively. Variables r_1 and r_2 are two random functions based on uniform probability distribution functions in the range [0, 1]. The inertia weight w is responsible for dynamically adjusting the velocity of the particles, so it is responsible for balancing between local and global searches and hence requiring less iteration for the algorithm to converge [18]. The following inertia weight is used in Eq. (1):

$$w = w_{\max} - \frac{w_{\max} - w_{\min}}{iter\_\max} \cdot iteration \tag{3}$$

where, *iter*_max is the maximum number of iterations and *iteration* is the current number of iteration. The Eq. (3) presents how the inertia weight is updated, considering $w_{\max}$ and $w_{\min}$ are the initial and final weights, respectively. Figure 2 shows the flowchart of the applied PSO algorithm.

2.2 Quantum Particle Swarm Optimization

In classical PSO technique, a particle is depicted by its position vector x_i and velocity vector v_i, which determine the trajectory of the particle. The dynamic behavior of the particle is widely divergent form that of the particle in CPSO systems in which exact values of x_i and v_i cannot be determined simultaneously. In quantum world, the term trajectory is meaningless, because x_i and v_i of a particle

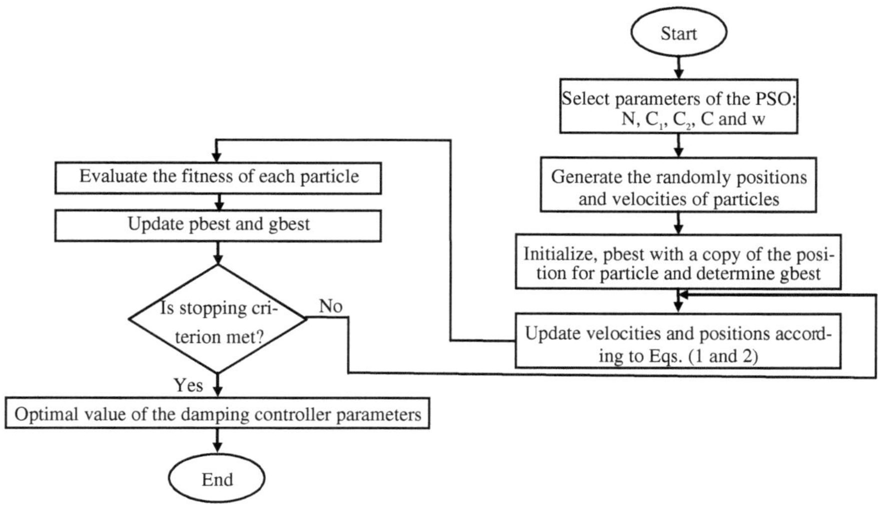

Fig. 2 Flowchart of the applied PSO technique

cannot be determined simultaneously according to uncertainty principle. Therefore, if individual particles in a PSO system have quantum behavior, the PSO algorithm is bound to work in a different fashion [23]. In the quantum model of a PSO called here QPSO, the state of a particle is depicted by wave function $\Psi(x, t)$ instead of position and velocity. Employing the Monte Carlo method, the particles move according to the following iterative equation:

$$\begin{aligned} x_i(t+1) &= p + \beta.|Mbest_i - x_i(t)|.\ln(1/u) \quad \text{if } k \geq 0.5 \\ x_i(t+1) &= p - \beta.|Mbest_i - x_i(t)|.\ln(1/u) \quad \text{if } k < 0.5 \end{aligned} \tag{4}$$

where, u and k are values generated according to a uniform probability distribution in range [0, 1], the parameter β is called contraction expansion coefficient, which can be tuned to control the convergence speed of the particle. In the QPSO, the parameter β must be set as $\beta < 1.782$ to guarantee convergence of the particle [24, 25]. Thus, the Eq. (4) is the fundamental iterative equation of the particle's position for the QPSO. Moreover, unlike the PSO, the QPSO needs no velocity vectors for particles at all, and also has fewer parameters to control (only one parameter β except population size and maximum iteration number), making it easier to implement. The experiment results on some well-known benchmark functions show that the QPSO described by the following procedure has better performance than the PSO [23].

Where Mbest called mean best position, is defined as the mean of the pbest positions of all particles. That is

$$\text{Mbest} = \frac{1}{\text{N}}\sum_{\text{d}=1}^{\text{N}} \text{p}_{\text{id}}(\text{t}) \tag{5}$$

Trajectory analyses in [25] demonstrated the fact that convergence of the PSO algorithm may be achieved if each particle converges to its local attractor, p defined at the coordinates

$$p = (c_1 p_{id} + c_2 P_{gd})/(c_1 + c_2) \tag{6}$$

The procedure for implementing the QPSO is given by the following steps:

Step 1 Initialization of swarm positions: Initialize a population (array) of particles with random positions in the n-dimensional problem space using a uniform probability distribution function.

Step 2 Evaluation of particle's fitness: Evaluate the fitness value of each particle.

Step 3 Comparison to pbest (personal best): Compare each particle's fitness with the particle's pbest. If the current value is better than pbest, then set the pbest value equal to the current value and the pbest location equal to the current location in n-dimensional space.

Step 4 Comparison to gbest (global best): Compare the fitness with the population's overall previous best. If the current value is better than gbest, then reset gbest to the current particle's array index and value.

Step 5 Updating of global point: Calculate the Mbest using Eq. (5).

Step 6 Updating of particles' position: Change the position of the particles according to Eq. (4), where c_1 and c_2 are two random numbers generated using a uniform probability distribution in the range [0, 1].

Step 7 Repeating the evolutionary cycle: Loop to Step 2 until a stop criterion is met, usually a sufficiently good fitness or a maximum number of iterations (generations).

3 Power System Modeling

In power systems, the problem of dynamic stability is primarily to keep the interconnected synchronous machines in synchronism [14]. The stability is also dependent on several other components such as the speed governors, excitation systems of the generators, the loads, the FACTS devices, etc. Therefore, an understanding of their characteristics and modeling of their performance are of fundamental importance for stability studies and control design. In this chapter, a quick overview of these models is given with respect to a particular study power system.

3.1 Overview of the Test Power Systems

Figure 3 shows a SMIB power system equipped with a UPFC. The synchronous generator is delivering power to the infinite-bus through a double circuit transmission line and a UPFC. The UPFC consists of an excitation transformer (ET), a boosting transformer (BT), two three-phase GTO based voltage source converters (VSCs), and a DC link capacitors. The four input control signals to the UPFC are m_E, m_B, δ_E, and δ_B, where, m_E is the excitation amplitude modulation ratio, m_B is the boosting amplitude modulation ratio, δ_E is the excitation phase angle, and δ_B is the boosting phase angle.

A four-machine, two-area power system, shown in Fig. 4, is considered for the damping control design for UPFC. This system is considered as one of the benchmark models for performing studies on inter-area oscillation because of its realistic structure and availability of system parameters [14]. Each area consists of two generator units. The rating of each generator is 900 MVA and 20 kV. Each of the units is connected through transformers to the 230 kV transmission line. There is a power transfer of 400 MW from Area 1 to Area 2. Each synchronous generator of the multi-machine power system is simulated using a third-order model. The loads are modeled as constant impedances.

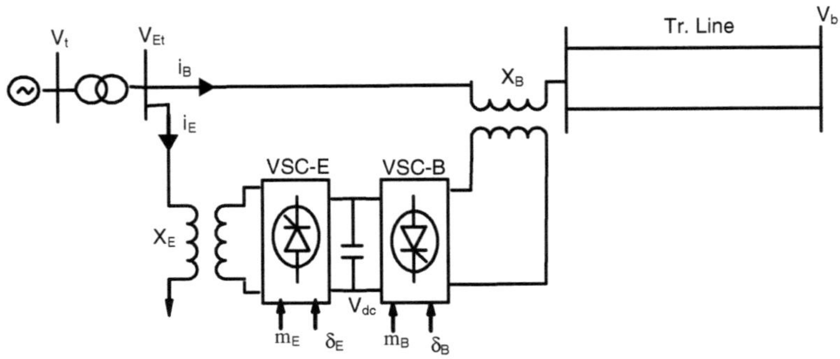

Fig. 3 SMIB power system equipped with UPFC

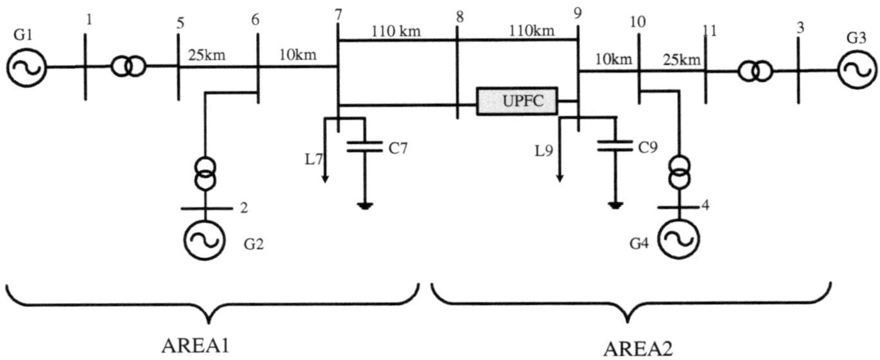

Fig. 4 Multi-machine power system with the UPFC

3.2 Modeling of Network Components

Different types of models have been reported in the literature for each of the power system components depending upon their specific application [14]. In this section, the relevant equations governing the dynamic behavior of only the specific types of models used in this research is described.

3.2.1 Generators

All four generators of the test systems are represented using the subtransient model with DC (IEEE-DCl A type) excitation system as described in [14]. The resultant dynamic equations are given by:

$$\dot{\delta}_i = \omega_b(\omega_i - 1) \tag{7}$$

$$\dot{\omega}_i = \frac{1}{M_i}(P_{mi} - P_{ei} - D_i(\omega_i - 1)) \tag{8}$$

$$\dot{E}'_{qi} = \frac{1}{T'_{doi}}(E_{fdi} - (x_{di} - x'_{di})i_{di} - E'_{qi}) \tag{9}$$

$$\dot{E}'_{di} = \frac{1}{T'_{q0.}}[-E'_{di} + (x_{qi} - x'_{qi})] \tag{10}$$

$$T_{ei} = E'_{qi}i_{qi} - (x_{qi} - x'_{di})i_{di}i_{qi} \tag{11}$$

where

δ rotor angle
ω rotor speed
P_m mechanical input power
P_e electrical output power
E'_q transient emf due to field flux-linkage
E'_d transient EMF due to flux-linkage in q-axis damper coil
E_{fd} equivalent excitation voltage
T_e electric torque
T'_{do} time constant of excitation circuit.

3.2.2 Excitation System

The differential equations governing the behavior of an IEEE-DClA type excitation system are given by:

$$\dot{E}_{fdi} = \frac{1}{T_{Ai}}(K_{Ai}(v_{\text{refi}} - v_i + u_i) - E_{fdi}) \tag{12}$$

where

T_A regulator time constant
K_A regulator gain
v_{ref} reference voltage
v_i terminal voltage.

4 UPFC Model

The UPFC device is installed in power systems to exert continuous control over the voltage profile or power flow pattern [4]. This enables the voltage profile and power flows to be changed in such a way that thermal limits are not exceeded, stability margins are increased, losses minimized, contractual requirements fulfilled, etc., without violating the economic generation dispatch schedule [26]. However, the sole presence of these devices does not improve the overall damping of the system appreciably. To enforce extra damping, supplementary control is required to be added to these FACTS devices and methods for designing such damping controller through FACTS-devices occupies a significant part of this book.

In this section, the steady-state and small signal dynamic models of UPFC device are briefly described. The current injection model is used for steady-state representation of this device as it is most suitable for incorporation into an existing power flow algorithm without altering the bus admittance matrix Y [26].

4.1 Modified Heffron-Phillips Model

The dynamic model of the UPFC is required in order to study the effect of the UPFC for enhancing the small signal stability of the single machine power system. By applying Park's transformation and neglecting the resistance and transients of the ET and BT transformers, the UPFC can be modeled as [8–10]:

$$\begin{bmatrix} v_{Etd} \\ v_{Etq} \end{bmatrix} = \begin{bmatrix} 0 & -x_E \\ x_E & 0 \end{bmatrix} \begin{bmatrix} i_{Ed} \\ i_{Eq} \end{bmatrix} + \begin{bmatrix} \frac{m_E \cos \delta_E v_{dc}}{2} \\ \frac{m_E \sin \delta_E v_{dc}}{2} \end{bmatrix} \tag{13}$$

$$\begin{bmatrix} v_{\mathrm{Btd}} \\ v_{\mathrm{Btq}} \end{bmatrix} = \begin{bmatrix} 0 & -x_B \\ x_B & 0 \end{bmatrix} \begin{bmatrix} i_{Bd} \\ i_{Bq} \end{bmatrix} + \begin{bmatrix} \frac{m_B \cos \delta_B v_{dc}}{2} \\ \frac{m_B \sin \delta_B v_{dc}}{2} \end{bmatrix} \tag{14}$$

$$\dot{v}_{dc} = \frac{3m_E}{4C_{dc}} [\cos \delta_E \quad \sin \delta_E] \begin{bmatrix} i_{Ed} \\ i_{Eq} \end{bmatrix} + \frac{3m_B}{4C_{dc}} [\cos \delta_B \quad \sin \delta_B] \begin{bmatrix} i_{Bd} \\ i_{Bq} \end{bmatrix} \tag{15}$$

where, v_{Et}, i_E, v_{Bt}, and i_B are the excitation voltage, excitation current, boosting voltage, and boosting current, respectively; C_{dc} and v_{dc} are the DC link capacitance and voltage.

From Fig. 2 we can have:

$$\bar{v}_t = jx_{tE}(\bar{i}_B + \bar{i}_E) + \bar{v}_{Et} \tag{16}$$

$$\bar{v}_{Et} = \bar{v}_{Bt} + jx_{BV}\bar{i}_B + \bar{v}_b \tag{17}$$

$$\begin{aligned} v_{td} + jv_{tq} &= x_q(i_{Eq} + i_{Bq}) + j(E'_q - x'_d(i_{Ed} + i_{Bd})) \\ &= jx_{tE}(i_{Ed} + i_{Bd} + j(i_{Eq} + i_{Bq})) + v_{Etd} + jv_{Etq} \end{aligned} \tag{18}$$

where, i_t and v_b, are the armature current and infinite bus voltage, respectively. From the above equations, we can obtain:

$$i_{Ed} = \frac{x_{BB}}{x_{d\sum}} E'_q - \frac{m_E \sin \delta_E v_{dc} x_{Bd}}{2x_{d\sum}} + \frac{x_{dE}}{x_{d\sum}} \left(v_b \cos \delta + \frac{m_B \sin \delta_B v_{dc}}{2} \right) \tag{19}$$

$$i_{Eq} = \frac{m_E \cos\delta_E v_{dc} x_{Bq}}{2x_{q\sum}} - \frac{x_{qE}}{x_{q\sum}}\left(v_b \sin\delta + \frac{m_B \cos\delta_B v_{dc}}{2}\right) \quad (20)$$

$$i_{Bd} = \frac{x_E}{x_{d\sum}} E_q' + \frac{m_E \sin\delta_E v_{dc} x_{dE}}{2x_{d\sum}} - \frac{x_{dt}}{x_{d\sum}}\left(v_b \cos\delta + \frac{m_B \sin\delta_B v_{dc}}{2}\right) \quad (21)$$

$$i_{Bq} = -\frac{m_E \cos\delta_E v_{dc} x_{qE}}{2x_{q\sum}} + \frac{x_{qt}}{x_{q\sum}}\left(v_b \sin\delta + \frac{m_B \cos\delta_B v_{dc}}{2}\right) \quad (22)$$

where,

$$\begin{aligned}
&x_{q\sum} = (x_q + x_T + x_E)(x_B + \frac{x_L}{2}) + x_E(x_q + x_T)\\
&x_{Bq} = x_q + x_T + x_B + \frac{x_L}{2}\\
&x_{qt} = x_q + x_T + x_E;\ x_{qE} = x_q + x_T;\ x_{d\sum} = (x_d' + x_T + x_E)(x_B + \frac{x_L}{2}) + x_E(x_d' + x_T)\\
&x_{Bd} = x_d' + x_T + x_B + \frac{x_L}{2};\ x_{Bd} = x_d' + x_T + x_E;\ x_{dE} = x_d' + x_T;\ x_{BB} = x_B + \frac{x_L}{2}
\end{aligned} \quad (23)$$

The x_E, x_B, x_d, x_d', and x_q are the ET, BT reactance, d-axis reactance, d-axis transient reactance, and q-axis reactance, respectively. A linear dynamic model is obtained by linearizing the nonlinear model round an operating condition. The Phillips-Herffron linearized model of power system with UPFC is given as follows:

$$\Delta\dot{\delta} = \omega_0 \Delta\omega \quad (24)$$

$$\Delta\dot{\omega} = (-\Delta P_e - D\Delta\omega)/M \quad (25)$$

$$\Delta \dot{E}_q' = (-\Delta E_q + \Delta E_{fd})/T_{do}' \quad (26)$$

$$\Delta\dot{E}_{fd} = (K_A(\Delta v_{ref} - \Delta v) - \Delta E_{fd})/T_A \quad (27)$$

$$\Delta\dot{v}_{dc} = K_7\Delta\delta + K_8\Delta E_q' - K_9\Delta v_{dc} + K_{ce}\Delta m_E + K_{c\delta e}\Delta\delta_E + K_{cb}\Delta m_B + K_{c\delta b}\Delta\delta_B \quad (28)$$

$$\begin{aligned}\Delta P_e = {} & K_1\Delta\delta + K_2\Delta E_q' + K_{pd}\Delta v_{dc} + K_{pe}\Delta m_E + K_{p\delta e}\Delta\delta_E\\ & + K_{pb}\Delta m_B + K_{p\delta b}\Delta\delta_B\end{aligned} \quad (29)$$

$$\begin{aligned}\Delta E_q' = {} & K_4\Delta\delta + K_3\Delta E_q' + K_{qd}\Delta v_{dc} + K_{qe}\Delta m_E + K_{q\delta e}\Delta\delta_E\\ & + K_{qb}\Delta m_B + K_{q\delta b}\Delta\delta_B\end{aligned} \quad (30)$$

$$\begin{aligned}\Delta V_t = {} & K_5\Delta\delta + K_6\Delta E_q' + K_{vd}\Delta v_{dc} + K_{ve}\Delta m_E + K_{v\delta e}\Delta\delta_E\\ & + K_{vb}\Delta m_B + K_{v\delta b}\Delta\delta_B\end{aligned} \quad (31)$$

where, K_1, $K_2 \ldots K_9$, K_{pu}, K_{qu}, and K_{vu} are linearization constants. The state-space model of power system is given by:

$$\dot{x} = Ax + Bu \tag{32}$$

where, the state vector x, control vector u, A and B are:

$$x = \begin{bmatrix} \Delta\delta & \Delta\omega & \Delta E_q' & \Delta E_{fd} & \Delta v_{dc} \end{bmatrix}; \; u = \begin{bmatrix} \Delta m_E & \Delta\delta_E & \Delta m_B & \Delta\delta_B \end{bmatrix}^T$$

$$A = \begin{bmatrix} 0 & w_0 & 0 & 0 & 0 \\ -\frac{K_1}{M} & 0 & -\frac{K_2}{M} & 0 & -\frac{K_{pd}}{M} \\ -\frac{K_4}{T_{do}'} & 0 & -\frac{K_3}{T_{do}'} & \frac{1}{T_{do}'} & -\frac{K_{qd}}{T_{do}'} \\ -\frac{K_A K_5}{T_A} & 0 & -\frac{K_A K_6}{T_A} & -\frac{1}{T_A} & -\frac{K_A K_{vd}}{T_A} \\ K_7 & 0 & K_8 & 0 & -K_9 \end{bmatrix}$$

$$B = \begin{bmatrix} 0 & 0 & 0 & 0 \\ -\frac{K_{pe}}{M} & -\frac{K_{p\delta e}}{M} & -\frac{K_{pb}}{M} & -\frac{K_{p\delta b}}{M} \\ -\frac{K_{qe}}{T_{do}'} & -\frac{K_{q\delta e}}{T_{do}'} & -\frac{K_{qb}}{T_{do}'} & -\frac{K_{q\delta b}}{T_{do}'} \\ -\frac{K_A K_{vc}}{T_A} & -\frac{K_A K_{v\delta e}}{T_A} & -\frac{K_A K_{vb}}{T_A} & -\frac{K_A K_{v\delta b}}{T_A} \\ K_{ce} & K_{c\delta e} & K_{cb} & K_{c\delta b} \end{bmatrix}$$

The block diagram of the linearized dynamic model of the SMIB power system with UPFC is shown in Fig. 5.

4.2 Power Injection Model

The UPFC consists of series and shunt voltage-source converters, which can be modeled as controllable voltage sources. The shunt converter is primarily used to provide active power demand of the series converter through the common DC link. The shunt converter can also generate or absorb reactive power, if it is desired, and thereby provides independent shunt reactive compensation for the line. The series converter provides the main function of the UPFC by injecting a voltage with controllable magnitude and phase angle in series with the line in the multi-machine power system case study, Fig. 4. The power balance between the series and the shunt converters is a prerequisite to maintain a constant voltage across the dc capacitor connected between the two converters [4].

The reactance x_s describes the reactance seen from terminals of the series transformer and is equal to:

$$x_s = x_k \cdot \frac{S_B}{S_s} \cdot r_{\max}^2 \tag{33}$$

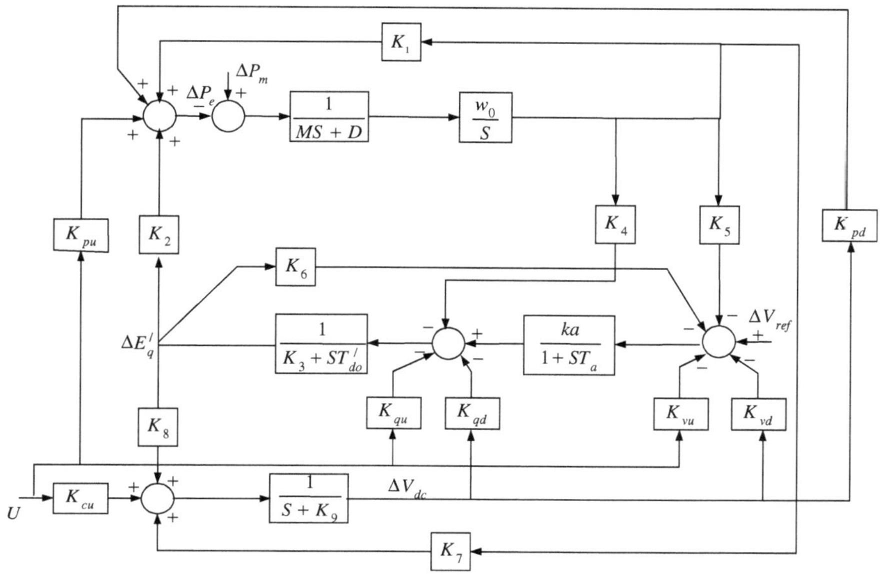

Fig. 5 Modified Heffron–Phillips transfer function model

where, x_k denotes the series transformer reactance, $r_{\max}$ is the maximum per unit value of injected voltage magnitude, S_B is the system base power, and S_S is the nominal rating power of the series converter.

The UPFC injection model is derived enabling three parameters to be simultaneously controlled [27]. They are namely the shunt reactive power, Q_{sh}, and the magnitude, r, and the angle, γ, of injected series voltage V_{se}. Figure 6 shows the circuit representation of a UPFC where the series connected voltage source is modeled by an ideal series voltage which is controllable in magnitude and phase and the shunt converter is modeled as an ideal shunt current source. In Fig. 6, we get:

$$I_{sh} = \bar{I}_t + \bar{I}_q = (I_t + jI_q)e^{j\theta_i} \tag{34}$$

where, I_t is the current in phase with V_i and I_q is the current in quadrature with V_i. In Fig. 6 the voltage source V_{se} is replaced by the current source $\bar{I}_{inj} = -jb_sV_{se}$ in parallel with x_s.

Fig. 6 The UPFC electric circuit

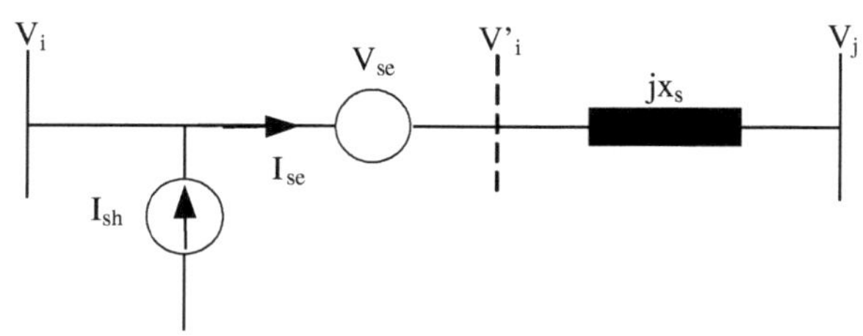

The active power supplied by the shunt current source can be calculated from:

$$P_{sh} = \mathrm{Re}[\bar{V}_i(-\bar{I}_{sh}^*)] = -V_i I_t \tag{35}$$

With the UPFC losses neglected, we get:

$$P_{sh} = P_{se} \tag{36}$$

The apparent power supplied by the series voltage source converter is calculated from:

$$S_{se} = \bar{V}_{se}.\bar{I}_{se}^* = re^{j\gamma}\bar{V}_i\left(\frac{\bar{V}' - V_j}{jx_s}\right)^* \tag{37}$$

Active and reactive powers supplied by series converter are distinguished as following:

$$P_{se} = rb_s V_i V_j \sin(\theta_i - \theta_j + \gamma) - rb_s V_i^2 \sin\gamma \tag{38}$$

$$Q_{se} = rb_s V_i V_j \cos(\theta_i - \theta_j + \gamma) - rb_s V_i^2 \cos\gamma \tag{39}$$

Substitution of Eqs. (38) and (35) into (36) gives:

$$I_t = -rb_s V_i V_j \sin(\theta_i - \theta_j + \gamma) + rb_s V_i \sin\gamma \tag{40}$$

The current of the shunt source is then given by:

$$I_{sh} = (-rb_s V_j \sin(\theta_{ij} + \gamma) + rb_s V_i \sin\gamma + jI_q)e^{j\theta_i} \tag{41}$$

From Fig. 7 the bus current injections can be defined as:

$$\begin{aligned} \bar{I}_i &= \bar{I}_{sh} - \bar{I}_{inj} \\ \bar{I}_j &= \bar{I}_{inj} \end{aligned} \tag{42}$$

where,

$$\bar{I}_{inj} = -jb_s V_{se} = -jb_s r\bar{V}_i e^{j\gamma} \tag{43}$$

Substituting (41) and (43) into (42) gives:

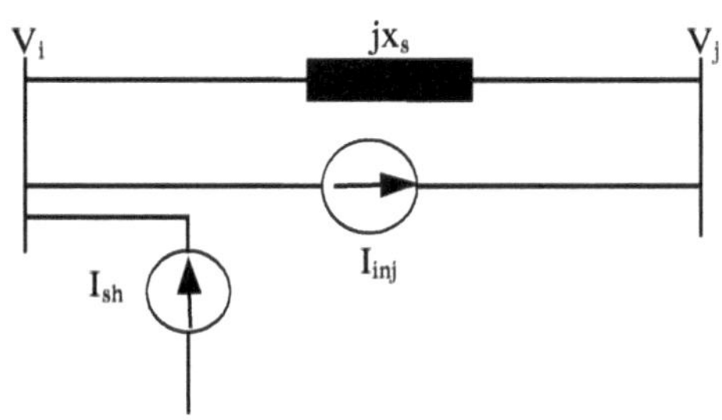

Fig. 7 Transformed series voltage source

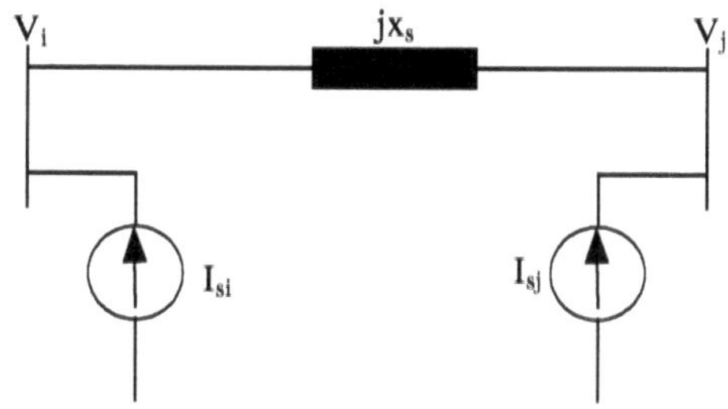

Fig. 8 UPFC current injection model

$$\begin{aligned}\bar{I}_{si} &= (-rb_sV_j\sin(\theta_{ij}+\gamma) + rb_sV_i\sin\gamma + jI_q)e^{j\theta_i} + jb_srV_ie^{j(\gamma+\theta)}\\ \bar{I}_{sj} &= -jb_srV_ie^{j(\gamma+\theta)}\end{aligned} \tag{44}$$

where, I_q is an independently controlled variable, representing a shunt reactive source. Based on Eq. (44), the current injection model can be presented as in Fig. 8. If there is a control objective to be achieved, the bus current injections are modified through changes of the UPFC control parameters r, γ, and I_q [27, 28].

5 Damping Controller for the UPFC

The damping controller is designed to produce an electrical torque in phase with the speed deviation according to phase compensation method. The four control parameters of the UPFC (m_B, m_E, δ_B, and δ_E) can be modulated in order to produce the damping torque. In this research δ_E and m_B are modulated in order to damping controller design in the SMIB. The structure of UPFC based damping controller is shown in Fig. 9. This controller may be considered as a lead-lag compensator [1]. However, an electrical torque in phase with the speed deviation is to be produced in order to improve damping of the system oscillations. It comprises gain block, signal-washout block, and lead-lag compensator. The parameters of the damping controller are obtained using CPSO and QPSO algorithms.

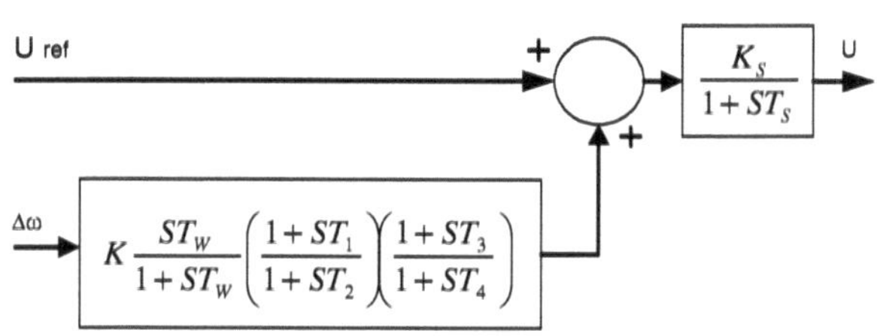

Fig. 9 UPFC with lead-lag controller

6 POD Controller Design Using PSO and QPSO

In the proposed methods, we must tune the UPFC controller parameters optimally to improve overall system dynamic stability in a robust way under different operating conditions and disturbances. To acquire an optimal combination, this research employs CPSO and QPSO to improve optimization synthesis and find the global optimum value of fitness function. For our optimization problem, two different objective functions are proposed for the PSSs design problem. The first objective function [29, 30] is eigenvalues based comprising the damping factor and the damping ratio of the lightly damped electromechanical modes while the second is time domain-based multi-objective function [1, 28] and are defined as follows:

$$\begin{aligned} f_1 &= J_1 + a \times J_2 \\ J_1 &= \sum_{j=1}^{NP} \sum_{\sigma_i \geq \sigma_0} (\sigma_0 - \sigma_i)^2, \quad J_2 = \sum_{j=1}^{NP} \sum_{\zeta_i \leq \zeta_0} (\zeta_0 - \zeta_i)^2 \end{aligned} \tag{45}$$

$$f_2 = \sum_{j=1}^{NP} \sum_{i=1}^{n} \int_0^{\mathrm{tsim}} t|\Delta\omega_i|dt \tag{46}$$

where, $\sigma_{i,j}$ and $\zeta_{i,j}$ are the real part and the damping ratio of the *i*th eigenvalue of the *j*th operating point. Also, speed deviation ($\Delta\omega$) of the machines is considered for the evaluation of the f_2. The value of α is chosen at 10. The t_{sim} is the time range of simulation, n is the number of machines, and NP is the total number of operating points for which the optimization is carried out. The value of σ_0 determines the relative stability in terms of damping factor margin provided for constraining the placement of eigenvalues during the process of optimization. By optimizing f_1, closed loop system poles are consistently pushed further left of the $j\omega$ axis with simultaneous reduction in real parts, too. Thus, enhancing relative stability and increasing the damping ratio over the ζ_0 are achieved. Finally, all the closed loop system poles should lay within a D-shaped sector are shown in Fig. 10 in the negative half plane of the $j\omega$ axis for which $\sigma_i < -2$, $\zeta_i > 0.3$. It is necessary to mention here that only the unstable or lightly damped electromechanical modes of the oscillations are relocated. For the second objective function calculation, the time-domain simulations of the power system model is carried out for the simulation period. It is aimed to minimize this fitness function in order to improve the system response in terms of the settling time and overshoots. The advantage of this selected objective function is that minimal dynamic plant information is needed. It is necessary to mention here that only the unstable or lightly damped electromechanical modes of oscillations are relocated. The design problem can be formulated as the following constrained optimization problem, where the constraints are the controller parameters bounds:

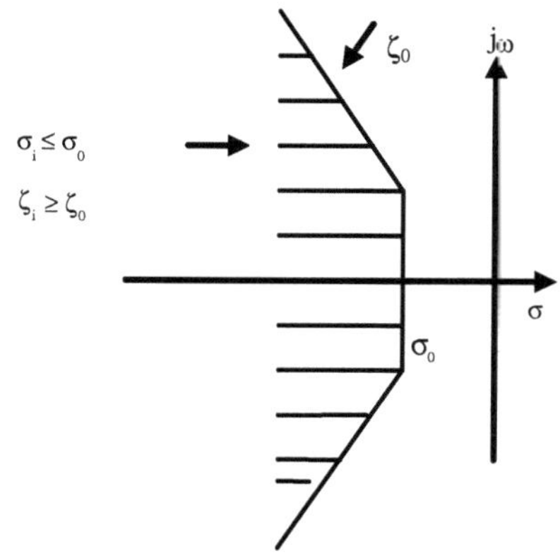

Fig. 10 Region of eigenvalue location for f_1 objective function

Minimize f_i *Subject to:*

$$K^{\min} \le K \le K^{\max}$$
$$T_i^{\min} \le T_i \le T_i^{\max} \tag{47}$$

It is emphasized that with this procedure, robust stabilizer, enable to operate satisfactorily over a wide range of the operating conditions, are obtained. Typical ranges of the optimized parameters are [0.01–100] for K and [0.01–1] for T_1, T_2, T_3, and T_4. The optimization of UPFC controller parameters is carried out by evaluating the multi-objective cost functions as given in Eqs. (45) and (46), which consider a multiple of operating conditions. The operating conditions are considered in the single machine and multi-machine power systems are given in the Tables 1 and 2, respectively.

In single machine power system, the values of σ_0 and ζ_0 are taken as −2 and 0.3, respectively. In order to acquire better performance, number of particle, particle size, number of iteration, c_1, c_2, and c is chosen as 30, 5, 50, 2, 2, and 1, respectively. Also, the inertia weight, w, is linearly decreasing from 0.9 to 0.4. In multi-machine power system, number of particle, particle size, number of iterations, c_1, c_2, and c are chosen as 50, 5, 200, 2, 2, and 1, respectively. It should be noted that the PSO algorithm is run several times and then optimal set of coordinated controller parameters is selected. The final values of the optimized parameters with both objective functions f_1 and f_2 with CPSO and QPSO algorithms in two case studies are given in Tables 3, 4, and 5.

Table 1 Operating conditions in single machine power system

Operating condition	P	Q	XL
Base case	0.8	0.114	0.3
Case 1	0.2	0.01	0.3
Case 2	1.2	0.4	0.3
Case 3	0.8	0.114	0.6
Case 4	1.2	0.4	0.6

Table 2 Operating conditions in multi-machine power system

Operating condition	P1	Q1	P2	Q2	P3	Q3	P4	Q4
Case 1	0.7778	0.2056	0.5556	0.2611	0.8020	0.0697	0.8889	0.2244
Case 2	0.5556	0.2056	0.5556	0.2611	1.3739	0.1502	0.5556	0.2244
Case 3	0.9911	0.1722	0.9444	0.3944	0.0095	0.0712	1.1111	0.2222

Table 3 The optimal parameter settings of the proposed controllers based on the f_1 objective function based on the δ_E and m_B in SMIB

Controller parameters	δ_E			m_B		
	J_1	J_2	f_1	J_1	J_2	f_1
K	64.45	94.2	100	100	79.34	68.65
T_1	0.4185	0.2566	0.1069	0.6	0.3235	0.01
T_2	0.5299	0.1563	0.2022	0.3655	0.1424	0.1105
T_3	0.3835	0.1361	0.4347	0.5262	0.4523	0.4704
T_4	0.3507	0.0965	0.4134	0.4489	0.5094	0.1235

Table 4 The optimal parameter settings of the proposed controllers based f_2 objective function based on the δ_E and m_B in SMIB

Controller parameters	QPSO based δ_E controller	PSO based δ_E controller	QPSO based m_B controller	PSO based m_B controller
K	97.75	74.35	96.70	89.12
T_1	0.7111	0.0665	0.0105	0.3227
T_2	0.6206	0.0177	0.1112	0.1408
T_3	0.5776	0.2565	0.4667	0.9457
T_4	0.6412	0.1908	0.1230	0.5011

Table 5 The optimal parameter settings of the proposed controllers in multi-machine power system

Type of algorithm	K	T_1	T_2	T_3	T_4
QPSO	37.05	1.5089	0.0855	0.0981	1.4550
CPSO	17.67	0.93125	0.0312	0.0288	0.4389

7 Results and Discussion

7.1 Eigenvalue Analysis

The electromechanical modes and the damping ratios obtained for all operating conditions both with and without proposed controllers in the SMIB are given in Tables 6 and 7. When UPFC is not installed, it can be seen that some of the modes are poorly damped and in some cases, are unstable (highlighted in Tables 6 and 7). It is also clear that the system damping with the proposed f_l based tuned UPFC controller are significantly improved. Moreover, it can be seen that electromechanical mode controllability via δ_E is higher than that m_B input.

7.2 Nonlinear Time Domain Simulation

7.2.1 Single Machine Power System

To assess the effectiveness and robustness of the proposed controllers, transient conditions are verified by applying a 6-cycle three-phase fault at $t = 1$ s, at the middle of the one transmission line. The fault is cleared by permanent tripping of the faulted line. The performance of the controllers when the multi-objective function (f_l) is used in the design is compared to that of the controllers designed using the single objective functions J_l and J_2. The speed deviation of generator at base case, case 2, and case 4 due to designed controller based on the δ_E and m_B are shown in Figs. 11 and 12. It can be seen that the PSO based UPFC controller tuned using the multi-objective function achieves good robust performance, provides superior damping in comparison with the other objective functions and enhance greatly the dynamic stability of power systems.

The performance of the controllers when the quantum-behaved particle swarm optimization is used in the design is compared to that of the controllers designed using the classical particle swarm optimization with second objective function. The speed deviation of generator at nominal, light, and heavy loading conditions due to designed controller based on the δ_E and m_B are shown in Figs. 13 and 14. It can be seen that the QPSO based UPFC controller achieves good robust performance, provides superior damping in comparison with the CPSO based UPFC controller and enhance greatly the dynamic stability of power system.

Table 6 Eigenvalues and damping ratios of electromechanical modes with and without CPSO based δ_E controller

Objective functions	Base case	Case 1	Case 2	Case 3	Case 4
Without controller	0.197 ± i4.51, −0.04	0.03 ± i5.32, −0.006	0.285 ± i4.49, −0.06	0.15 ± i4.03, −0.036	0.23 ± i3.88, −0.059
	−2.99 ± i0.17, 0.99	−2.7951, −3.1728	−3.1878, −2.9126	−3.18 ± i0.043, 0.99	−3.3868, −3.0673
	−96.582	−96.268	−96.643	−96.407	−96.48
J_1	−2.012 ± i6.994, 0.27	−2.071 ± i6.984, 0.28	−2.112 ± i7.901, 0.25	−2.079 ± i7.256, 0.27	−2.160 ± i7.022, 0.29
	−2.772 ± i0.144, 0.98	−2.788 ± i0.092, 0.99	−2.763 ± i0.125, 0.99	−2.894 ± i0.078, 0.99	2.887 ± i0.059, 0.99
	−3.97, −1.8499	−3.3636, −1.7719	−4.3202, −1.8653	−4.2825, −1.7806	−4.7938, −1.8063
	−96.544	−96.262	−96.598	−96.377	−96.443
J_2	−1.031 ± i2.830, 0.34	−1.840 ± i2.946, 0.54	−0.9004 ± i2.801, 0.31	−1.284 ± i2.075, 0.52	−1.130 ± i2.011, 0.49
	−6.9885, −2.7324	−2.7044, −3.2538	−7.016, −2.7103	−24.963, −7.0502	−25.868, −7.0744
	−3.3804, −20.766	−22.579, −6.9473	−3.5115, −21.66	−2.9761, −3.4973	−3.6079, −2.965
	−96.403	−96.24	−96.427	−96.259	−96.294
f_1	−2.961 ± i4.392, 0.56	−3.173 ± i6.196, 0.45	−3.337 ± i4.141, 0.63	−4.206 ± i5.034, 0.64	−4.507 ± i4.918, 0.67
	−2.783 ± i2.264, 0.77	−3.137 ± i0.736, 0.97	−2.473 ± i2.638, 0.68	−2.188 ± i1.333, 0.85	−2.949 ± i1.510, 0.89
	−2.460 ± i0.121, 0.99	−2.428 ± i0.279, 0.98	−2.468 ± i0.089, 0.99	−2.5341, −2.7465	−2.4526, −2.8285
	−96.546	−96.262	−96.6	−96.378	−96.445

Table 7 Eigenvalues and damping ratios of electromechanical modes with and without CPSO based m_B controller

Objective functions	Base case	Case 1	Case 2	Case 3	Case 4
Without controller	0.197 ± i4.51, −0.04	0.03 ± i5.32, −0.006	0.285 ± i4.49, −0.06	0.15 ± i4.03, −0.036	0.23 ± i3.88, −0.059
	−2.99 ± i0.17, 0.99	−2.7951, −3.1728	−3.1878, −2.9126	−3.18 ± i0.043, 0.99	−3.3868, −3.0673
	−96.582	−96.268	−96.643	−96.407	−96.48
J_1	−2.075 ± i6.844, 0.28	−2.197 ± i7.063, 0.29	−2.015 ± i7.801, 0.25	−2.060 ± i6.744, 0.289	−2.067 ± i7.454, 0.27
	−2.019 ± i1.779, 0.75	−2.806 ± i1.318, 0.9	−2.969 ± i1.873, 0.84	−2.749 ± i1.736, 0.84	−2.794 ± i1.951, 0.82
	−2.227 ± i0.026, 0.99	−2.5071, −2.1509	−2.217 ± i0.069, 0.99	−2.6419, −2.1348	−2.6123, −2.1367
	−96.764	−96.351	−96.842	−96.504	−96.587
J_2	−1.323 ± i4.027, 0.31	−1.103 ± i3.008, 0.34	−1.406 ± i4.207, 0.32	−1.786 ± i3.847, 0.42	−1.1739 ± i3.684, 0.30
	−3.121 ± i0.249, 0.99	−7.5057, −1.9445	−6.548, −1.947	−7.173, -1.9377	−7.1849, −1.9378
	−1.9436, −6.6017	−3.1945, −2.8223	−3.0151, −3.477	−3.135, −3.3676	−3.5778, −3.0725
	−96.737	−96.339	−96.812	−96.489	−96.571
f_1	−2.668 ± i4.105, 0.55	−2.69 ± i5.159, 0.462	−3.928 ± i4.463, 0.66	−2.098 ± i3.795, 0.48	−2.027 ± i3.595, 0.49
	−3.647 ± i3.652, 0.70	−6.791 ± i3.813, 0.87	−2.409 ± i3.613, 0.55	−6.841 ± i3.024, 0.91	−6.725 ± i3.282, 0.89
	−2.6635, −7.8836	−3.7891, −2.7154	−2.643, −7.8342	−4.7565, −2.8963	−5.0951, −2.8801
	−96.581	−96.267	−96.642	−96.406	−96.479

Fig. 11 Dynamic responses for $\Delta\omega$ in scenario 1 with δ_E controller at (**a**) base case (**b**) case 2 and (**c**) case 4 loading; *Solid* (f_1), *Dashed* (J_2), and *Dotted* (J_1)

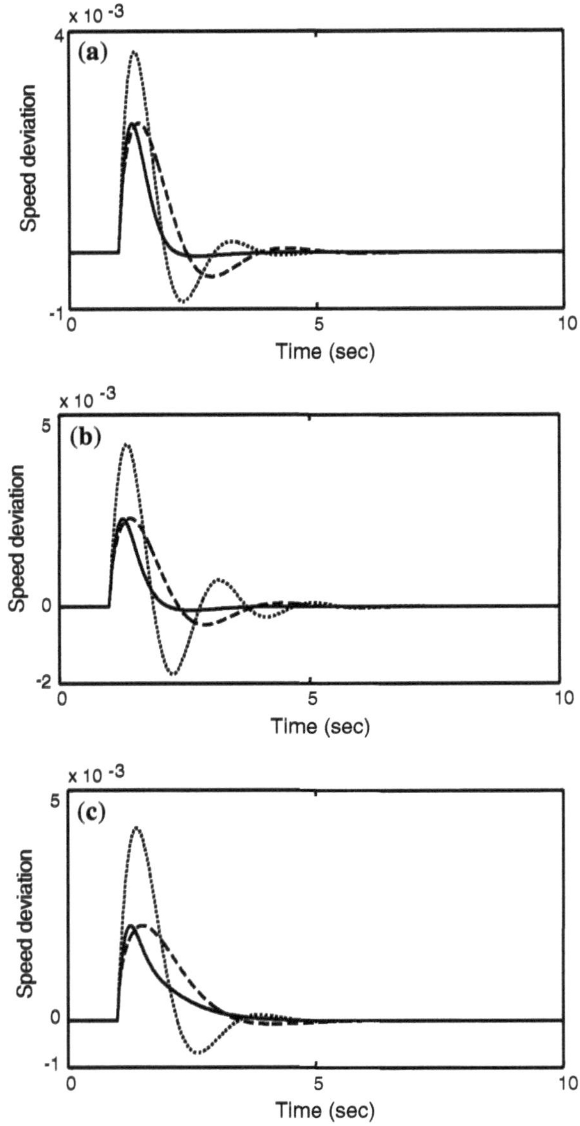

7.2.2 Multi-Machine Power System

To test the validity and robustness of the performance of the proposed control algorithm, a three-phase fault of 100 ms duration at the middle of one of the transmission lines between bus-7 and bus-8 is applied. The inter-area and local mode of oscillations are shown in Figs. 15, 16, 17, respectively. The performance

Fig. 12 Dynamic responses for $\Delta\omega$ in scenario 1with m_B controller at (**a**) base case (**b**) case 2 and (**c**) case 4 loading; *Solid* (f_1), *Dashed* (J_2), and *Dotted* (J_1)

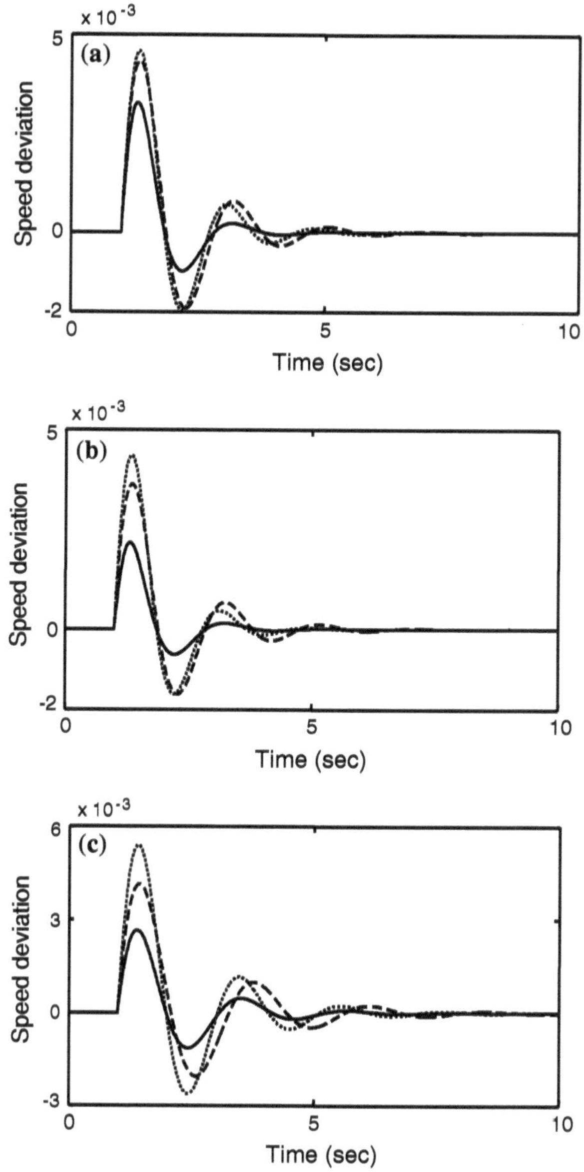

of the QPSO based damping controller is quite prominent in comparison with the CPSO based damping controller and the overshoots and settling time are significantly improved for the proposed controller.

Fig. 13 Dynamic responses for $\Delta\omega$ in scenario 1 at (**a**) nominal (**b**) light (**c**) heavy loading conditions; *Solid* (QPSO based δ_E controller) and *Dashed* (PSO based δ_E controller)

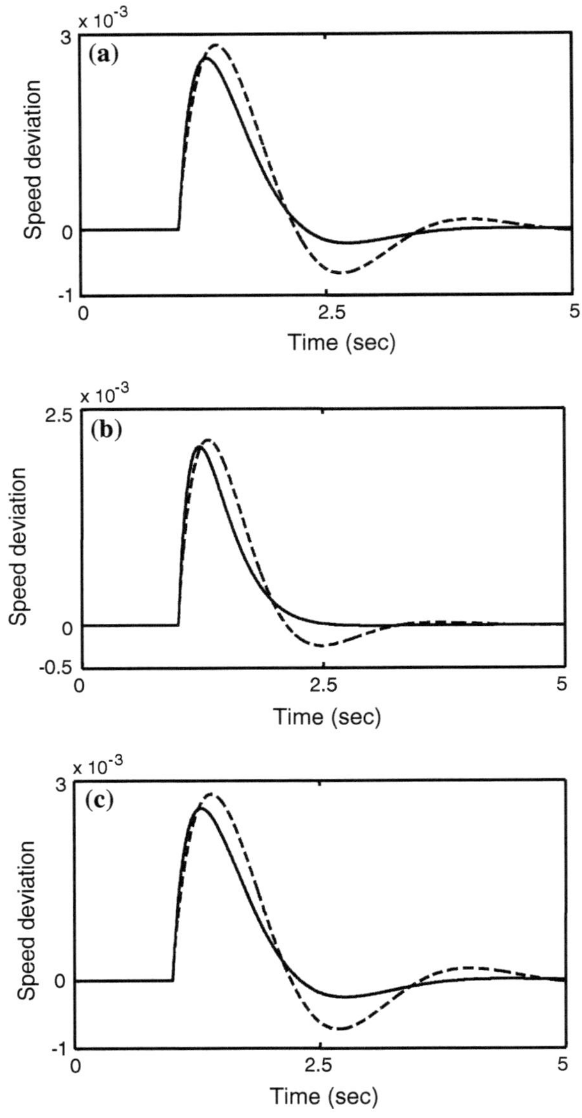

Fig. 14 Dynamic responses for $\Delta\omega$ in scenario 1 at (**a**) nominal (**b**) light (**c**) heavy loading conditions; *Solid* (QPSO based m_B controller) and *Dashed* (PSO based m_B controller)

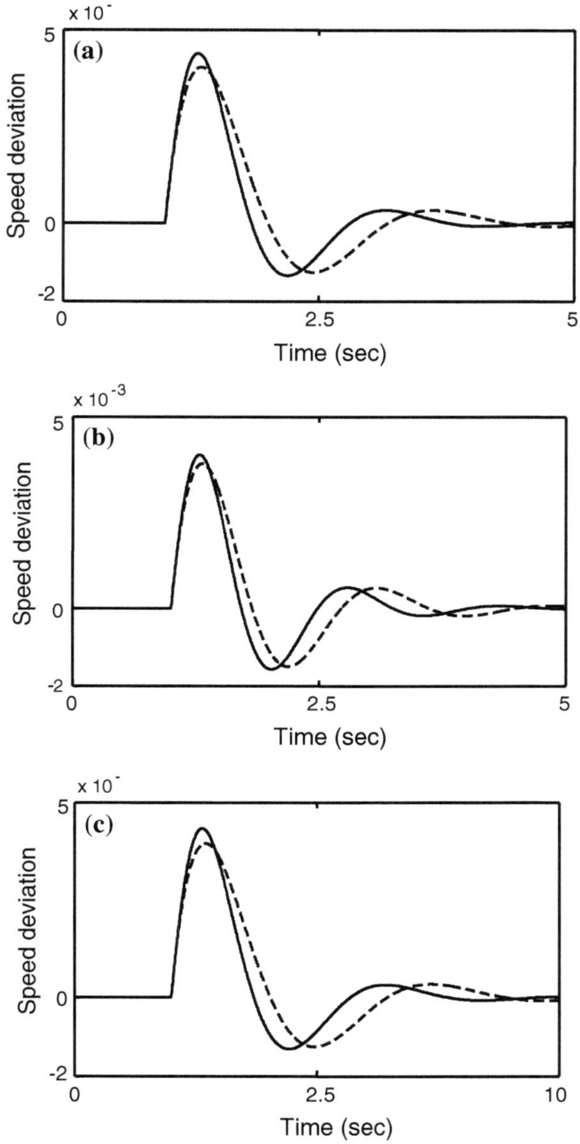

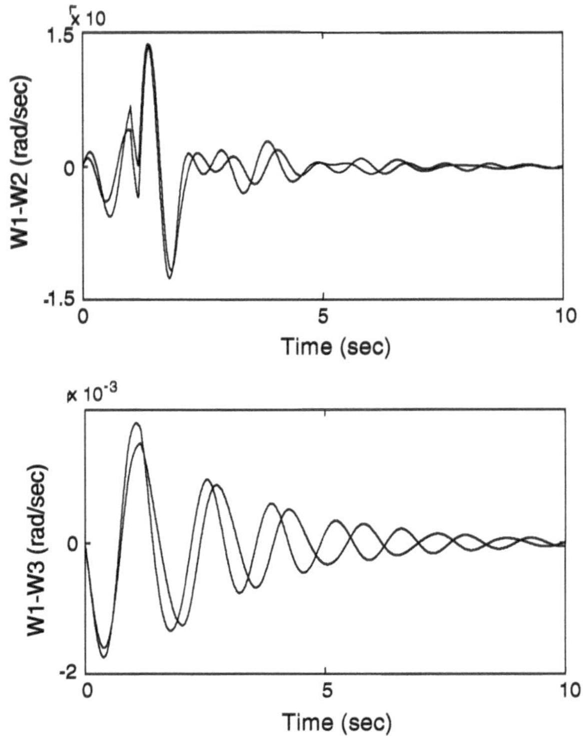

Fig. 15 Inter-area and local mode of oscillations for case 1; *Solid* (PSO) and *Dashed* (QPSO)

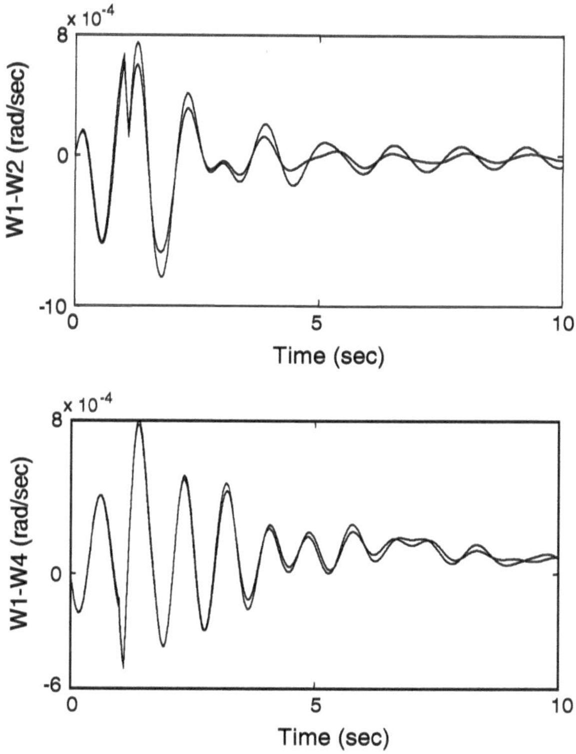

Fig. 16 Inter-area and local mode of oscillations for case 2; *Solid* (PSO) and *Dashed* (QPSO)

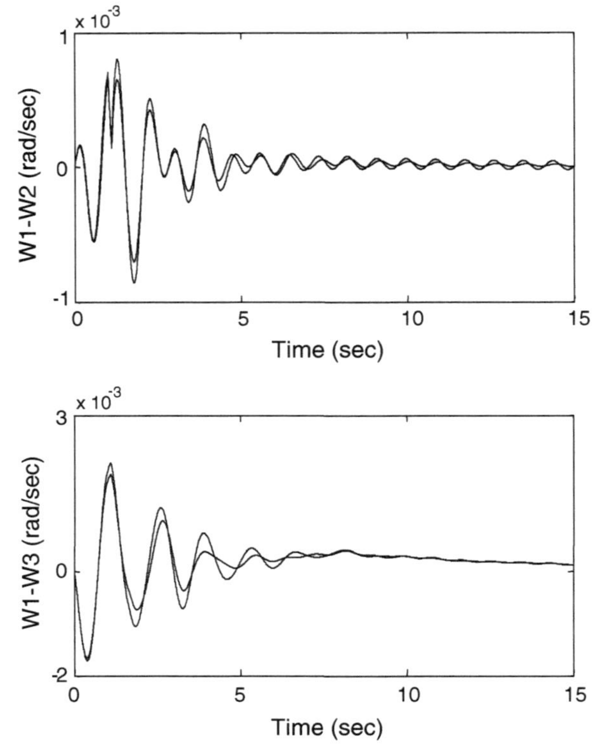

Fig. 17 Inter-area and local mode of oscillations for case 3; *Solid* (PSO) and *Dashed* (QPSO)

8 Conclusions

In this chapter, in order to maintain the transient stability and damp the power system oscillation after fault effectively, a robust damping controller for the UPFC has been designed. Controller design problem is formulated as an optimization problem for different operating conditions. Then, the CPSO and QPSO algorithms have been successfully applied to find the optimal solution of the design problem. Two different objective functions based eigenvalue analysis and time simulation are proposed in this work for the design of damping controller. The effectiveness of the proposed controllers has been tested on single machine and multi-machine power systems through the simulation studies under different operating conditions. The simulation results show the effectiveness of the proposed controllers and their ability to provide good damping of low frequency oscillations. Compared with the CPSO method, the QPSO technique demonstrates its superiority in computational complexity, success rate, and solution quality.

References

1. Shayeghi H, Shayanfar HA, Jalilzadeh S, Safari A (2009) Design of output feedback UPFC controllers for damping of electromechanical oscillations using PSO. Energy Conver Manage 50:2554–2561
2. Anderson PM, Fouad AA (1977) Power system control and stability. Iowa State University Press, Ames
3. Keri AJF, Lombard X, Edris AA (1999) Unified power flow controller: modeling and analysis. IEEE Trans Power Deliv 14(2):648–654
4. Hingorani NG, Gyugyi L (1999) Understanding FACTS: concepts and technology of flexible AC transmission systems. Wiley-IEEE Press, New York
5. Gyugyi L (1992) Unified power-flow control concept for flexible ac transmission systems. IEE Proc Gen Transm Distrib 139 (4): 323–331
6. Song YH, Johns AT (1999) Flexible ac transmission systems (FACTS). IEE Press, London
7. Nabavi-Niaki A, Iravani MR (1996) Steady-state and dynamic models of unified power flow controller (UPFC) for power system studies. IEEE Trans Power Syst 11(4):1937–1943
8. Wang HF (1999) Damping function of unified power flow controller. IEE Proc Gen Transm Dist 146(1):81–87
9. Wang HF (1999) Application of modeling UPFC into multi-machine power systems. IEE Proc Gen Transm Dist 146(3):306–312
10. Wang HF (2000) A unified model for the analysis of FACTS devices in damping power system oscillations - Part III: unified power flow controller'. IEEE Trans Power Deliv 15(3):978–983
11. Rouco L (2001) Coordinated design of multiple controllers for damping power system oscillations. Elect Power Energy Syst 23:517–530
12. Dash PK, Mishra S, Panda G (2000) A radial basis function neural network controller for UPFC. IEEE Trans Power Syst 15(4):1293–1299
13. Kazemi A, Vakili Sohrforouzani M (2006) Power system damping controlled facts devices. Elect Power Energy Syst 28:349–357
14. Kundur P (1994) Power system stability and control. McGrew Hill, New York
15. Khon L, Lo KL (2006) Hybrid micro-GA based FLCs for TCSC and UPFC in a multi machine environment. Elec Power Syst Res 76:832–843
16. Mok TK, Liu H, Ni Y, Wu FF, Hui R (2005) Tuning the fuzzy damping controller for UPFC through genetic algorithm with comparison to the gradient descent training. Elect Power Energy Syst 27:275–283
17. Tripathy M, Mishra S, Venayagamoorthy GK, Bacteria foraging: a new tool for simultaneous robust design of UPFC controllers. In: Proceedings of the international joint conference on neural networks, pp 2274–2280
18. Shayeghi H, Jalili A, Shayanfar HA (2008) Multi-stage fuzzy load frequency control using PSO. Energy Conver Manage 49:2570–2580
19. Eberhart R, Kennedy J (1995). A new optimizer using particle swarm theory. In: Proceedings of the sixth international symposium on micro machine and human science, pp 39–43
20. Clerc M, Kennedy J (2002) The particle swarm-explosion, stability, and convergence in a multidimensional complex space. IEEE Trans Evolutionary Comput 6(1):58–73
21. Poli R, Kennedy J, Blackwell T (2007) Particle swarm optimization: an overview. Swarm Intell 1:33–57
22. Shayeghi H, Shayanfar HA, Jalilzadeh S, Safari A (2010) Tuning of damping controller for UPFC using quantum particle swarm optimizer. Energy Conver Manage 51:2299–2306
23. Coelho LS (2008) A quantum particle swarm optimizer with chaotic mutation operator. Chaos Soliton Fractals 37:1409–1418
24. Sun J, Fang W, Chen W, Xu W (2008) Design of two dimensional IIR digital filters using an improved quantum-behaved particle swarm optimization algorithm. In: Proceedings of the American control conference, pp 2603–2608

25. Sun J, Fang W, Chen W, Xu W (2005) Adaptive parameter control for quantum-behaved particle swarm optimization on individual level. In: Proceedings of the IEEE international conference on systems, man and cybernetics, Big Island, USA, pp 3049–3054
26. Noroozian M, Angquist L, Ghandhari M, Andersson G (1997) Improving power system dynamics by series connected FACTS devices. IEEE Trans Power Deliv 12(4):1635–1641
27. Sadikovic R (2006) Use of FACTS devices for power flow control and damping of oscillations in power systems. Ph. D. dissertation, University of Tuzla, Zurich, Swiss
28. Mishra S (2006) Neural network based adaptive UPFC for improving transient stability performance of power system. IEEE Trans Neural Netw 17(2):461–470
29. Abdel-Magid YL, Abido MA (2003) Optimal multiobjective design of robust power system stabilizers using genetic algorithms. IEEE Trans Power Syst 18(3):1125–1132
30. Nguyen TT, Gianto R (2007) Optimisation based control coordination of PSSs and FACTS devices for optimal oscillations damping in multi-machine power system. IET Proc Gener Transm Distrib 1(4):564–573

Application of PSO and GA for Transmission Network Expansion Planning

Hossein Shayeghi and Meisam Mahdavi

Abstract TNEP is one of the important parts of power system planning which determines the number, time, and location of new lines for adding to transmission network. It is a hard, large-scale and highly nonlinear combinatorial optimization problem that can be solved by classic, nonclassic or heuristic methods. Classic methods like linear programming and Bender decomposition are only based on mathematical principles, but their difficulty is that if the scale of problem is large, it is very difficult to find accurate and reasonable solutions. Contrary to classic methods, nonclassic ones such as evolutionary algorithms like GAs are not based on mathematical rules and simply can be applied for solution of complex problems. GA is a random search method that has demonstrated the ability to deal with nonconvex, nonlinear, integer-mixed optimization problems like the STNEP problem. Although global optimization techniques like GA to be good methods for the solution of TNEP problem, however, when the system has a highly epistatic objective function and number of parameters to be optimized is large, then they have degraded efficiency to obtain global optimum solution and also simulation process use a lot of computing time. Heuristic methods like PSO can improve speed and accuracy of the solution program. PSO is a novel population-based heuristic that is a useful tool for engineering optimization. Unlike the other heuristic techniques, it has a flexible and well-balanced mechanism to enhance the global and local exploration abilities. In this chapter, first we review some research in the field of TNEP. Then, the method of mathematical modeling for TNEP problem is presented. Afterward, GA and PSO algorithms are described completely. Finally, effective parameters on network losses with a few examples are introduced.

H. Shayeghi (✉)
Technical Engineering Department, University of Mohaghegh Ardabili, Ardabili, Iran
e-mail: hshayeghi@gmail.com

M. Mahdavi
School of ECE, College of Engineering, University of Tehran, Tehran, Iran
e-mail: me.mahdavi@ut.ac.ir

N. Bizon et al. (eds.), *Analysis, Control and Optimal Operations in Hybrid Power Systems*, Green Energy and Technology, DOI: 10.1007/978-1-4471-5538-6_6,

1 Introduction

TNEP is an important part of power system planning. Its task is to determine an optimal network configuration according to load growth while meeting imposed technical, economic, and reliability constraints. The basic principle of TNEP is to minimize the network construction and operational cost while satisfying the requirement of delivering electric power safely and reliably to load centers along the planning horizon [1–3]. In majority of power systems, generating plants are located far from the load centers. In addition, the planned new projects are still so far from completion. Due to these situations, the investment cost for transmission network is huge. Thus, the TNEP problem acquires a principal role in power system planning and should be evaluated carefully, because any effort to reduce the cost of the transmission system expansion by some fraction of a percent allows saving of a significant amount of capital. On the other hand, determination of investment cost for power system expansion is very difficult work, because we should accept the fact that a costumer plays important role in the competitive electricity market, so, this cost should be determined from grid owners with agreement of customer and considering the various reliability criteria [1, 4].

Generally, TNEP could be classified as static or dynamic. Static expansion determines where and how many new transmission lines should be added to the network up to the planning horizon. If in the static expansion the planning horizon is categorized in several stages, then it becomes dynamic planning [5, 6]. In other words, in static planning, the goal is finding an appropriate number of new circuits that should be added to the transmission network. In this case, the time of lines construction is not specified by planner and the total expansion investment is carried out at the beginning of the planning horizon [7]. But, in dynamic one, multiple years are considered and an optimal expansion design is outlined over the whole planning period. In this case, the model includes time restrictions for considering the coupling among the years. So, the expansion cost along the planning horizon is minimized while the imposed restrictions are observed [8]. The long-term TNEP is a hard, large-scale, and highly nonlinear combinatorial optimization problem. Therefore, in expansion planning, approximate models such as DC, hybrid, and transportation models are often used for representation of transmission network. After mathematical modeling of the problem, it can be solved by classic, nonclassic, or heuristic methods [5]. Classic methods are only based on mathematical principles, so, they cannot be applied directly for solution of nonlinear complicated problems like TNEP. For this reason, first, such problems are converted to a linear problem using linearization technique and then are solved. But its difficulty is that if the scale of problem is large, it is very difficult to find accurate and reasonable solutions. Linear programming and Bender decomposition are common classical methods for solution of TNEP problem.

Contrary to classic methods, nonclassic ones are not based on mathematical rules and their foundation is random search. Complication of these methods is less than classic ones and simply can be applied for solution of complex problems without linearization. Random search algorithms such as evolutionary algorithms, SA, and Tabu search are examples of nonclassical methods. Heuristic methods like GRASP and HIPER[1] are combination of different approaches. These methods can improve speed and accuracy of the solution program. In this chapter, first we review some research in the field of transmission network expansion planning. Then, the method of mathematical modeling for TNEP problem is presented. Afterward, GA and PSO algorithms are described completely. Finally, effective parameters on network losses with a few examples are introduced.

2 TNEP: A Review

For a long time, the only tools available to solve the long-term transmission expansion planning problem were analysis tools, such as power flow [3]. One of the first approaches for solving the TNEP problem was proposed by Garver in 1970. He formulated the problem as a power flow problem and used a linear programming algorithm to find the most direct routes from generation to loads. After Garver's paper that was published in 1970 [9], much research has been done on the field of TNEP problem. Some of this is related to problem solution method. Some others, irrespective of solution method, proposed different approach for solution this problem considering various parameters such as uncertainty in demand, reliability criteria, and economic factors. Also, some of them investigated this problem and generation expansion planning together. However, most of them have involved static planning and a few works are about dynamic one. The dynamic models are currently in an underdeveloped status, and they have excessive limitations concerning the system size and the system modeling complexity level.

Chanda and Bhattacharjee [10] solved STNEP problem in order to obtain a maximum reliable network. Reliability criteria are relating to actual systems that considering them help to maintain a higher degree of reliability of the system. Later, they [11] proposed a new method for designing a maximum reliable network when failure probabilities of the lines are fuzzy in nature instead of deterministic as mentioned in [10]. Sohtaoglu [12] has studied the effect of economic factors on STNEP using load flow methods and the objective is minimization of the capital, variable, and power loss costs. The effects of economic parameters have been incorporated as a variable coefficient in objective function. Granville et al. [13] were formulated the STNEP problem by a linearized power flow model and used the Benders decomposition for its solution. However, classical decomposition approaches, e.g., Benders decomposition may fail to converge to optimal solutions due to the nonconvex nature of the STNEP problem. To handle with these nonconvexities difficulties, HIPER was proposed by Romero and Monticelli

[1] Benders hierarchical decomposition approach.

[14], where the power network constraints were represented by a chain of three model. The two first models relax all the nonconvexities constraints, which results in the optimal solution. Then, the nonconvexities are introduced (third model). Nevertheless, the nonconvexities still exist in the mathematical model used and application of this approach to networks with a large number of candidate circuits is limited by computational limitations.

Binato et al. [3] presented a heuristic approach, called GRASP to solve the static transmission expansion planning problem. The objective is an expansion of a network with the minimum cost and loss of load. GRASP is an expert iterative sampling technique that due to its generality and simplicity, is a useful alternate approach that can be applied to many other kinds of decision problems. However, this technique is the most time consuming and the local search procedure used in this approach leads to some difficulties related with pruning by comparison.

Lee et al. [15] adopted branch and bound algorithm in a way to preserve the discrete nature of investments for solution of STNEP problem. However, some problems that a planner may face with them exist when uses this algorithm. These problems are: (1) depending on the problem complexity the convergence of the algorithm is too slow; (2) there are memory storage problems when the fathoming rate of the current candidate problems is low; and (3) the difficulty in making an efficient implementation. Periera and Pinto [16] proposed a technique based on sensitivity analysis for static expansion of transmission network. But difficulty of proposed method is that if the number of nodes or number of participants is large, the planning for expansion is combinatorial complicated and that makes it very difficult to find reasonable solutions within short computational time. Romero et al. [17] presented SA for optimizing the investment cost and loss of load of the network in STNEP. SA mimics the physical process of annealing in solids (i.e., heating up a solid, and cooling it down until it crystallizes). It is a point-to-point search method with a strong theoretical base that its ability to reach global, or near global, optimal solutions under certain circumstances (slow cooling schedules) makes it a robust optimization algorithm. However, in the hard combinational problems such as STNEP problem, both the number of alternatives to be analyzed and the number of local minimum points increase with the dimension of the network. This fact can negatively affects on computing time and solution quality of the problem. Later, Gallego et al. [18] in order to improve the performance of the SA, proposed PSA approach. The objective function is the same one of Ref. [17]. The simulation results show that the proposed method give not only faster solutions but better ones as well. But, implementation of this method for solving large-scale, hard, and highly nonlinear combinational problems like long-term STNEP problem is so hard. Al-Saba and El-Amin [19] proposed a neural network-based method for solution of the STNEP problem with considering both the network losses and construction cost of the lines. Also, Contreras and Wu [20] included the network expansion costs and transmitted power through the lines in objective function and the goal is optimization of both expansion costs and lines loading.

Recently, Silva et al. [21] used a GA for solving the proposed problem of Ref. [17]. GA is a random search method that has demonstrated the ability to deal with nonconvex, nonlinear, integer-mixed optimization problems like the STNEP problem. Later, Silva et al. [22] introduced a TS-based method for optimization of investment cost in transmission expansion planning. TS is an iterative search procedure that moving from one solution to another looks for improvements on the best solution visited. The basic concepts of TS are movements and memory. A movement is an operation to jump from one solution to another while memory is used with different objectives such as to guide the search to avoid cycles. The simulation results for two real-world case studies (Brazilian Southern and Brazilian Southeastern network) have been shown that TS is a feasible and powerful technique to be applied to STNEP problem. Also, the authors have been shown that the performance of TS for finding the best solutions is almost similar to GA. Thus, it can be concluded that the most important advantage of GA is its simple implementation in addition to reach the good solutions, respectively.

So, in [23, 24], the expansion cost of substations with the network losses have been considered for the solution of STNEP problem using DCGA. The results evaluation in [23] indicated that the network with considering higher voltage level save capital investment in the long-term and become overload later. In [24], it was shown that the total expansion cost of the network was calculated more exactly considering effects of the inflation rate and load growth factor and therefore the network satisfies the requirements of delivering electric power more safely and reliably to load centers. Although global optimization techniques like GA and TS seem to be good methods for the solution of TNEP problem, however, when the system has a highly epistatic objective function (i.e., where parameters being optimized are highly correlated), and number of parameters to be optimized is large, then they have degraded efficiency to obtain global optimum solution and also simulation process use a lot of computing time. In order to overcome these drawbacks, Shayeghi et al. [25] proposed the DPSO approach for solving TNEP problem. They concluded that ability of DPSO algorithm for finding the more optimal solutions is more than GA. Also, it is shown that the convergence speed of the DPSO is more than GA method. DPSO is a novel population-based metaheuristic that is a useful tool for engineering optimization. Unlike the other heuristic techniques, it has a flexible and well-balanced mechanism to enhance the global and local exploration abilities. However, during the running of the algorithm, the particles become more similar, and cluster into the best particle in the swarm, which make the swarm premature convergence around the local solution. This phenomenon is caused that the precision of obtained solutions decreases and the best solution is not resulted. In order to resolve this problem, they [26] introduced an IDPSOMS algorithm. The lines loading parameter has been involved in fitness function of STNEP and investment cost in problem constraints. The proposed improved DPSO is a new conception, collectivity, which is based on similarity between the particle and the current global best particle in the swarm that can prevent the premature convergence of DPSO around the local solution. The results evaluation shows this fact.

Contreras and Wu [27] included the network expansion costs and transmitted power through the lines in objective function and the goal is optimization of both expansion costs and lines loading. Braga and Saraiva [28] presented a multicriteria formulation for DTNEP problem. The objective function only includes expansion and generation costs and one of the reliability criteria, i.e., PNS. In [29], authors investigated DTNEP problem and generation scheduling together using a new constructive heuristic approach. The goal is minimizing investment and fuel supply requirement cost of power plants, expansion cost of transmission lines and network losses. Bulent Tor et al. [30] solved the DTNEP problem considering the transmission system security and congestion in a competitive electricity market using Benders decomposition. They showed that annual evaluation of transmission investments and congestion along with local generation investment costs enables more realistic assessments of generation and transmission investment decisions. Mahdavi et al. [31] investigated the effect of bundle lines on static expansion planning of a multiple voltage level transmission network considering the network losses by DCGA. For this reason, the expansion cost of related substations from the voltage level point of view and network losses were included in objective function. They concluded that considering the effect of bundle lines on static transmission expansion planning of a network with different annual losses is caused that the total expansion cost of network (expansion and operational costs) is considerably decreased and therefore the capital investment significantly saved. Moreover, it was shown that construction of bundle lines in transmission network with different voltage levels considering the network losses is caused that the network lines is overloaded later and the network would have higher adequacy.

3 Mathematical Modeling of the TNEP Problem

As said, TNEP can be classified as STNEP and DTNEP. So, we describe modeling of both cases separately. General modeling of these two problems will be described in following sections.

3.1 Classical STNEP Problem Modeling

Normally, DCPF model is used for formulation of the TNEP problem. As known, the main goal of TNEP is minimizing the investment cost while satisfying the operational and economic constraints. So, STNEP problem can be formulated as follows using classical DCPF model.

$$TC = \sum_{i,j=1}^{ns} \beta . cl_{ij} n_{ij} \tag{1}$$

where TC is the construction cost of lines along the planning horizon; ns is the number of substations (buses); cl_{ij} is the construction cost of each line in corridor *i-j* and n_{ij} is the number of new circuits added to corridor *i-j*. It should be mentioned that β is a coefficient for converting the costs to present worth and can be acquired from Eq. (2).

$$\beta = \frac{1}{1+d} \tag{2}$$

In Eq. (2), d declares the discount rate. Equation (1) represents the construction cost of new lines which should be added to the network for delivering safe and reliable electric power to load centers after the planning horizon. Several restrictions have to be modeled in a mathematical representation to ensure that the mathematical solutions are in line with the planning requirements. These constraints are as follows (see Ref. [32] for more details):

- **DCPF node balance constraint**

This linear equality constraint represents the conservation of power at each node.

$$Sf + g - D = 0 \tag{3}$$

In the above equation, S is branch-node incidence matrix; f is active power matrix in each corridor. Also, g and D are generation and demand vectors.

- **Power flow limit on transmission lines constraint**

The following inequality constraint is applied to STNEP in order to limit the power flow for each path.

$$|f_{ij}| \leq (n_{ij}^0 + n_{ij}) \times \overline{f_{ij}} \tag{4}$$

where f_{ij} is total active power flow in branch *i-j* that can be calculated by Eq. (5). $\overline{f_{ij}}$ and n_{ij}^0 are the maximum transmissible power and number of initial circuits of corridor *i-j*, respectively. Also, θ_i and θ_j declare voltage phase angles of buses i and j.

$$f_{ij} = \gamma_{ij}(n_{ij}^0 + n_{ij}) \times (\theta_i - \theta_j) \tag{5}$$

- **Right-of-way constraint**

It is clear that for accurate expansion planning of a transmission system, exact place, and capacity of new circuits should be considered in TNEP problem. Mathematically, this constraint defines maximum number of circuits that can be installed into each corridor after planning horizon. Equation (6) represent that number of circuits installed in each corridor is limited by $n_{ij}^{\max}$.

$$0 \leq n_{ij} \leq n_{ij}^{\max} \tag{6}$$

- **Bus voltage phase angle limit constraint**

In DCPF model, the bus voltage magnitude is neglected, and only the voltage phase angle is considered. This parameter is considered as a STNEP constraint and the calculated phase angle should be less than a predefined value.

$$|\theta_i| \leq \left|\theta_i^{\max}\right| \tag{7}$$

Because of network losses effect on TNEP and its important role in determination of network configuration and arrangement [23], we should consider this parameter in solving of TNEP problem. Thus, in next section, we express formulation of the STNEP problem in the presence of active power losses.

3.1.1 STNEP Formulation Considering Network Losses

With considering active losses of the network in STNEP problem, the proposed objective function is defined as follows:

$$C_T = TC + LC \tag{8}$$

where C_T is total expansion cost of the network. LC is total cost of the network losses that its calculation method will be explained in following. Equation (8) should be minimized subjecting to constraints (3)–(7), which have been detailed in previous section.

Calculation Method of the Network Losses Cost

Regarding the fact that DCLF model has been used for solution of TNEP problem, network losses cannot be obtained directly using load flow. So, first, active power flow of each branch (corridor) is calculated by DCLF (Eq. (5)). Then, flow current of each branch is acquired using this power flow.

$$I_{ij} = \frac{|f_{ij}|}{\sqrt{3} \times V_L} \tag{9}$$

In Eq. (9), I_{ij} and V_L are flow current of branch *i-j* and voltage level of the related line, respectively. Assuming r_{ij} as resistance of branch *i-j* and regarding Eq. (9), network losses can be calculated as follows:

$$Loss = \sum_{i,j=1}^{ns} r_{ij} \times (I_{ij})^2 \tag{10}$$

If we represent generation cost of one MWh by C_{MWh}, total cost of the losses can be obtained by Eq. (11).

$$LC = \beta \times k_{Loss} \times C_{MWh} \times 8760 \times Loss \tag{11}$$

In the above equation, k_{Loss} is the losses coefficient. This coefficient that simulates changes of load is equal to the square of the under the LDC [23]. Usually, expansion planning of transmission network without expansion of substations and generation units is not possible practically because of annual load growth. Expansion of substation can be included in solution of STNEP, but for considering generation costs in TNEP we should solve DTNEP problem. Since we always face to dynamic load growth, it is more practical if generation costs are considered in DTNEP problem. Accordingly, in the next section, we include only expansion cost of substations in STNEP formulation.

3.1.2 STNEP Formulation Considering Network Losses and Expansion Cost of Substations

Expansion of substations can include construction of new substations due to increasing the transformers loading and expansion of substations from voltage level point of view. Since, construction of new substations is related to SEP problem, so, here, we only consider expansion cost of substations from voltage level view point, i.e., if the voltage levels of related substations (the placed substations on both of the candidate lines) are not equal to the voltage level of its candidate line, these substations must be expanded from the voltage level point of view. Therefore, the aim of calculating the expansion cost of the substations is calculating the expansion cost of substations that their voltage levels are not matched to the voltage levels of their related candidate lines. For calculating this cost, the DCFL program is run with the presence of the candidate lines. Then, according to the transmitted power trough the lines and using the KCL law, the power of the transmission substations is calculated. In accordance with this obtained power and, on the other hand, the standard capacities of transformers, the number of required transformers is determined. Therefore, the total expansion cost of the substations can be calculated. Consequently, objective function of STNEP problem considering network losses and expansion of substations from voltage level point of view is defined as follows:

$$C_T = TC + LC + SC \tag{12}$$

where

$$SC = \sum_{k \in \psi} SC_k \tag{13}$$

In the above equation, SC_k is expansion cost of kth substation and ψ is set of all substations. Equation (12) should be minimized subjecting to constraints (3)–(7). But, it should be noted that, here, $\bar{f}_{ij}$ which is expressed in Eq. (4) have two different rates according to voltage level of candidate line.

3.2 Classical Modeling of DTNEP Problem

DTNEP problem can be formulated as follows using classical DCPF model:

$$DTC = \sum_{i,j=1}^{ns} \sum_{t=1}^{T} \beta^t cl_{ij} n_{ij}^t \tag{14}$$

In the above equation, n_{ij}^t and β^t are the number of new circuits added to corridor *i-j* and inflation rate at year *t*. *T* is the planning horizon year. This equation indicates the construction cost of new added lines to the network during the planning horizon. Similar to STNEP problem, restrictions have to be modeled in a mathematical representation. These constraints are as follows [32]:

- **Power flow node balance constraint**

This linear equality constraint represents the conservation of power at each node.

$$S^t f^t + g^t - D^t = 0 \tag{15}$$

In this equation, S^t and f^t are branch-node incidence and each corridor's active power matrixes at year *t*, respectively. Also, g^t and D^t are generation and demand vectors at year *t*.

- **Power flow limit on transmission lines constraint**

The following inequality constraint shows the limitation of the power flow for each corridor at year *t*.

$$\left|f_{ij}^t\right| \leq (n_{ij}^0 + \sum_{k=1}^{t} n_{ij}^k)\overline{f_{ij}} \tag{16}$$

where

$$f_{ij}^t = \gamma_{ij}(n_{ij}^0 + \sum_{k=1}^{t} n_{ij}^k) \times (\theta_i^t - \theta_j^t) \tag{17}$$

In the above-mentioned equation, f_{ij}^t, θ_i^t and θ_j^t are respectively total active power flow in branch *i-j* and voltage phase angles of buses *i* and *j* at year *t*.

- **Power generation limit constraint**

Since, in DTNEP, load growth of the system is as dynamical and generation must be increased with load growth, cost of generating plant units should be considered in solution of DTNEP problem. So, the real power generation of bus *i* at year *t* must be restricted by $g_i^{t,\min}$ and $g_i^{t,\max}$ as follows:

$$g_i^{t,\min} \leq g_i^t \leq g_i^{t,\max} \tag{18}$$

- **Right-of-way constraint**

Mathematically, this constraint defines maximum number of circuits that can be installed into each corridor along planning horizon. Since, here, the planning horizon is divided into multiple stages; this constraint can be represented by two following restrictions. Equation (19) represents that number of circuits installed in each corridor at year t is limited by $n_{ij}^{t,\max}$. Also, Eq. (20) shows that sum of new constructible circuits in each corridor along the planning horizon is restricted to $n_{ij}^{t,\max}$.

$$0 \leq n_{ij}^t \leq n_{ij}^{t,\max} \tag{19}$$

$$\sum_{t=1}^{T} n_{ij}^t \leq n_{ij}^{\max} \tag{20}$$

- **Bus phase angle limit constraint**

In DTNEP, this constraint can be defined by Eq. (21). In this equation, $\theta_j^{t,\max}$ is maximum voltage phase angle of bus i at year t.

$$\left|\theta_i^t\right| \leq \left|\theta_i^{t,\max}\right| \tag{21}$$

3.2.1 DTNEP Formulation Considering Network Losses, Expansion Cost of Substations and Generation Costs

Due to considering network losses, expansion cost of substations and generation costs in dynamic planning of a transmission network, following objective function is proposed:

$$C_T = DTC + DLC + DSC + GC \tag{22}$$

$$DLC = \sum_{t=1}^{T} \beta^t \times k_{\text{Loss}} \times C_{\text{MWh}} \times 8760 \times \text{DLoss} \tag{23}$$

$$\text{DLoss} = \sum_{i,j=1}^{ns} r_{ij}^t \times (I_{ij}^t)^2 \tag{24}$$

$$I_{ij}^{t} = \frac{\left|f_{ij}^{t}\right|}{\sqrt{3} \times V_L} \tag{25}$$

where r_{ij}^{t} and I_{ij}^{t} are resistance and flow current of branch *i-j* at year *t*. DSC is the expansion cost of substations from voltage level point of view in dynamic state. *GC* is the cost of generation plant units along the planning horizon that consists of two components. The first component is related to the construction cost of the units along the planning horizon (Eq. (26)) and second one includes operation cost of the units over this time (Eq. (27)).

$$CC = \sum_{i=1}^{ns} \sum_{t=1}^{T} \beta^{t} C_{i}^{t}.n_{i}^{t} \tag{26}$$

$$OC = \sum_{i=1}^{ns} \sum_{t=1}^{T} \beta^{t} \times 8760 \times (a_i P_{it}^{2} + b_i P_{it} + c_i) \tag{27}$$

In Eq. (27), C_i^t and n_i^t are construction cost and number of generation plant units constructed in substation i at year t, respectively. Also, a_i, b_i, and c_i are cost coefficients of the unit i. *OC* is formulated as an OPF problem. So, *GC* is equal to sum of these two components as follows:

$$GC = CC + OC \tag{28}$$

But, due to the policies of the decision makers, the generation amount and number of units which should be constructed on each bus may be limited to some specific values. The generation limit constrain was mentioned earlier as Eq. (18), where g_i^t represented real power generation on each bus. But, here, since real power generation of each unit is shown by P_{it}, this constraint is reformed as follows:

$$P_{it}^{\min} \le P_{it} \le P_{it}^{\max} \tag{29}$$

Also, number of generation plant units which should be constructed in each bus is restricted to $n_{ij}^{t,\max}$.

$$0 \le n_i^t \le n_i^{t,\max} \tag{30}$$

In order to meet the required adequacy of the lines for delivering safe and reliable electric power to load centers, reliability constrains should be considered for solution of the problem. Normally, in TNEP, *N-1 safe criterion* is used for this purpose. This constraint expresses the robustness of transmission network in ordinary state and *N*-1 contingency (when a line fails) for not overloading and supplying the loads. Shayeghi and Mahdavi [33] defined a new constraint instead of *N-1 safe criterion*. They presented this new condition as follows:

$$LL \leq LL_{\max} \tag{31}$$

where LL is lines loading rate of the network and LL_{max} is maximum of this amount. This inequality constraint can be included in the DTNEP problem restrictions. It should be noted that LL_{max} is an experimental parameter that is determined according to load growth coefficient and its rate is between 0 and 1. Reducing rate of this parameter is caused that added lines to the network, the network adequacy (increasing of overload duration time) and expansion cost are increased. Also, network losses and lines loading average in operational time are decreased. Equation (22) should be minimized subjecting to constraints (14)–(21) and (31). For solving such large-scale problems, there are various methods such as classic, nonclassic, and heuristic methods. However, only nonclassic and heuristic methods can be applied to such hard problems. As said in introduction, among the nonclassic methods DCGA and among heuristic methods IDPSO are good methods for solution of the long-term and large-scale TNEP problems due to flexibility, simple implementation, and the advantages which were mentioned in [23, 24, 26] In following, these two algorithms are described.

4 Genetic Algorithm

GA is part of a class of what is known as evolutionary algorithms that mimics the metaphor of natural biological evolution [34, 35]. It were first described by John Holland in the 1960s and further developed by Holland and his students and colleagues at the University of Michigan in the 1960s and 1970s [36]. This algorithm is a random search method that can be used to solve nonlinear system of equations and optimize many practical complex problems such as TNEP. The base of this algorithm is the selection of individuals. It doesn't need a good initial estimation for sake of problem solution. In other words, the solution of a complex problem can be started with weak initial estimations and then be corrected in evolutionary process of fitness. GA is based on population of chromosomes which each of them consists of a linear string of bits. Each chromosome's organ represents an unknown parameter of a problem. A bit can include binary or decimal codes that regarding the codification type are divided into two following categories: 1-BCGA, 2-DCGA.

4.1 BCGA (Standard Genetic Algorithm)

BCGA that is usually called standard genetic algorithm or GA manipulates the binary strings, which may be the solutions of the problem. Each chromosome or individual is exhibited by an L-bit binary string. The number of bits (L) depends upon the desired resolution and number of problem's parameter. In every optimization problem, there is a space which is referred to as the search or state space. This space comprises all possible solutions to the optimization problem at hand. At every evolutionary step also known as generation, the individuals in the current population are decoded and evaluated according to a fitness function set for a given problem [37]. In reproduction process, every member of a population (individual) cooperates in optimization with respect to its fitness value. In other words, individuals with higher fitness value have a high probability of being selected for mating whereas less fit individuals have a correspondingly low probability of being selected [35]. Also, mutation operator is applied as random variations during reproduction. Finally in iterative algorithm, the population moves toward evolution and the best individual is accepted as problem solution. The GA generally includes the three fundamental genetic operators of selection, crossover and mutation. These operators conduct the chromosomes toward better fitness. In following, these operators are completely described.

4.1.1 Selection Operator

Selection operator selects the chromosome in the population for reproduction. The more fit the chromosome, the higher its probability of being selected for reproduction. Thus, selection is based on the survival-of-the-fittest strategy, but the key idea is to select the better individuals of the population, as in tournament selection, where the participants compete with each other to remain in the population. The most commonly used strategy to select pairs of individuals is the method of roulette-wheel selection, in which every string is assigned a slot in a simulated wheel sized in proportion to the string's relative fitness. This ensures that highly fit strings have a greater probability to be selected to form the next generation through crossover and mutation. After selection of the pairs of parent strings, the crossover operator is applied to each of these pairs.

4.1.2 Crossover Operator

The crossover operator involves the swapping of genetic material (bit-values) between the two parent strings. Based on predefined probability, known as crossover probability, an even number of chromosomes are chosen randomly. A random position is then chosen for each pair of the chosen chromosomes. The two chromosomes of each pair swap their genes after that random position. Crossover may be applied at a single position or at multiple positions.

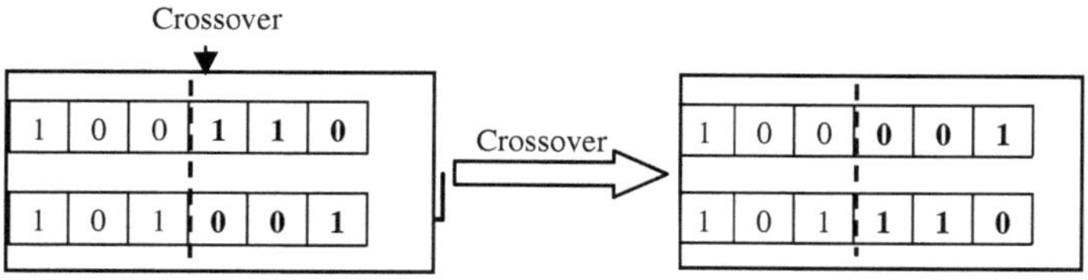

Fig. 1 Single position crossover

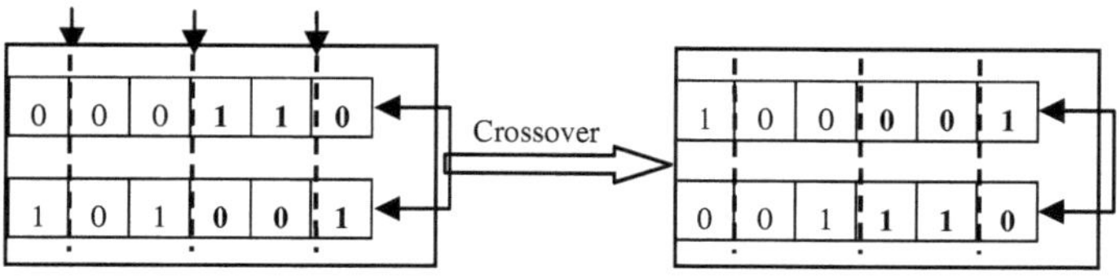

Fig. 2 Multiple position crossover

Single Position Crossover

As shown in Fig. 1, this operator randomly chooses a locus (a bit position along the two chromosomes) and exchange subsequences before and after that locus between two chromosomes to create two offspring [37].

Multiple Position Crossover

Multiple position crossover which is known as uniform crossover randomly chooses multiple locus (more than one bit position along the two chromosomes) and exchange subsequences between two chromosomes [37]. Multiple position crossover for three points is shown in Fig. 2.

4.1.3 Mutation Operator

Each individuals (children) resulting from each crossover operation will now be subjected to the mutation operator in the final step to forming the new generation. The mutation operator enhances the ability of the GA to find a near optimal solution to a given problem by maintaining a sufficient level of genetic variety in the population, which is needed to make sure that the entire solution space is used in the search for the best solution. In a sense, it serves as an insurance policy: it helps prevent the losses of genetic material. This operator randomly flips or alters one or more bit values usually with very small probability known as a mutation probability (typically between 0.001 and 0.01). It is done by changing the gene from 1 to 0 or vice versa. Similar to crossover, this operator can be classified to

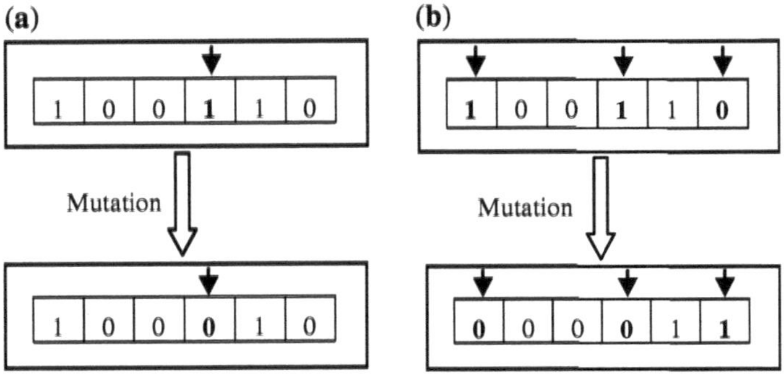

Fig. 3 Mutation operator

single and multiple mutations. These two types of mutation are illustrated in Fig. 3. Practical experience has shown that in the transmission expansion planning application the rate of mutation has to be larger than ones reported in the literature for other application of the GA.

4.2 DCGA

Although binary codification is conventional in GA but it has some features that we use decimal codification [31]: (1) Avoiding difficulties which are happened at coding and decoding problem, (2) Preventing the production of completely different offspring from their parents and subsequent occurrence of divergence in mentioned algorithm. In this method crossover can take place only at the boundary of two integer numbers. Mutation operator selects one of existed integer numbers in chromosome and then changes its value randomly, i.e., the gene value is randomly increased or decreased by one providing not to cross its limits. Reproduction operator, similar to standard form, reproduces each chromosome proportional to value of its objective function. Thus, the chromosomes which have better objective functions will be selected more probable than other chromosomes for the next population. After mutation, the production of new generation is completed and it is ready to start the process all over again with fitness evaluation of each chromosome. The process continues and it is terminated by either setting a target value for the fitness function to be achieved, or by setting a definite number of generations to be produced. A typical chromosome for a network with six corridors is shown in Fig. 4. In the first corridor, one transmission circuit, in the second corridor and finally in the sixth corridor, two transmission circuits have been predicted. The flowchart of DCGA has been shown in [23].

Fig. 4 A typical chromosome

1	2	3	1	1	2	1	0	0	1	1	0

5 PSO Algorithm

PSO algorithm, which is tailored for optimizing difficult numerical functions and based on metaphor of human social interaction, is capable of mimicking the ability of human societies to process knowledge [38]. It has roots in two main component methodologies: artificial life and, evolutionary computation. It lies somewhere in between evolutionary programming and the GAs [24]. As in evolutionary computation paradigms, the concept of fitness is employed and candidate solutions to the problem are termed particles or sometimes individuals, each of which adjusts its flying based on the flying experiences of both itself and its companion. Vectors are taken as presentation of particles since most optimization problems are convenient for such variable presentations. In fact, the fundamental principles of swarm intelligence are adaptability, diverse response, proximity, quality, and stability [39]. It is adaptive corresponding to the change of the best group value. The allocation of responses between the individual and group values ensures a diversity of response. The population is responding to the quality factors of the previous best individual values and the previous best group values. As it is reported in [38], this optimization technique can be used to solve many of the same kinds of problems as GA and does not suffer from some of GAs difficulties. It has also been found to be robust in solving problem featuring nonlinearing, nondifferentiability, and high-dimensionality. It is the search method to improve the speed of convergence and find the global optimum value of fitness function. PSO starts with a population of random solutions ''particles'' in a D-dimension space. The *i*th particle is represented by $X_i = (x_{i1}, x_{i2},\ldots,x_{iD})$. Each particle keeps track of its coordinates in hyperspace, which are associated with the fittest solution it has achieved so far. The value of the fitness for particle i is stored as $P_i = (p_{i1}, p_{i2},\ldots,p_{iD})$ that its best value is represented by (*pbest*). The global version of the PSO keeps track of the overall best value (*gbest*), and its location obtained thus far by any particle in the population. PSO consists of, at each step, changing the velocity of each particle toward its *pbest* and *gbest* according to Eq. (31). The velocity of particle i is represented as $V_i = (v_{i1}, v_{i2}\ldots v_{iD})$. Acceleration is weighted by a random term, with separate random numbers being generated for acceleration toward *pbest* and *gbest*. The position of the *i*th particle is then updated according to Eq. (32) [38]:

$$v_{id}(t'+1) = \omega \times v_{id}(t') + c_1 r_1 (P_{id} - x_{id}(t')) + c_2 r_2 (P_{gd} - x_{id}(t')) \tag{31}$$

$$x_{id}(t'+1) = x_{id}(t') + c v_{id}(t'+1) \tag{32}$$

where P_{id} and P_{gd} are *pbest* and *gbest*. It is concluded that *gbest* version performs best in terms of median number of iterations to converge. However, *pbest* version with neighborhoods of two is most resistant to local minima. The results of past experiments about PSO show that ω was not considered at an early stage of PSO algorithm. However, ω affects the iteration number to find an optimal solution. If the value of ω is low, the convergence will be fast, but the solution will fall into the local minimum. On the other hand, if the value will increase, the iteration number will also increase and therefore the convergence will be slow. Usually, for running the PSO algorithm, value of inertia weight is adjusted in training process. It was shown that PSO algorithm is further improved via using a time decreasing inertia weight, which leads to a reduction in the number of iterations [39]. In Eq. (31), term of c_1r_1 (P_{id} -x_{id} (t')) represents the individual movement and term of c_2r_2 (P_{gd} -x_{id} (t')) represents the social behavior in finding the global best solution. Regarding the fact that parameters of the TNEP problem are discrete time type and the performance of standard PSO is based on real numbers, this algorithm cannot be used directly for solution of the TNEP problem. There are two methods for solving the TNEP problem based on the PSO technique [40]: (1) BPSO, (2) DPSO. In the following, details of both algorithms are presented.

5.1 BPSO

Regarding the fact that parameters of the TNEP problem are discrete time type and the performance of standard PSO is based on real numbers, this algorithm cannot be used directly for solution of the TNEP problem. Thus, in order to overcome this drawback a BPSO algorithm is used for solution of the TNEP problem. In this method, in a D-dimensional binary solution space, the position of ith particle can be expressed by an D-bit binary string as $X_i = (x_{i1}, x_{i2},\ldots,x_{iD})$, where X_i ε {0,1}. Since each bit X_i is binary-valued, the term of (P_{id}−x_{id} (t')) or (P_{gd}−x_{id} (t')) has only three possible values 0, 1, and −1. Moreover, it should be mentioned that although PSO algorithm is further improved via using a time decreasing inertia weight but the results of past research about BPSO show that the time-varying inertial weight does not improve the convergence of BPSO [41] and a constant weight of 1.0 is suggested. Therefore, velocity of particle i that is represented by circuit's change of each corridor is updated by the following equation:

$$v_{id}(t'+1) = v_{id}(t') + c_1r_1(P_{id} - x_{id}(t')) + c_2r_2(P_{gd} - x_{id}(t')) \tag{33}$$

$$\begin{aligned} P_{id} - x_{id}(t') &= 1; \quad && if\ P_{id} = 1, x_{id} = 0 \\ P_{id} - x_{id}(t') &= 0; && if\ P_{id},\ x_{id} = 0\ or\ P_{id}, x_{id} = 1 \\ P_{id} - x_{id}(t') &= -1; && if\ P_{id} = 0,\ x_{id} = 1 \end{aligned}$$

In Eq. (33), t' is the number of algorithm iterations and the velocity v_{id} ($t' + 1$) is a real number in $[-V_{max}, V_{max}]$. According to Eq. (32), for updating the position of the ith particle, the real value v_{id} ($t' + 1$) must be added to the binary value x_{id} (t'), but this is not possible mathematically. So an intermediate variable S (v_{id} ($t' + 1$)) via the sigmoid limiting transformation is defined as Eq. (34) [41]:

$$S(v_{id}(t'+1)) = \frac{1}{1+e^{-v_{id}(t'+1)}} \tag{34}$$

Equation (34) maps the domain of $[-V_{max}, V_{max}]$ into the range of $[1/(1+e^{V_{\max}}), 1/(1+e^{-V_{\max}})]$, which is a subset of (0, 1). The value of S (v_{id} (t' + 1)) can be therefore interpreted as a probability threshold. A random number with a uniform distribution in (0, 1), R, is then generated and compared to S (v_{id} (t' + 1)). Thus, the position of the particle i can be updated as follows:

$$\begin{aligned} x_{id}(t'+1) &= 1; \quad if\ R\langle S(v_{id}(t'+1)) \\ x_{id}(t'+1) &= 0; \quad if\ R \geq S(v_{id}(t'+1)) \end{aligned} \tag{35}$$

The probability that x_{id} (t' + 1) equals to 1 is $S(v_{id}$ (t' + 1)) and the probability that it equals to 0 is $1-S(v_{id}$ (t' + 1)). The flowchart of the proposed algorithm has been given in [42].

5.2 IDPSOMS

DPSO is better than BPSO due to avoid difficulties which are happened at coding and decoding problem, increasing convergence speed and simplification. In this approach, the each particle is represented by three arrays: starting bus ID, ending bus ID and number of transmission circuits (the both of constructed and new circuits) at each corridor. In the DPSO iteration procedure, only the number of transmission circuits needs to be changed while starting bus ID and ending bus ID are unchanged in calculation. So, the particle can omit the starting and ending bus ID. Thus, particle can be represented by one array. A typical particle with 12 corridors is shown in Fig. 5.

In Fig. 5, in the first, second, third corridor and finally 12th corridor, one, two, three and two transmission circuits have been predicted, respectively. Also, the particle's velocity is represented by circuit's change of each corridor. ω is considered as a time decreasing inertia weight that its value is determined by Eq. (36).

$$\omega = \frac{1}{\ln(t')} \tag{36}$$

Finally, position and velocity of each particle is updated by the following equations:

$$v_{id}(t'+1) = Fix[\omega \times v_{id}(t') + c_1 r_1 (P_{id} - x_{id}(t')) + c_2 r_2 (P_{gd} - x_{id}(t'))] \tag{37}$$

$$x_{id}(t'+1) = x_{id}(t') + v_{id}(t'+1) \tag{38}$$

Fig. 5 A typical particle

$$X_{typical} = (1, 2, 3, 1, 0, 2, 1, 0, 0, 1, 1, 2)$$

where $v_{min} \leq v_{id} \leq v_{max}$, and *fix* (.) is getting the integer part of *f*. When v_{id} is bigger and smaller than v_{max} and v_{min}, make $v_{id} = v_{max}$ and $v_{id} = v_{min}$, respectively. While, x_{id} is bigger than upper bound of circuit number allowed to be added to a candidate corridor for expansion, then make x_{id} equal the upper bound. While $x_{id} < 0$, make $x_{id} = 0$. The other variables are the same to Eqs. (31) and (32). In the DPSO, particle is attracted toward the global best particle and particles become more and more similar as iteration go on which make the algorithm lack of the diversity fast. In order to measure how much similar two particles are, the following formula is used to measure the *similarity* between two particles [43]:

$$s(i,j) = \begin{cases} 1, & d(i,j) < d_{\min} \\ 1 - [\frac{d(i,j)}{s_{\max}}]^{\beta} & d_{\min} \leq d(i,j) < d_{\min} \\ 0 & d(i,j) \geq d_{\max} \end{cases} \tag{39}$$

where $s(i, j)$ represents the *similarity* of two particles *i* and *j*. $d(i, j)$ shows the distance of two particle *i* and *j* in the space. $d_{\max}$ and $d_{\min}$ are specified parameters which are determined according to the function optimized. β is a constant parameter that is considered as 1 here. It should be noted that $s(i, j) \in [0,1]$ and its value is considered 1 for every particle *i*. From Eq. (39), it can be concluded that the smaller distance of two particles leads to their larger *similarity*. So, the conception *similarity* can be used to measure how much similar two particles are.

If *gbest* and *pbest* of the *i*th particle is fixed and $\sqrt{2(1 + \omega - (c_1 + c_2)^2 - 4\omega}$ <2 $x_i(t')$ converges to the weighted centre of *gbest* and *pbest* $(c_1 r_1 P_{id} + c_2 r_2 P_{gd})/(c_1 + c_2)$ [44]. Thus, while *pbest* keeps on updating, *pbest* and subsequent $x_i(t')$ converges *to gbest*. It is concluded from the above analysis that particles in the basic DPSO tend to cluster too closely as the iteration goes on. When *gbest is* found by one particle, the other particle will be drawn toward it. So to measure the cluster degree of the particle swarm (swarm diversity), a conception *collectivity* is defined as follow [43]:

$$C(t') = \frac{1}{n} \sum_{i=1}^{n} s(i, g) \tag{40}$$

where $s(i, g)$ denotes the *similarity* of the particle *i* to *gbest* at the t'th generation; $C(t')$ is used to measure the cluster degree of t'th generation of the whole particle swarm. If the swarm is going to converge to *gbest* found so far, the evolution of DPSO will be stagnated as iteration goes on and be lack of the diversity. If all particles end up in *gbest*, they will stay at the *gbest* without much chance to escape. If the identified *gbest* is only local, it would be disadvantageous to let some of the particles explore other areas of the search space while the remaining particles stay at this optimum to finetune the solution. In order to keep the diversity of the particle swarm when particles started to cluster, our model is designed that the position x_i of the *i*th particle is randomly mutated according to the swarm *collectivity* and its *similarity* to the *gbest*. So, the following formula is designed [43].

$$\text{if } rand < \alpha \times C(t') \times s(i,j) \quad \text{then} \quad x_{id} = N + x_{id} \tag{41}$$

where *rand* and N are random numbers between 0 and 1. α is a coefficient which can be considered constant or variable. Here, in order to keep the tradeoff of the exploration and the exploitation, the parameter α is considered variable, i.e., its value is decreased linearly with iteration going on, and equal to zero ultimately, which makes the algorithm to improve the exploration in global domain in the early times, and keep the exploitation in local fields in the later times $\alpha = 1 + 1/(1 + \ln t)$. Also, it can be known from Eq. (41): larger *collectivity* of the swarm and *similarity* of the particle to *gbest* are, more possible it is for the particle to randomly mutate and increase the diversity of particle swarm. The flowchart of the proposed IDPSOMS algorithm has been described in [26]. Now, we present examples of TNEP problem that have been solved using GA and PSO methods.

Example 1. Consider Garver's 6-bus system which all details are given in [24]. Solve the STNEP problem for this network if only the construction cost of lines is to be a minimum. Assume that the units have no generation limits and the reliability constraint is *N-1 safe criterion*. The planning horizon is year 2021 (15 years ahead: according to Ref. [23], it has been accounted with respect to 2006, so, 15 years after 2006 is 2021).

Solution: So, we modeled the problem mathematically as Eq. (1) that is subjected to constraints (3)–(6) and *N-1 safe criterion*. DCGA method is applied for solution of this problem with following input data (Table 1):

Due to the stochastic nature of the GA, there is no guarantee that different executions of the program converge to the same solution. Thus, in this study, the program has been executed for four times as continual, i.e., after running of the genetic program, obtained results are inserted in initial population of next run and this process is iterated for three times. In addition to this continual run, a more suitable criteria termination has accomplished that is production of predefined generations after obtaining the best fitness and finding no better solution. In this work, a maximum number of 1,000 generations has been chosen ($N = 1000$). After applying DCGA, the results are obtained in Table 2.

Also, total expansion cost of the network (C_T) is 96.175 Million \$ (M\$).

Table 1 Input data

Parameter	Value
Load growth factor (LGF)	1.07
The number of initial population	5
β, n_{ij}^{max}, P_C, P_M and C_{MWh} (\$/MWh)	0.1, 4, 0.3, and 1 and 36.1

Table 2 Configuration of the network

Corr.	VL (kV)	NC	Corr.	VL (kV)	NC
2–6	230	4	4–6	230	4
3–5	400	2	5–6	230	1

Table 3 Configuration of the network considering losses

Corr.	VL (kV)	NC	Corr.	VL (kV)	NC	Corr.	VL (kV)	NC
2–6	400	4	3–5	400	2	4–6	230	3

Table 4 Configuration and expansion costs of network neglecting losses

Corr.	VL (kV)	NC	Corr.	VL (kV)	NC	TC (M$)	SC (M$)	C_T (M$)
2–6	230	4	4–6	230	4	89.18	31.475	120.655
3–5	230	3	5–6	230	1	–	–	–

Table 5 Configuration and expansion costs of the network considering losses

Corr.	VL (kV)	NC	Corr.	VL (kV)	NC	TC (M$)	SC (M$)	C_T (M$)
2–6	400	3	4–6	230	3	99.809	31.257	131.066
3–5	230	3	5–6	230	1	–	–	–

Example 2. Let us consider the same Example 1. Now, solve the STNEP problem with considering network losses.

Solution: The mathematical model of the problem is as Eq. (8) with constraints of given in Example 1. Due to considering the different voltage levels for the transmission lines and also for simplicity in programming the selected chromosome is divided into two parts [23]. After running the DCGA problem, the results are obtained as Table 3 and 108.415 M$ is obtained for C_T.

Example 3. Consider the same system of Example 1 with changes applied to voltage levels of lines and substations. Assume that voltage level of all lines is 230 kV and voltage level of substations is as follows: substations 1–5 are 230/63 and substation 6 is 400/63. Please find the optimal expansion design considering expansion cost of substations from voltage level view point under two following cases: neglecting and considering network losses.

Solution: For both cases, the constraints are exactly similar to previous examples but the objective function is different.

Case 1. *Regardless of the network losses*

$$C_T = TC + SC$$

Here, DCGA is applied to the problem and the results are acquired according to Table 4.

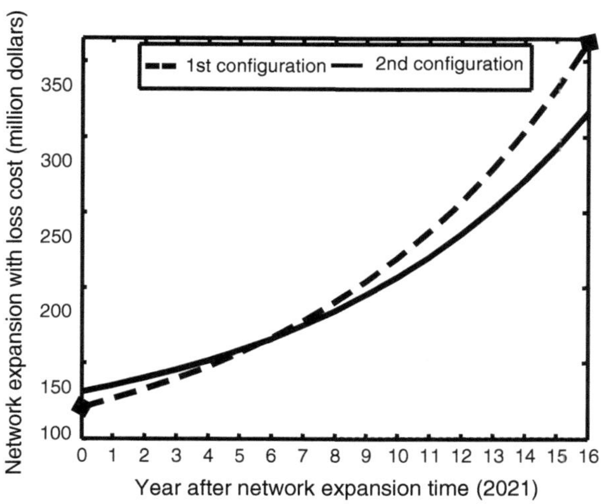

Fig. 6 Total expansion cost of the network for two proposed configurations in cases 1 and 2

Case 2. *Considering the network losses*

$$C_T = TC + LC + SC$$

The simulated results are given in Table 5.

As seen, solution of STNEP problem with consideration of network losses is caused that C_T value and the number of 400 kV lines constructed are increased in the network. For more investigation, the total expansion cost of the network for two proposed configurations (considering and neglecting network losses) after planning horizon is shown in Fig. 6. As shown in Fig. 6, the total expansion cost of the network for the second configuration (most of its lines are 400 kV) is more than that of the first one until about 6 years after the planning horizon (year 2021), but then the total expansion cost for the first configuration (all the lines are 230 kV) becomes more than that of the other one. In the beginning, it can be seen that regardless of the loss, the first configuration is suitable. However, 6 years after the expansion time (when network losses is considered), the second configuration is more economic and saves capital in the long term. Also, from the transmitted power point of view through the lines, the second configuration is more suitable than the first one, because it is overloaded 2 years later than first configuration. As seen, we solve the STNEP problem with different objective functions and considering different parameters such as losses and voltage level of lines and substations. In these examples, it was showed that network losses has important role for determining configuration and arrangement of the network with different voltage levels of lines. Considering this parameter in transmission expansion planning decreases the operational costs considerably and the network satisfies the requirement of delivering electric power more safely and reliably to load centers. So, studying effective parameters on network losses can be helpful. So, in next

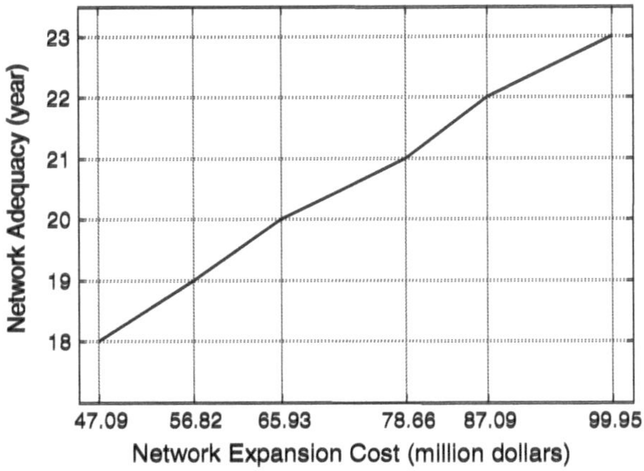

Fig. 7 Adequacy curve with respect to network expansion cost

section, we will introduce important parameters which can be affected on network losses and subsequent STNEP problem solution.

6 Effective Parameters on Network Losses in TNEP

The effect of three important parameters, i.e., IRF, LGF and bundle lines on network losses for solution of STNEP problem have been investigated in this section. For this reason, the losses cost and also the expansion cost of related substations from the voltage level point of view are included in the objective function. We describe influence of these three parameters on STNEP solution in the following sections.

6.1 IRF

One of important parameters that affected on network losses is IRF. So, for evaluating effect of this parameter on the network losses in STNEP problem with various voltage levels and subsequent adding expansion cost of substations to expansion costs, the problem is formulated as Eq. (12) subjecting to constraints (3)–(6) and (31). DCGA method with input data given in Table 6 is performed

Table 6 Input data of proposed method

Parameter	Value
LGF, C_{MWh}, n_{ij}^{max}, P_C, P_M, N, LL_{Max}	1.07, 36.1, 4, 0.1, 1, 1000, 0.5
The number of initial population	5

Table 7 First and second configurations in case 1 for IRF = 0

	First configuration		Second configuration	
Corr.	VL (kV)	NC	VL (kV)	NC
2–6	230	3	230	3
4–6	230	3	230	3
3–5	400	1	400	1

under two cases. In case 1, the IRF is considered zero, but in case 2, value of this parameter is assumed 5 %. The load growth is considered 7 % for both cases. Chromosome structure is divided into two parts [23]. It should be noted that planning horizon for both cases is 10 years ahead when the paper has been written (Shayeghi et al. 2008a) (from 2007 to 2017).

Case 1. In this case, *IRF* is 0 and the DCGA is applied to the same case study system of Example 1 and the results (lines which must be added to the network up to planning horizon year) are given in Table 7. The first and second configurations are obtained neglecting and considering the network losses, respectively (Table 7).

TC and *SC* for first and second configuration in case 1 and considering $IRF = 0$ are 0 and 55.79 M$, respectively. Considering and neglecting the losses in objective function of the STNEP problem, is resulted in similar configurations for expansion of the network. The reason is that, when the inflation rate is zero the network losses cannot compete with expansion costs (expansion cost of substations and lines) well. Also, expansion cost of substations has obtained zero. Because the voltage level of the proposed lines for network expansion has been existed in their both first and end substations and therefore substations have not required expansion from the voltage level point of view.

Case 2. Here, the *IRF* is assumed 5 % and similar to the previous case, the results are given in Table 8. Total expansion cost of expanded network with the two proposed configurations has been shown in Fig. 4 of Ref. [24].

TC and *SC* for first and second configurations in case 2 and considering $IRF = 5$ % are 0 and 81.28 M$, respectively. Due to the obtained results, it can be concluded that inflation rate has important role in determining of the network configuration. Also, considering this economic parameter in transmission expansion planning is caused that the total expansion cost of the network (expansion costs and the operational cost) is calculated more exactly and network losses effect

Table 8 First and second configurations for IRF = 5 %

Corr.	First configuration		Second configuration	
	VL (kV)	NC	VL (kV)	NC
2–6	230	3	430	3
4–6	230	3	230	3
3–5	400	1	400	1

Table 9 The First and second configurations for LGF = 4 %

Corr.	First configuration		Second configuration	
	VL (kV)	NC	VL (kV)	NC
2–6	230	3	400	2
4–6	230	2	400	1

Table 10 Expansion cost of network with first and second configurations for LGF = 4 %

	First configuration	Second configuration
SC	0	6.99 M$
TC	37.16 M$	47.79 M$
C_T	37.16 M$	54.78 M$

Table 11 The first and second configurations for LGF = 7 %

Corr.	First configuration		Second configuration	
	VL (kV)	NC	VL (kV)	NC
2–6	230	3	400	3
4–6	230	3	230	3
3–5	400	1	400	1

on transmission expansion planning is increased with growing rate of this parameter.

6.2 LGF

Second important parameter that affected on network losses is LGF. So, for evaluating effect of this parameter on the network losses, we consider the same objective function and constraints is presented in Sect. 6.1. Effects of LGF on the network losses are evaluated in this section under three cases. In case 1, 2, and 3 the LGF is considered 4, 7, and 10 %, respectively. For three cases, the *IRF* and planning horizon year is 10 % and 2017 [24].

Case 1. In this case, the LGF is 4 %. Tables 9 and 10 show the results with applying of the DCGA method on the Garver's network described in previous section.

Case 2. Here, the LGF is considered 7 % and the desired method is applied to the case study system and the results are given in Table 11.

Also, *TC* and *SC* for first and second configurations in are 55.79 and 81.28 M$, respectively.

Table 12 The First and second configurations for LGF = 10 %

Corr.	First configuration		Second configuration	
	VL (kV)	NC	VL (kV)	NC
2–6	230	4	400	3
4–6	230	3	230	4
3–5	400	2	400	2
3–6	230	1	–	–

Case 3. In this case, LGF is assumed 10 %. The results are shown in Table 12 with applying of the proposed method on the Garver's network. *TC* and *SC* for first and second configuration in this case are 85.99 and 99.92 M$, respectively.

As expected and from Figs 6, 8, and 10 of Ref. [24], regardless to the network losses, growing of the load growth factor is caused that more 230 and 400 kV lines are added to the network. The reason is that, the power flow capability through the lines is improved with growing this factor. Also, it can be said that the LGF has important effect on the network losses. Because the curve of the second configuration cuts the curve of first one earlier and subsequent more 400 kV lines are added to the network with growing this factor. Thus, the LGF plays important role in determining of the network configuration in solving of the TNEP problem.

6.3 Bundle Lines

Bundle lines can affect on STNEP problem directly and indirectly, i.e., number of bundles has role in both reductions of network losses and lines construction costs separately. Therefore, here, we investigate this parameter under two sections. In Sect. 6.3.1, we study role of this parameter only on STNEP problem while its effect on network losses is investigated in Sect. 6.3.2. In following, we describe details of them. For studying the effect of this parameter on the network losses and STNEP, we should introduce a real case study system which includes bundle lines. The transmission network of the Azerbaijan regional electric system is used to test and evaluation of the proposed idea. This actual network has been located in northwest of Iran. All details of the test system are given in [31].

6.3.1 Effect of Bundle Lines on STNEP

In this section, the goal is evaluation of role of bundle lines only in STNEP.

So, it's the mathematical model is defined as objective function of case 1 in Example 3. Constrains are defined as Eqs. (3)–(6) and:

$$0 \leq nb_{230} \leq nb_{230}^{\max}$$

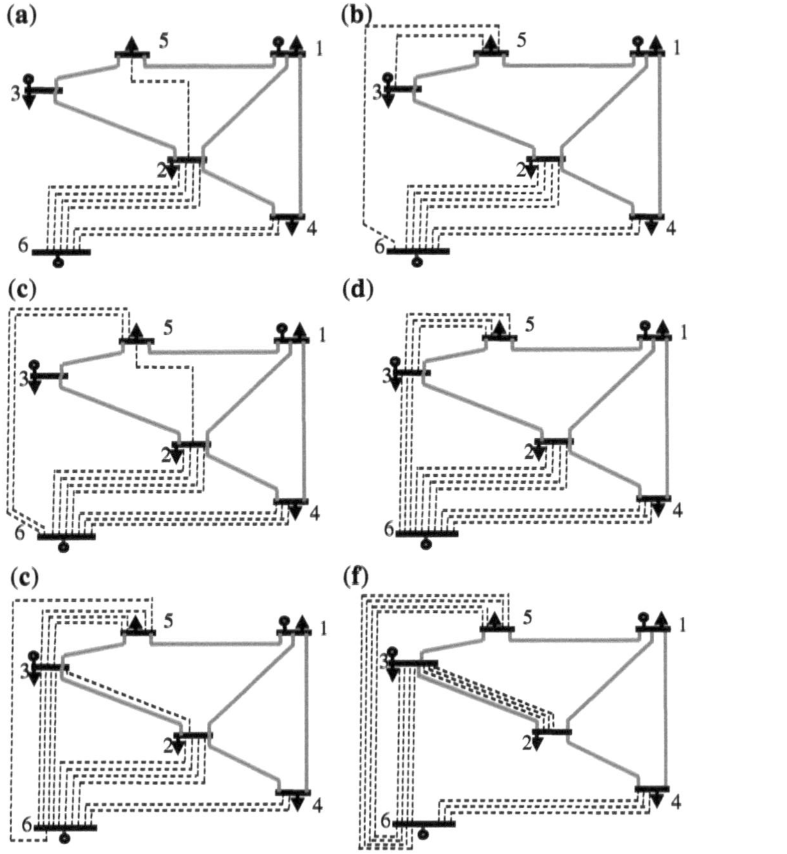

Fig. 8 Proposed configuration for: **a** $C_{\max} = 30$, **b** $C_{\max} = 40$, **c** $C_{\max} = 50$, **d** $C_{\max} = 60$, **e** $C_{\max} = 70$, and **f** $C_{\max} = 80$

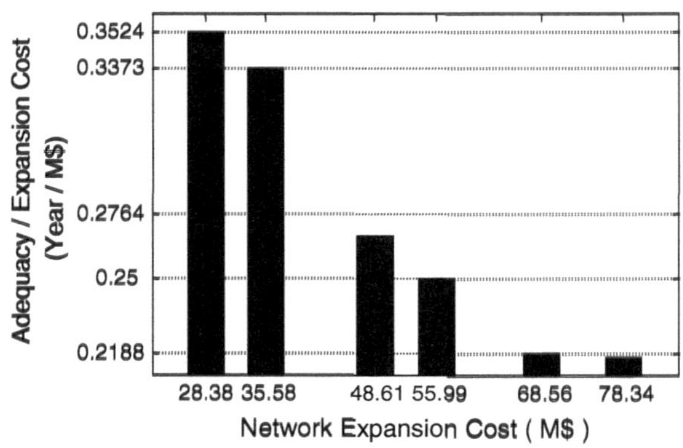

Fig. 9 The curve of adequacy index on the expansion cost for $C_{\max} = 30–80$

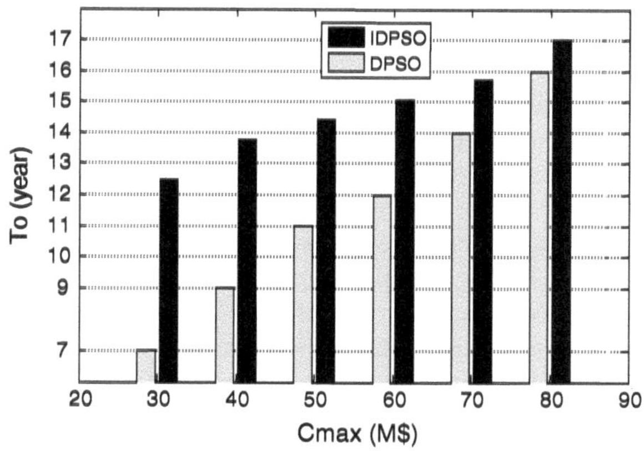

Fig. 10 Diagram of year of missing the network adequacy versus maximum investment for $C_{max} = 30–80$

Table 13 Input data of solution method

Parameter	Value
LGF, C_{MWh}, β, $n_{ij}^{t,max}$, nb_{230}^{max}, nb_{400}^{max}, P_C, P_M, N	1.08, 36.1,1.15, 4, 3, 2, 0.3, 0.1, 3500
The number of initial population	5

$$0 \leq nb_{400} \leq n_{400}^{max}$$

where nb_{230} and nb_{400} are number of bundles for new 230 and 400 kV lines added to the network respectively that their values are limited to nb_{230}^{max} and nb_{400}^{max}. In order to study the effect of bundle lines on STNEP using DCGA, with input data given in Table 13, the chromosome structure is represented based on two cases. In case 1, the bundle of lines has not been considered in chromosome structure, but in case 2, the bundle of lines has been included in chromosome. In the following, the details of these cases are specified.

Case 1. In this case, with respect to neglecting the effect of bundle lines for solution of STNEP problem in a multiple voltage level transmission network and also simplicity in programming, selected chromosome is divided into two parts (Fig. 1 of Ref. [31].

Case 2. Here, unlike case 1, for considering the effect of bundle lines, another new part which describes the number of bundle lines is added to selected chromosome in case 1 (Fig. 2 of Ref. [31]). In order to evaluate the role of bundle lines and planning horizon year in an actual transmission network, the proposed idea is

Table 14 First configuration (case 1) and second configuration (case 2) in scenario 1

	First configuration		Second configuration		
Corr.	VL (kV)	NC	VL (kV)	NC	NB
8–9	230	2	230	2	2
2–8	230	2	230	1	2

test on the case study system, considering and neglecting the bundle lines based on two scenarios.

Scenario 1:

In this scenario, the planning horizon year and the maximum loading are considered as follows: (1) planning horizon year is 10 (year 2018: according to [31], the reference year is 2008). (2) $LL_{Max} = 50\ \%$. The proposed method is applied to the case study system and the results (lines which must be added to the network during the planning horizon year) are given in Table 14. The first and second configurations are obtained neglecting and considering the bundle lines, respectively Also, SC is zero for both configurations, and TC is 11.222 and 9.665 M$ for first and second configurations, respectively. Due to given results, although the bundle lines are more expensive than lines which have no bundle conductors, it can be seen that the second configuration is more economic than first one. In other words, the proposed GA has been decreased the expansion cost of lines and subsequent the total expansion cost of network by constructing of bundle lines instead of increasing the number of circuits.

Moreover, the capital investment saving for second configuration in comparison with first one is about 1.5 million dollars (this cost is equal to subtraction of total network expansion costs of two mentioned configurations). Thus, expansion of the network by second configuration is caused that about 13.4 % of total expansion cost of network with first configuration is saved. Accordingly, considering the bundle lines in transmission expansion planning is caused that total expansion cost of the network is calculated more precisely. It is well known, ability of transmitting the power of 400 kV lines is more than the lines with voltage level of 230 kV, i.e., construction of these lines can prevent useless increasing the number of 230 kV lines. On the other hand, expansion cost of 400 kV lines is more than other lines (230 kV) and construction of them in corridors which their sending and receiving substations have not voltage level of 400 kV, which would be caused substations are expanded and subsequent total expansion cost of the network is increased. Consequently, if expansion cost of these lines and substations can compete with the expansion cost of 230 kV lines, construction of them is suggested by the proposed method. However, expansion of the network by 400 kV lines is not economic and it is rejected by the DCGA based method. Its reason is that, in mid-term, expansion cost of the lines with higher voltage levels (here, 400 kV) and substations cannot compete with another expansion cost (expansion cost of 230 kV lines).

Scenario 2:

In this scenario, in order to more studying behavior of the network in long term from bundle and voltage level of the lines point of view, planning horizon year and maximum loading of lines and substations have been increased and decreased, respectively. Thus, planning horizon and maximum loading are considered as follows: (1) Planning horizon year is 15 [31] (year 2023: the reference year is 2008), (2) $LL_{Max} = 0.3$ (30 %). The proposed method is tested on the above-mentioned case study and the results are given in Tables 15. Also, the expansion

Table 15 Proposed configurations for scenario 2

	First configuration		Second configuration		
Corr.	VL (kV)	NC	VL (kV)	NC	NB
1–7	–	–	230	1	2
1–9	230	2	230	1	1
1–11	230	1	–	–	–
1–8	230	1	–	–	–
2–8	400	2	400	2	3
4–8	230	2	230	1	2
6–8	230	2	–	–	–
7–8	400	1	–	–	–
9–8	–	–	230	2	2
8–10	230	2	–	–	–
8–18	1	2	–	–	–
5–15	230	1	–	–	–
10–18	230	1	–	–	–
11–18	230	2	400	1	3
11–15	–	–	230	1	2

costs are listed in Table 16. Similar to for $LL_{Max} = 0.5$, it can be seen that with considering the bundle lines the second configuration is more economic. Moreover, DCGA has decreased the total expansion cost of network by constructing the bundle lines instead of increasing the number of circuits. Thus, although the cost of lines which have bundle conductors is more than unbundled lines, construction of them reduces the both expansion cost of lines and substations and therefore is caused that the total expansion cost of network decreases considerably. According to Table 16, the capital investment saving of second configuration in comparison with first one is about 26 million dollars. This cost is about 27 % of total expansion cost of network with first configuration that is considerable rate.

According to Table 15, the number of added candidate corridors to the network for second configuration is less that first one. Consequently, it can be said that construction of bundle lines in transmission network prevents useless expansion of unbundled lines in separate corridors and subsequent is caused that the size of the proposed network configuration for expansion decreased and therefore the transmission expansion planning is optimized. It is well known that, the loading of lines can be decreased when planning horizon year is increased. On the other hand, ability of transmitting the power and loading of 400 kV lines are more than and less than 230 kV lines, respectively. Due to the above discussion, although

Table 16 Expansion costs of network in scenario 2 for $LL_{Max} = 0.3$

	First configuration	Second configuration
Expansion Cost of Substations	25.6 million $US	22.4 million $US
Expansion Cost of Lines	71.019 million $US	48.378 million $US
Total Expansion Cost of Network	96.619 million $US	70.778 million $US

construction cost of 400 kV lines is more than 230 kV and construction of them would be caused substations are expanded, increasing of planning horizon years and decreasing of maximum loading of lines is caused 400 kV lines are expanded in addition to expansion of 230 kV lines, too. From voltage level of the added lines point of view, it can be seen that considering of bundle lines in long-term transmission expansion planning with different voltage levels is caused that from among expansion plans, configuration which its ratio of 400 lines to 230 kV lines is more than the first configuration will be selected. Its reason is that bundle lines with voltage level of 400 kV are overloaded later than the bundle lines with lower voltage levels (in here, 230 kV). Moreover, it can be seen that nearly all the mentioned lines (230 and 400 kV) have been added to the network with their maximum the number of bundle conductors (2 and 3), respectively.

The results analysis of both scenarios reveals that considering the bundle lines in transmission expansion planning is caused the both expansion cost of lines and substations and therefore the total expansion cost of network are decreased. Moreover, it can be said that although cost of bundle lines are more than those which have not bundle conductor, constructing this type of lines in transmission network with different voltage levels prevents useless expansion of unbundled lines in separate corridors and subsequent is caused that the size of the proposed network for expansion is reduced and therefore the transmission expansion planning is optimized. Thus, with considering the effect of these lines in expansion planning of a transmission network the capital investment cost is considerably saved and then the total expansion planning is calculated more exactly. Consequently, bundle lines play important role in transmission expansion planning and subsequent determination of network arrangement and configuration. In addition, from voltage level of the added lines point of view, considering the bundle lines in long-term TNEP problem with different voltage levels is caused that, from among expansion plans, configuration which its ratio of 400 lines to 230 kV lines is more than the first configuration will be selected.

6.3.2 Effect of Bundle Lines on Network Losses in STNEP

Bundle lines can affect indirectly on STNEP problem. In other words, it can affect on network losses and subsequent influence on STNEP problem. So, the objective function is Eq. (12) with the constraints described in Sect. 6.3.1. Similar to previous section, in order to evaluate the effect of bundle lines on the network losses and subsequent transmission expansion planning, DCGA method is tested on Azerbaijan network with the same input data considering and neglecting the network losses for two cases. It should be noted that the planning horizon year and maximum loading of lines and substations are 15 (year 2023: according to [33], the start time is 2008) and 30 % for both cases, respectively.

Case 1. Proposed method neglecting the bundle lines is applied to above-mentioned test network and the results are given in Tables 17 and 18.

Table 17 Proposed configurations in case 1

Corr	First configuration		Second configuration	
	VL (kV)	NC	VL (kV)	NC
1–9	230	2	230	2
2–8	400	2	400	2
4–8	230	2	230	2
6–8	230	2	230	2
7–8	400	1	400	1
8–10	230	2	230	2
5–15	230	1	230	2
1–11	230	1	230	1
5–18	230	1	230	1
1–18	230	1	230	1
10–18	230	2	230	2

Table 18 Expansion costs of proposed configurations in case 1

	First configuration	Second configuration
Expansion Cost of Substations	25.6 million $US	25.6 million $US
Expansion Cost of Lines	72.019 million $US	73.679 million $US
Total Expansion Cost of Network	96.619 million $US	99.279 million $US

Table 19 Proposed configurations in case 2

Corr.	First configuration		Second configuration		
	VL (kV)	NC	VL (kV)	NC	NB
1–9	230	1	–	–	–
8–9	230	2	230	2	2
4–8	230	1	230	2	2
6–8	–	–	230	2	2
2–8	400	2	400	2	3
5–15	–	–	230	2	2
8–18	230	1	230	2	2
11–15	–	–	230	1	2
11–18	400	1	400	1	3
1–7	230	1	230	1	2
10–18	-	–	400	2	3

Table 20 Expansion costs of proposed configurations in case 2

	First configuration	Second configuration
SC	25.6 million $US	19.2 million $US
TC	48.378 million $US	64.998 million $US
TC	70.778 million $US	84.198 million $US

Case 2. In this case, the proposed idea considering the effect of bundle lines is tested on case study, and the results are given in Tables 19 and 20.

In this case, from the transmitted power through the lines point of view, the first configuration is overloaded 16 and second one is overloaded 19 years after expansion time, too. Construction of bundle lines in transmission network with various voltage levels is caused the total expansion cost is decreased and calculated more exactly. Moreover, it can be seen that most of the mentioned lines (400 and 230 kV) have been added to the network with their maximum the number of bundle conductors (2 and 3), respectively. Finally, it can be concluded, considering the network losses, with considering the bundle lines is caused the network lines are overloaded later. Thus, construction of these lines increases the network adequacy and consequently network satisfies the requirement of delivering electric power more safely and reliably to load centers. We presented various examples for STNEP which solved by DCGA. As said, PSO algorithms like DPSO and IDPSO are better methods than DCGA for solution of such problems. So, we present new examples for showing more robustness and efficiency of these methods. In these examples, it is tried that a new mathematical model is introduced for STNEP and then solved by DPSO and IDPSO in comparison with and DCGA.

Example 4. Assume a STNEP problem has been formulated as follows:

$$Fitness = T_{\text{overload}}$$

S.t.: Eqs (3)–(6), *N-1 safe criterion* and $C \leq C_{\max}$.

where T_{overload} is required time for missing the expanded network adequacy (year) and $C_{\max}$ is maximum construction cost of lines (maximum investment) along the planning horizon for expanding the network.

Remark. The lines adequacy of network is necessary to provide load demands when the network is expanding because its lack (i.e., lines overloading) caused to load interrupting. Consequently, if expanded network is more reliable and therefore its lines overloaded later, will be more economic and caused to utilize favorably. It should be noted that the lines adequacy of transmission network is proportional to the investment cost. In fact, the lines adequacy is increased by increasing the investment cost and using the exact planning and the proper solution method. On the other hand, with a low costing, the network operates weakly to support load demand and becomes overloaded early. Thus, with compromising between two parameters, i.e., investment cost and network adequacy rate and finally defining a total index, STNEP can be implemented in order to have a network with maximum efficiency technically and economically. By defining the foregoing fitness function, a design for transmission network expansion could be acquired to represent a maximum probabilistic adequacy according to a maximum value of specified investment cost ($C_{\max}$). The goal of solving mentioned model is obtaining number of required circuits for appending to the network until it is brought to a maximum adequacy with minimum cost during one specified horizon year. Maximize above-mentioned objective function using DPSO.

Table 21 Value of parameters for DPSO algorithm

Parameter	Value
Problem dimension	15
Number of particles	30
Number of iterations	1000
C_1, C_2, v_{max}, v_{min}	1.7, 2.3, 3, −3

Table 22 Proposed configurations for $C_{max} = 50-100$ M$

C	50	60	70	80	90	100
Corridor	NC	NC	NC	NC	NC	NC
2–6	4	4	–	–	–	–
3–5	2	2	1	–	1	–
3–6	2	1	4	1	4	4
4–6	3	3	3	3	4	3
5–6	–	2	2	4	3	4
2–4	–	–	–	1	1	–
2–5	–	–	–	4	1	–
1–3	–	–	–	–	–	1
1–5	–	–	1	–	1	–
1–6	–	–	–	–	–	2
2–3	–	–	2	2	2	3

Solution: To prove the validity of the proposed planning technique, it was applied to the same system described in Example 1, but, here, all the existed lines and substations are 230 and 230/63 kV, respectively. It should be mentioned that the planning horizon year is 2014 (5 years ahead: the reference year is 2009 [25]). In following, test results of proposed algorithm on mentioned network will be described in comparison with DCGA approach. To acquire better performance and fast convergence of the proposed algorithm, parameters which are used in DPSO algorithm have been initialized according to Table 21.

By implementing the proposed method (DPSO) on the network according to various investment costs, the results are given in Table 22.

Network adequacy versus network expansion cost has been depicted in Fig. 7. As shown in Fig. 7, increasing in higher investment cost (C_{max} = 70–80 and 90–100), changes the network adequacy more slightly than other investment costs. Thus, a parameter, named adequacy index on expansion cost rate, is defined for obtaining best design according to the investment cost and the network adequacy. This parameter is the network adequacy rate (year) per the investment cost. Thus, a high value is desirable for this index. According to Fig. 7, the optimized point is 47.09 million dollars for the investment cost (C_{max} = 50) [25].

Convergence curves of DPSO method for different cases which illustrated in [25] show the fitness function is optimized more and faster than DCGA one. Thus, it can be concluded that optimization of lines loading in TNEP by DPSO algorithm is more precise, faster and finally better than DCGA method.

Table 23 Value of parameters for IDPSOMS algorithm

Parameter	Value
Problem dimension	15
Number of particles	20
Number of iterations	300
C_1, C_2, v_{max}, v_{min}, d_{min}, d_{max}, S_{max}	1.7, 2.3, 2, −2, 3, 20, 198

Example 5. Solve the STNEP problem described in Example 4 using IDPSO.

Solution: For having better a good form in convergence curves, we convert the maximization problem of previous example to minimization problem. In other words, objective function is defined as $Fitness = 1/T_o$. where T_O is the same $T_{overload}$. IDPSO is applied to the same test systems of Example 4 with following parameters (Table 23).

By applying IDPSOMS on the network according to various investment costs (C_{max} changes between 30 and 80 million dollars by 10 million steps), the results are obtained as follows (the dash lines into figures are number of required circuits for adding to the network until planning horizon year) [26].

It is noted that, by investment cost limit increasing, required lines which could be appended to the network is increased and overloaded later. However, it seems that the network adequacy may be acquired with lower relative investment cost. Thus, the parameter of adequacy index on expansion cost rate is used for obtaining best design (see [25] for more details), as shown in Fig. 9. A high value is desirable for this index. According to Fig. 9, the optimized point is 28.38 million dollars for the investment cost ($C_{max} = 30$) [26].

In order to show the efficiency and correctness of the proposed IDPSOMS, DPSO algorithm is applied to the desired STNEP problem and years of missing the network adequacy versus maximum investments for both methods are shown in Fig. 10.

Regarding the simulation results of both case study systems, it can be concluded that optimization of the lines loading problem in STNEP by IDPSOMS is caused that for the same expansion costs the network adequacy is more increased than DPSO. IDPSOMS by preventing the premature convergence of the algorithm around local solutions is caused that the fitness function is optimized more than

Table 24 Input data for IDPSO algorithm

Parameter	Value
Problem dimension	15
Number of particles	20
Number of iterations	300
C_1, C_2, v_{max}, v_{min}, d_{min}, d_{max}, S_{max}, $n_1^{l,max}$, $n_3^{l,max}$, $n_6^{l,max}$	1.7, 2.3, 2, -2, 3, 20, 198, 6, 8, 6
LL_{max}	30 %
a_i ($/MW^2-hr), b_i ($/MW-hr), c_i, P_{li}^{min}	0.009, 25, 0, 25 % of ith unit capacity

Table 25 The number of generation units constructed along the planning horizon

Year / Substation	1	2	3	4	5
1	0	1	0	0	0
3	1	1	1	0	0
6	1	0	0	0	0

Table 26 The generation amount of power plant units along the planning horizon

Year / Substation	1	2	3	4	5
1	100	125	125	125	125
3	300	336	391	373	399
6	303	293	291	366	399

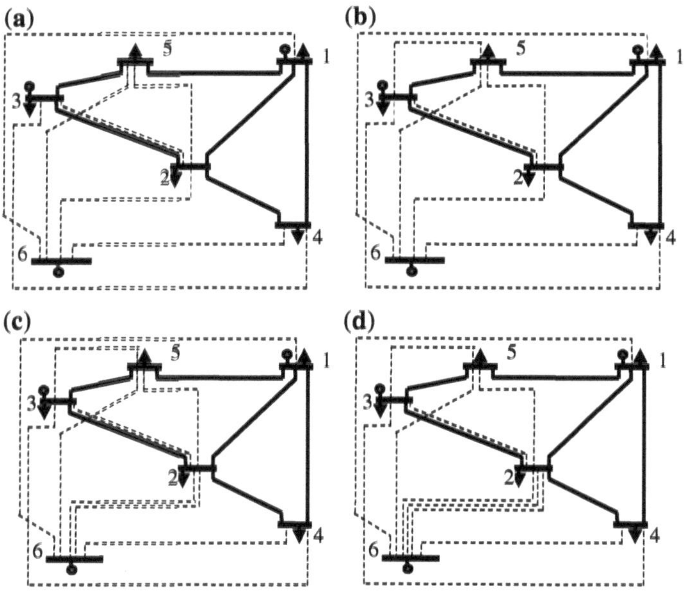

Fig. 11 Proposed configuration during the planning horizon for: **a** first year, **b** second and third year, **c** fourth year, and **d** fifth year

DPSO one. Thus, it can be concluded that optimization of lines loading in TNEP by IDPSOMS algorithm is more precise and finally better than DPSO method.

Example 6: Solve the DTNEP problem for Garver's network Using IDPSO method if the construction cost of lines, network losses and generation costs is to be minimum. So, the problem is formulated as follows:

$$C_T = DTC + DLC + GC$$

S.t. (15)–(17), (19)–(21), (29)–(31).

In this network, the existing generation plants on buses 1, 3, and 6 are considered as four units of 25 MW, five units of 50 MW and three units of 25 MW, respectively. Also, construction cost of each power plant is assumed 800000 $/MW. IDPSO is applied to test case study system with following input data (Table 24):

The results are obtained as follows. Also, the number of generation plant units are constructed along the planning horizon is listed in Table 25. Moreover, generation amount of these power plant units are given in Table 26 (Fig. 11).

References

1. Abdelaziz AR (2000) Genetic algorithm based power transmission expansion planning. 7th IEEE Int Conf Electron Circuits and Syst 2:642–645
2. Levi VA, Calovic MS (1993) Linear-programming-based decomposition method for optimal planning of transmission network investments. IEE Proc Gener Transm Distrib 140:516–522
3. Binato S, de Oliveira GC, Araujo JL (2001) A greedy randomized adaptive search procedure for transmission expansion planning. IEEE Trans Power Syst 16:247–253
4. Choi J, Mount T, Thomas R (2006) Transmission system expansion plans in view point of deterministic, probabilistic and security reliability criteria. The 39th Hawaii Int Conf Syst Sci 10:1–10
5. Silva IDJ, Rider MJ, Romero R, Murari CA (2005) Transmission network expansion planning considering uncertainness in demand. Power Eng Soc Gen Meet 2:1424–1429
6. Binato S, Periera MVF, Granville S (2001) A new Benders decomposition approach to solve power transmission network design problems. IEEE Trans Power Syst 16:235–240
7. Escobar AH, Gallego RA, Romero R (2004) Multistage and coordinated planning of the expansion of transmission systems. IEEE Trans Power Syst 19:735–744
8. Dodu JC, Merlin A (1981) Dynamic model for long-term expansion planning studies of power transmission systems: The Ortie model. Int J Electr Power & Energy Syst 3:2–16
9. Garver LL (1970) Transmission net estimation using linear programming. IEEE Trans Power Appar Syst PAS-89:1688–1696
10. Chanda RS, Bhattacharjee PK (1995) A reliability approach to transmission expansion planning using minimal cut theory. Electr Power Syst Res 33:111–117
11. Chanda RS, Bhattacharjee PK (1998) A reliability approach to transmission expansion planning using fuzzy fault-tree model. Electr Power Syst Res 45:101–108
12. Sohtaoglu NH (1998) The effect of economic parameters on power transmission planning. 9th Mediterr Electrotech Conf 2:941–945
13. Granville S, Pereira MVF et al (1988) Mathematical decomposition techniques for power system expansion planning-analysis of the linearized power flow model using the benders decomposition technique. EPRI, Technical Report, RP, pp 2473–2476
14. Romero R, Monticelli A (1994) A hierarchical decomposition approach for transmission network expansion planning. IEEE Trans Power Syst 9:373–380
15. Lee STY, Hocks KL, Hnyilicza H (1970) Transmission expansion of branch and bound integer programming with optimal cost capacity curves. IEEE Trans Power Appar Syst PAS-93:1390–1400
16. Periera MVF, Pinto L (1985) Application of sensitivity analysis of load supplying capacity to interactive transmission expansion planning. IEEE Trans Power Appar Syst PAS-104:381-389

17. Romero R, Gallego RA, Monticelli A (1996) Transmission system expansion planning by simulated annealing. IEEE Trans Power Syst 11:364–369
18. Gallego RA, Alves AB, Monticelli A, Romero R (1997) Parallel simulated annealing applied to long term transmission network expansion planning. IEEE Trans Power Syst 12:181–188
19. Al-Saba T, El-Amin I (2002) The application of artificial intelligent tools to the transmission expansion problem. Electr Power Syst Res 62:117–126
20. Contreras J, Wu FF (2000) A kernel-oriented algorithm for transmission expansion planning. IEEE Trans Power Syst 15:1434–1440
21. Silva EL, Gil HA, Areiza JM (2000) Transmission network expansion planning under an improved genetic algorithm. IEEE Trans Power Syst 15:1168–1174
22. Silva EL, Oritz JMA, Oleveria GC, Binato S (2001) Transmission network expansion planning under a Tabu search approach. IEEE Trans Power Syst 16:62–68
23. Jalilzadeh S, Kazemi A, Shayeghi H, Mahdavi M (2008) Technical and economic evaluation of voltage level in transmission network expansion planning using GA. Energy Convers Manag 49:1119–1125
24. Shayeghi H, Jalilzadeh S, Mahdavi M, Haddadian H (2008) Studying influence of two effective parameters on network losses in transmission expansion planning using DCGA. Energy Convers Manag 49:3017–3024
25. Shayeghi H, Mahdavi M, Bagheri A (2010) Discrete PSO algorithm based optimization of transmission lines loading in TNEP problem. Energy Convers Manag 51:112–121
26. Shayeghi H, Mahdavi M, Bagheri A (2010) An improved DPSO with mutation based on similarity algorithm for optimization of transmission lines loading. Energy Convers Manag 51:2715–2723
27. Contreras J, Wu FF (2000) A kernel-oriented algorithm for transmission expansion planning. IEEE Trans Power Syst 15:1434–1440
28. Braga ASD, Saraiva JT (2005) A multiyear dynamic approach for transmission expansion planning and long-term marginal costs computation. IEEE Trans Power Syst 20:1631–1639
29. Sepasian MS, Seifi H, Foroud AA, Hatami AR (2009) A multiyear security constrained hybrid generation-transmission expansion planning algorithm including fuel supply costs. IEEE Trans Power Syst 24:1609–1618
30. Bulent Tor O, Guven AN, Shahidehpour M (2008) Congestion-driven transmission planning considering the impact of generator expansion. IEEE Trans Power Syst 23:781–789
31. Mahdavi M, Shayeghi H, Kazemi A (2009) DCGA based evaluating role of bundle lines in TNEP considering expansion of substations from voltage level point of view. Energy Convers Manag 50:2067–2073
32. Sum-Im T, Taylor GA, Irving MR, Song YH (2009) Differential evolution algorithm for static and multistage transmission expansion planning. IET Gener Transm Distrib 3:365–384
33. Shayeghi H, Mahdavi M (2009) Genetic algorithm based studying of bundle lines effect on network losses in transmission network expansion planning. J Electr Eng 60:237–245
34. Eldershaw C (2000) A brief Survey of genetic algorithms as applied to non-linear optimization. Technical Report, University of Queensland, pp 1–14
35. Chipperfield A, Fleming P, Pohlheim H, Fonseca C (1994) Genetic algorithms toolbox for use with MATLAB®. Users Guide, University of Sheffield, p 1–94
36. Mitchell M (1995) Genetic algorithms: an overview, Technical Report, vol. 1. Santa Fe Institute, pp 31–39
37. Townsend AAR (2003) Genetic algorithms, A tutorial, pp 1–52
38. Shayeghi H, Jalili A, Shayanfar HA (2008) Multi-stage fuzzy load frequency control using PSO. Energy Convers Manag 49:2570–2580
39. Clerc M, Kennedy J (2002) The particle swarm-explosion, stability, and convergence in a multidimensional complex space. IEEE Trans Evol Comput 6:58–73
40. Jin YX, Cheng HZ, Yan JYM, Zhang L (2007) New discrete method for particle swarm optimization and its application in transmission network expansion planning. Electr Power Syst Res 77:227–233

41. Jin N, Samii YR (2007) Advances in particle swarm optimization for antenna designs: real-number, binary, single-objective and multiobjective implementations. IEEE Trans Antennas Propag 55:556–567
42. Mahdavi M, Monsef H, Bagheri A (2012) Lines loading optimization in transmission expansion planning based on binary PSO algorithm. i-manag J Inf Technol 1:24–32
43. Liu J, Fan X, Qu Z (2007) An improved particle swarm optimization with mutation based on similarity. Third Int Conf Nat Comput 4:824–828
44. Trelea IC (2003) The particle swarm optimization algorithm: convergence analysis and parameter selection. Inf Process Lett 85:317–325

Applications in Control of the Hybrid Power Systems

Nicu Bizon, Mihai Oproescu and Marian Raducu

Abstract This chapter analyses the control of the Hybrid Power Sources (HPS) based on some applications performed. Usually, a HPS combines two or more energy sources that work together with the Energy Storage Devices (ESD) to deliver power continuously to the DC load or to the AC load via the inverter system. In the automotive applications, the ESD stack can be charged from the regenerative braking power flow or from other power sources like the thermoelectric generator or the renewable source. The last may have a daily variable power flow such as the photovoltaic panels integrated into a car's body or into buildings. In the first section, an efficient fuel cell/battery HPS topology is proposed for high power applications to obtain both performances in energy conversion efficiency and fuel cell ripple mitigation. This topology uses an inverter system directly powered from the appropriate Polymer Electrolyte Membrane Fuel Cell (PEMFC) stack that is the main power source and a buck Controlled Current Source (buck CCS) supplied by a batteries stack, which is the low power auxiliary source. The buck CCS is connected in parallel with the main power source, the PEMFC stack. Usually, the FC HPS supply inverter systems and PMFC current ripple normally appear in operation of the inverter system that is grid connected or supply the AC motors in vehicle applications. The Low Frequency (LF) ripple mitigation is based on the active nonlinear control placed in the tracking control loop of the fuel cell current ripple shape. So the buck CCS will generate an anti-ripple current that tracks the FC current shape. This anti-ripple current is injected into the output node of the HPS to mitigate the inverter current ripple. Consequently, the buck CCS must be designed in order to assure the dynamic requested in the control loop. The ripple mitigation performances are evaluated by some

---

N. Bizon (✉) · M. Oproescu · M. Raducu
Faculty of Electronics, Communication and Computers, University of Pitesti, Pitesti, Romania
e-mail: nicu.bizon@upit.ro

N. Bizon
Doctoral School, University 'Politehnica' of Bucharest, Bucharest, Romania

N. Bizon et al. (eds.), *Analysis, Control and Optimal Operations in Hybrid Power Systems*, Green Energy and Technology, DOI: 10.1007/978-1-4471-5538-6_7,

indicators related to the LF harmonics mitigation. It is shown that good performances are also obtained with the hysteretic current—mode control, but the nonlinear control has better performances. The nonlinear control of the buck CCS is implemented based on a piecewise linear control law. This control law is simply designed based on the inverse gain that is computed to give a constant answer for all levels of the LF current ripple. The control performances are shown by the simulations performed. Finally, the designed control law will be validated using a Fuzzy Logic Controller (FLC). In the second and the third section is proposed and analysed a nonlinear control for FC HPS based on bi-buck topology that further improves FC performance and its durability in use in the low and medium power applications. The nonlinear voltage control is analysed and designed in the second section using a systematic approach. The design goal is to stabilise the HPS output voltage. This voltage must have a low voltage ripple. Additionally, the power spectrum of this ripple must be spread in a wide frequencies band using an anti-chaos control. All the results have been validated with several simulations.

Keywords Fuel cell · Hybrid power sources · Inverter systems · Ripple mitigation · Spread power spectrum · Energy efficiency · Nonlinear control

1 Introduction

The PEMFC stack represents one of the most used solutions as main energy source in the Energy Generation Systems (EGS) and vehicle applications. This is due to its new performances in applications: Small size, ease of construction, good energy efficiency, fast start-up and low operating temperature. Even if there are a lot of advantages in their using, the extensive use in such applications is unfortunately still limited due to their relatively short lifetime [1]. As it is known, the inverter current ripple is one of the main factors for low performances regarding the PEMFC energy efficiency [2, 3] and the PEMFC life cycle [4–6]. Also, it is known that the LF FC current ripple affects in much measure the PEMFC life cycle, causes hysteretic losses and subsequently more fuel consumption. The LF inverter current ripple contributes with up to 10 % reduction in the available output power [7, 8]. Consequently, some limits for the LF FC current ripple or other slower load transients on different frequencies bands are specified. The FC stack has a rather large capacitance that can mitigate the High Frequency (HF) current ripple if the limits are not exceeded. Usually, only one limit for the HF ripple is specified.

The USA National Energy Technology Laboratory (NETL) published the first guidelines for the FC current ripple limits that can assure PEMFC stack operation without degradation of its performance [9]. The FC current ripple limits are given experimentally as values of the Ripple Factor (RF) measured for different frequency bands (for example, LF RF must be up to 5 % from 10 to 100 % load, but should not exceed 0.5 A for lighter loads; HF RF must be up to 40 % from 10 to

100 % load, but should not exceed 2 A for lighter loads). Lower values for the RF are certainly recommended to be obtained by an appropriate active control to further increase the PEMFC performances. For example, the interleaved control technique used in the parallel topology of the power converters supplied from the same PEMFC stack may be a solution for the FC current ripple mitigation [10, 11].

On the other hand, the slow FC power profiles (variations below grid frequency) represent the "load following" action of the EGS control and two control loops must be used: (1) the load following control loop will set the fuelling values according to the load requirements, and (2) the energy harvesting loop will use a Maximum Power Point (MPP) tracking technique to increase the energy amount extracted from the PEMFC stack in a real-time optimisation manner. The MPP signal from the PEMFC should be tracked within 1 % with a current-mode controller for purposes of both PEMFC reliability and efficiency [12–14].

In vehicle applications usually appear high energy demands that will cause high current slopes and obviously voltage drops, which are recognised as fuel starvation phenomenon. Consequently, it is necessary to add ESDs in the vehicles supplied by the PEMFC stack [15–17]. The ESDs having the short time response (for example, the ultracapacitors stack) could be used as a Power Dynamic Compensators (PDC). Batteries and ultracapacitors are usually used as ESDs and PDCs, respectively. Those devices are used in the hybrid ESDs stack to compensate the fast power demand, reducing the FC starvation phenomenon by improving the dynamic performance of the HPS [18]. For the above considerations, it is obviously that the hybrid ESDs and fuel cell stacks need to be merged technologies in the HPS topologies. Usually, a HPS topology uses two or more energy sources and a hybrid ESD/PDC stack that work together as an embedded power unit (named as power hub) to deliver or store energy. Consequently, the challenge for the power management of the HPS is to enhance the performance of the entire HPS through these technologies working together [19]. The current slopes are given experimentally for different levels of the PEMFC stacks power (about 10 A/s per each kW of the rated power) [20–22]. Recently, some HPS topologies of FC/ultracapacitor type have been reported for vehicle applications [23], such as FC/battery HPS [24] and FC/ultracapacitor/battery HPS [8, 25]. In this chapter a FC/battery HPS topology is analysed from the control point of view and behaviour under dynamic load. The control goal is to mitigate the PEMFC current ripple as much as possible. Some ripple models for the PEMFC stack and appropriate power interfaces used to mitigate the FC current ripple are analysed in [26–29]. The state-of-art for the FC HPS topologies is presented in one of the chapters of this book, too.

In the Sect. 1, the analysis will be focused on modelling, designing and operating of the FC HPS with active control used to mitigate the inverter ripple based on a nonlinear control law. A high power FC/ESD/PDC HPS topology is proposed to obtain both performances in energy conversion and in ripple mitigation. This topology uses an inverter system that is directly powered from the appropriate FC stack and a buck CCS, which is also powered from the FC stack. In the next sections, a FC/ESD/PDC HPS structure based on the bi-buck topology is considered for medium power applications. The control goal for the buck CCS is the same: high mitigation of the FC current ripple based on an active control. The necessity of a nonlinear gain in the control loop is shown by simulation. After that, this control law is validated through a FLC that generates a 2-D control surface based on two input control variables: (1) the output voltage error and the FC current ripple. One of the contour projections for this 2-D control surface can be chosen as a nonlinear control law. On the other hand, the control goal for the buck Controlled Voltage Source (CVS) is to stabilise the HPS output voltage having a low voltage ripple that is spread in wide frequencies band. The simulation results successfully show that nonlinear voltage control performs good performances in the frequency-domain (high spreading level of the power spectrum) and in the time-domain (low RF level of the output voltage), too. Conclusions are given in the Sect. 6.

2 Problem Statement and Used Models

In order to estimate the inverter current ripple using the Matlab-Simulink® toolboxes, some simulations are made for the inverter systems of mono-phase and three-phase type. The obtained results are shown in one of the chapters of this book. The LF current ripple of the input inverter current is back propagated from load to the DC output of the HPS via the inverter system. Obviously, the power spectrum of the current ripple has the HF harmonics situate at multiples of the carrier frequency, but their levels are much smaller than the levels of the LF harmonics. The HF harmonics were generated by the switching action of the inverter system. Usually a PWM pure sine command is used, having the carrier frequency in range of 10–100 kHz.

As it was known, the main harmonic for the grid-connected mono-phase inverters is situated at twice of the grid frequency and the significant LF harmonics are situates at multiples of this harmonic. If the grid frequency is of 50 Hz, then the significant LF harmonics are situates at frequencies of 100 Hz, 200 Hz and 300 Hz. Also, for the grid-connected three-phase inverters, the main harmonic is situated at triple of the grid frequency. Thus, the significant LF harmonics are situates at frequencies of 150 Hz, 300 Hz and 600 Hz. If the mono- and three-phase inverter supply an AC load (for example, an electrical AC machine), then the main harmonic is situated at twice or triple of the working frequency and the significant LF harmonics are situated at multiples of it. Those results obtained (namely, the spectral distribution observed for the LF harmonics of the current ripple, and the observed levels of these harmonics in simulations and experiments performed) will be considered in designing an equivalent load for the inverter system behaviour. The equivalent load is implemented by a CCS having the control signal a superposition of three rectified sine wave with frequency of 50 Hz, 150 Hz and 300 Hz. Also, the levels for these harmonics can be set independently. Those levels are set to 30, 5 and 5 A for case shown in Fig. 1. The HF current ripple is not considered in the load model because the PEMFC stack has a high

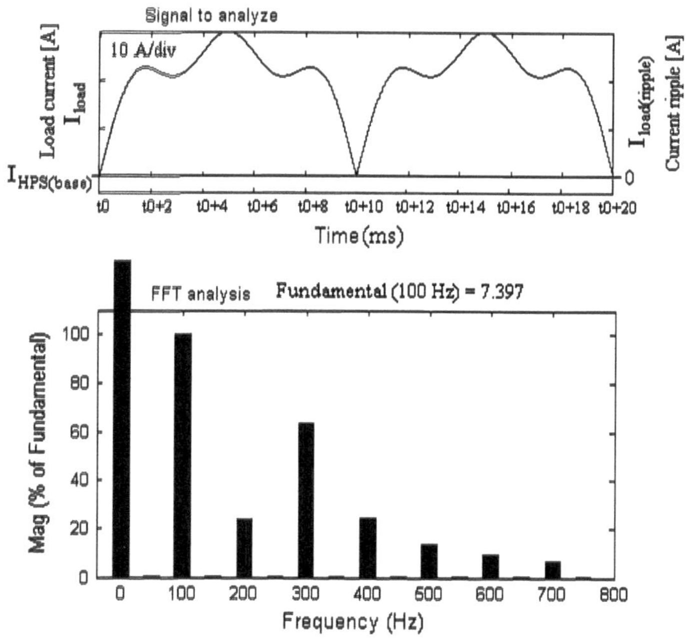

Fig. 1 The equivalent load current (*top*) and its power spectrum (*bottom*); adapted from [35]

tolerance to it. The DC reference current, $I_{HPS(base)}$, will be chosen in correlation with the MPP of the used PEMFC stack (see Figs. 1 and 2).

Three types of PEMFC stack will be used in simulation: one for high power applications and two for medium power applications. Also, two types of batteries stack will be used in accordance with the power level of application implemented. This statement is also valid for used type of ultracapacitors stack.

Next subsections will briefly present the HPS models and the parameters values set for the PEMFC model used.

2.1 Fuel Cell Model

Some detailed PEMFC models are now available in the literature [30–32]. One that combines in a well manner, the PEMFC operating relationships are now available in Matlab—Simulink®.

Three preset PEMFC models are used in this chapter, having the main parameters specified below [33–35]:

- 1.26 kW PEMFC stack that for a FuelFr = 10.5 lpm will have the MPP approximately at 45 A and 27 V (Fig. 3a);

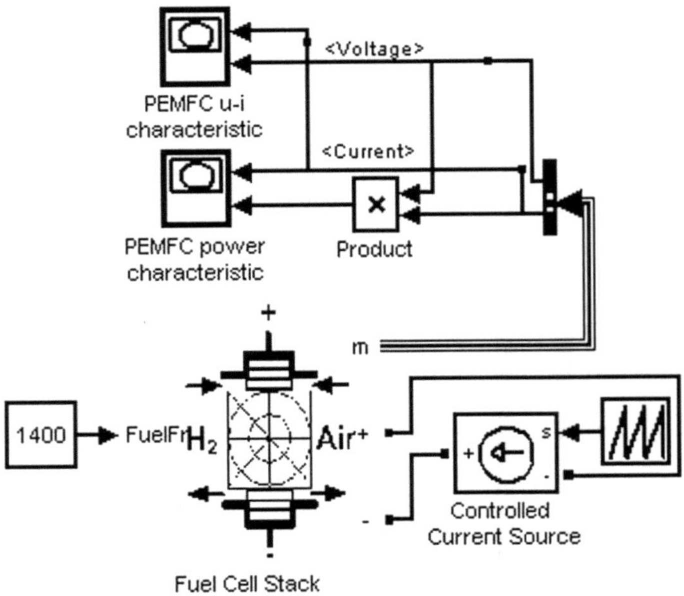

Fig. 2 The Simulink diagram used to obtain the static characteristics of the PEMFC stack used

- 6 kW PEMFC stack that for a FuelFr = 47 lpm will have the MPP approximately at 120 A and 50 V (Fig. 3b).
- 50 kW PEMFC stack that for a FuelFr = 1400 lpm will have the MPP approximately at 240 A and 560 V (Fig. 3c).

Other parameters of the preset model for the used Fuel flow rate (named FuelFr and measured in litres per minute, lpm) are shown in Fig. 3, too.

2.2 Energy Storage Devices Models

Usually, the medium power FC HPS topologies use a power interface based on the MPP tracking control to extract the maximum energy from the PEMFC stack via a grid-connected inverter. Off-grid power systems also use MPP tracking controller to harvest the energy from the PEMFC stack. The power delivered by the FC HPS must always be bigger than the needed load power. The flow rate regulator (which is driven by the FC current) control the power delivered by the PEMFC to assure this requirement until the maximum available FC power. In the transitory regime, when the load power requirements are less or greater than the power currently available (which must always be near the MPP), the power difference is delivered by an auxiliary energy source (wind turbine or/and photovoltaic panel) or an ESD

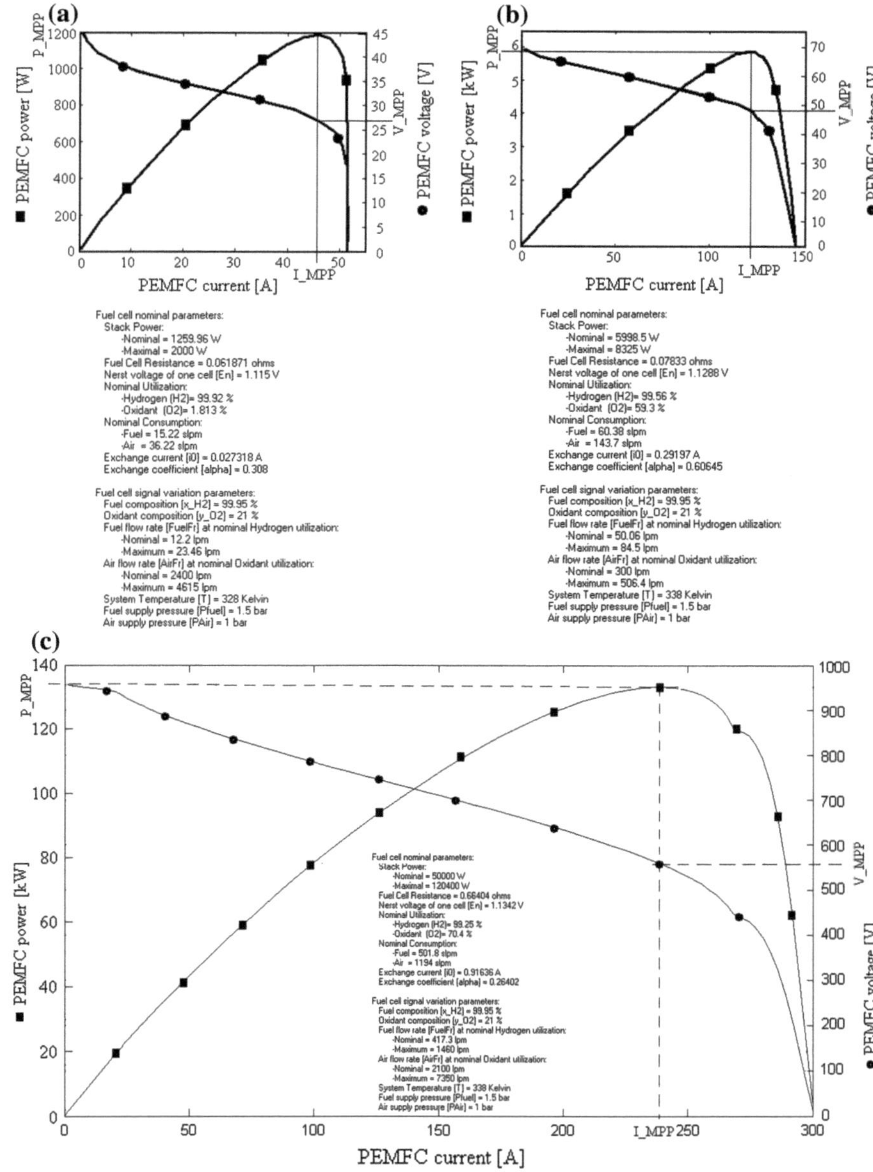

Fig. 3 The PEMFC static characteristics; adapted from [33, 35], **a** 1.26 kW PEMFC static characteristics at the fuel flow rate of 10.5 lpm, **b** 6 kW PEMFC static characteristics at the fuel flow rate of 47 lpm, **c** 50 kW PEMFC static characteristics at the fuel flow rate of 1400 lpm

(usually a batteries stack). If the CVS controller sets a constant voltage in the HPS output node, then the energy management based on the power balance implies an ESD current controlled by the CCS controller. The buck CCS extracts the necessary energy from the ESD in order to compensate the sharp power profiles for the dynamic loads (the power difference that appears for short time in output node). A PDC stack (usually an ultracapacitors stack) is used directly (in parallel with the PEMFC stack) or via a bidirectional DC–DC power converter.

If a bi-buck topology is used, then the HPS output voltage will be lower than the PEMFC voltage at MPP. This value was chosen to be 25 and 40 Vdc for the case of using the 1.26 and 6 kW PEMFC stack, respectively. The batteries stacks were chosen in relation with those voltage values. Detailed models for battery are now available in the literature [35] and one generic is now available in Matlab—Simulink®, too. A preset NiMH battery model will be used. For the preset model the model parameters based on the battery type, nominal voltage value and the rated capacity are used. The initial State-Of-Charge (SOC) is set to 80 % in all simulations. The used parameters are specified below for each PEMFC's stack:

- For the 1.26 kW PEMFC: The NiMH battery parameters are set to 40 V and 20 Ah for the nominal voltage and rated capacity, respectively (see discharge characteristics on Fig. 4a);
- For the 6 kW PEMFC: The NiMH battery parameters are set to 60 V and 200 Ah for the nominal voltage and rated capacity, respectively (see discharge characteristics on Fig. 4b).
- For the 50 kW PEMFC: The NiMH battery parameters are set to 800 V and 100 Ah for the nominal voltage and rated capacity, respectively (see discharge characteristics on Fig. 4c).

The set values are chosen to obtain a reasonable value for the duty cycle of the PWM command applied to the buck CCS.

Batteries technology represents a good option to be used as ESD in different power applications, while the ultracapacitors technology represents an attractive option to be used as PDC in burst power applications.

The ultracapacitors stack provides the difference between the load demand and the power delivered by the PEMFC/battery hybrid system. In this chapter a PEMFC/ultracapacitors/battery hybrid system is adopted. The capacitance value used for the ultracapacitors stack depends on the imposed HF voltage ripple, the switching frequency and the load power level. A first order model is used to model the ultracapcitors stack.

2.3 Load Test Model

It is obvious that the LF ripple current appears on the HPS DC voltage bus in the same way for all multi-phase inverter systems topologies. Consequently, the equivalent load for the inverter system was implemented by a CCS that is controlled to cover all these cases. The control signal can be a superposition of three rectified LF sine waves having different levels for these LF harmonics.

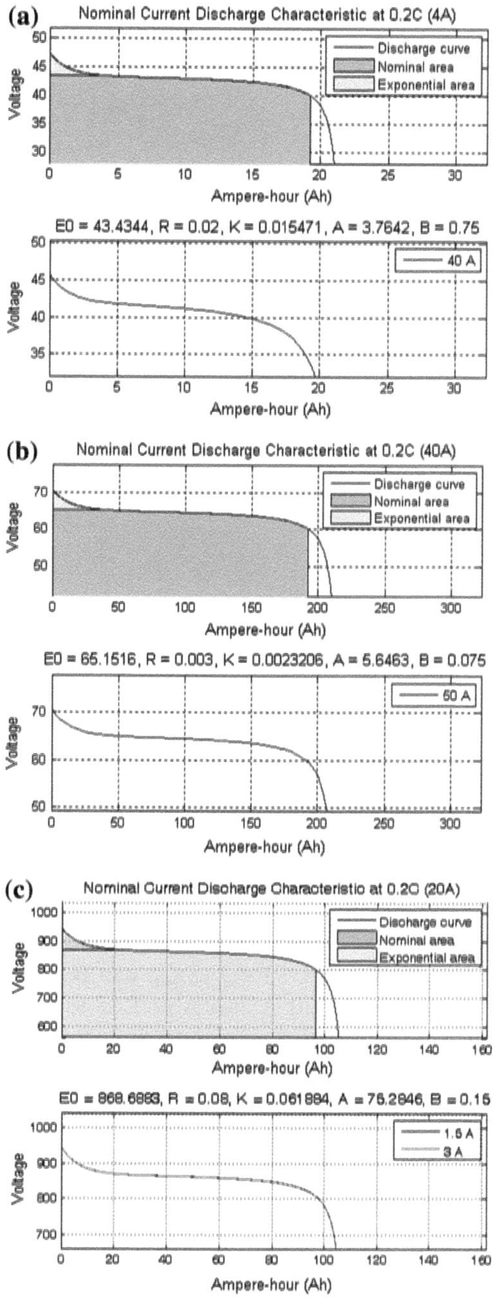

Fig. 4 The NiMH battery discharge characteristics; adapted from [33], **a** 20 Ah (40 V) NiMH battery, **b** 200 Ah (60 V) NiMH battery, **c** 100 Ah (800 V) NiMH battery

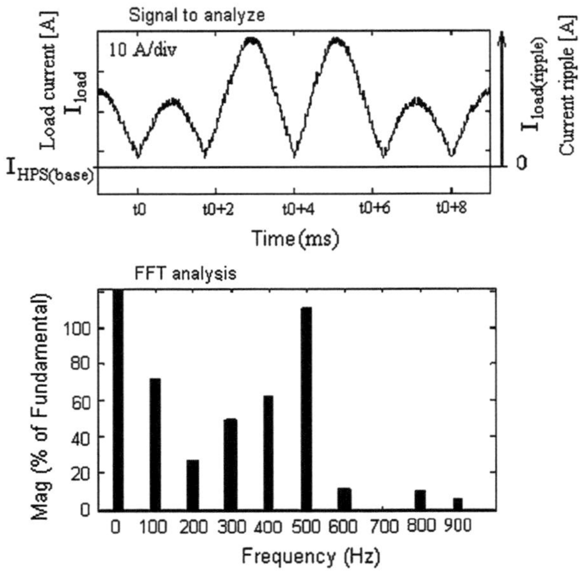

Fig. 5 The load current (*top*) and its power spectrum (*bottom*); adapted from [33]

The levels for the LF harmonics of 50 Hz, 150 Hz and 300 Hz are set at 30 A, 30 A and 5 A for the example shown in Fig. 5. The DC reference current, $I_{HPS(base)}$, defines the base level from which the current ripple shall be deemed (see Figs. 1 and 5).

As it was mentioned before, the current slopes are given experimentally for different power of the PEMFC stacks and the recommended value is 10 A/s per each kW power. So the recommended limit for the PEMFC stacks considered is of 12, 60 and 500 A/s, respectively. The maximum current slope is higher than 20 A/ms (2,000 A/s) for the both current load shapes shown in Figs. 1 and 5. So it is much higher than the admissible FC current slope. If a ripple of 30 $A_{p\text{-}p}$ (peak-to-peak value) is considered for the load current, then the RF value for the PEMFC stacks considered, $RF_{load\_current}$, will be about of 30 A/45 A = 66.7 %, 30 A/120 A = 25 % and 30 A/240 A = 12.5 %, respectively. Thus, the RF for the load current was set high enough to show the performances of the ripple mitigation for the HPS under proposed control. Of course, all the above can be easily changed by using a new set of parameters for the equivalent load. Values used in the simulation will be mentioned in each case.

It can be noted that if the ripple mitigation loop operates, then the buck CCS will compensate the main part of this ripple and the rest will propagate back to the PEMFC stack. Moreover, this remaining ripple can be spread in a large HF band through the anti-chaos control of the CVS (see the Sect. 6 of this chapter). Consequently, the both LF and HF $RF_{PEMFC\_current}$ values will be lower than the recommended limits. Considering a current slope higher than 20 A/ms, it is obvious that these parameters of the load model will ensure an unacceptable dynamic for the PEMFC stack. So the dynamic of the HPS power flows must be compensated via a buck CCS converter. Further details about the HPS operation will be shown in the modelling section of the bi-buck HPS topology and the appropriate control section.

3 Nonlinear Control of the High Power FC HPS

This section is organised as follows. The issues of high-power HPS topologies based on PEMFC stack as main energy source are presented in the first subsection. Some simulation results for the FC HPS that supplies an inverter system or an equivalent load are shown in the second subsection. The necessity to have a nonlinear gain in the control loop is analysed in the third subsection. Its design is shown, too. The possibility to design this nonlinear control law by a fuzzy logic controller is shown in the fourth subsection. Last subsection concludes this section.

3.1 The High-Power HPS Topology

The EGS architecture with the mitigation control for the FC current ripple is shown in Fig. 6. The topology of the buck CCS and the structure of its controller are shown in Figs. 7 and 8. The modelling of the FC/battery HPS topology and the design of the nonlinear law control that can replace the linear gain, G_{Ifc}, are presented in [35]. The low-pass filter inductance, L_f, is used to connect in parallel the PEMFC stack and the buck CCS. Because the ripple of the FC current, I_{FC}, without use of the buck CCS, and the anti-ripple generated by the buck CCS current, I_{CCS}, will have almost the same magnitude, L_f value could be equal to the buck CCS inductance, L_{buck} (for example, $L_f = L_{buck} = 100$ μH). The L_f inductance and the internal capacitance of the PEMFC stack forms a low-pass filter that mitigates the HF current ripple. The C_f capacitor is used to obtain the imposed RF voltage on the HPS DC bus (usually, $C_f > 100$ μF).

The value of the L_{buck} inductance must be chosen small to assure a short time response of Buck CCS, but large enough to operate it in the current continuous mode (for example, $50 < L_{buck}$ [μH] < 150 if a 6 kW PEMFC stack is used and the load ripple is 30 $A_{p\text{-}p}$) [26]. The mitigation control loop must have a short time response to better track the shape of the inverter current ripple. So a value close to the minimal value (also named as critical value) must be used. However, note that a too low value could increase the HF ripple over the imposed limits. The HF ripple magnitude dependents on the hysteresis value that is set for the relay block (see Fig. 8). So the hysteresis value was chosen to obtain a HF ripple up to the imposed limits. Considering the simplicity of the circuit design, the hysteretic control was chosen to be used as a current-mode control method. The switching frequency for the buck CCS will be in range of 5–50 kHz if 10 Amps is used for

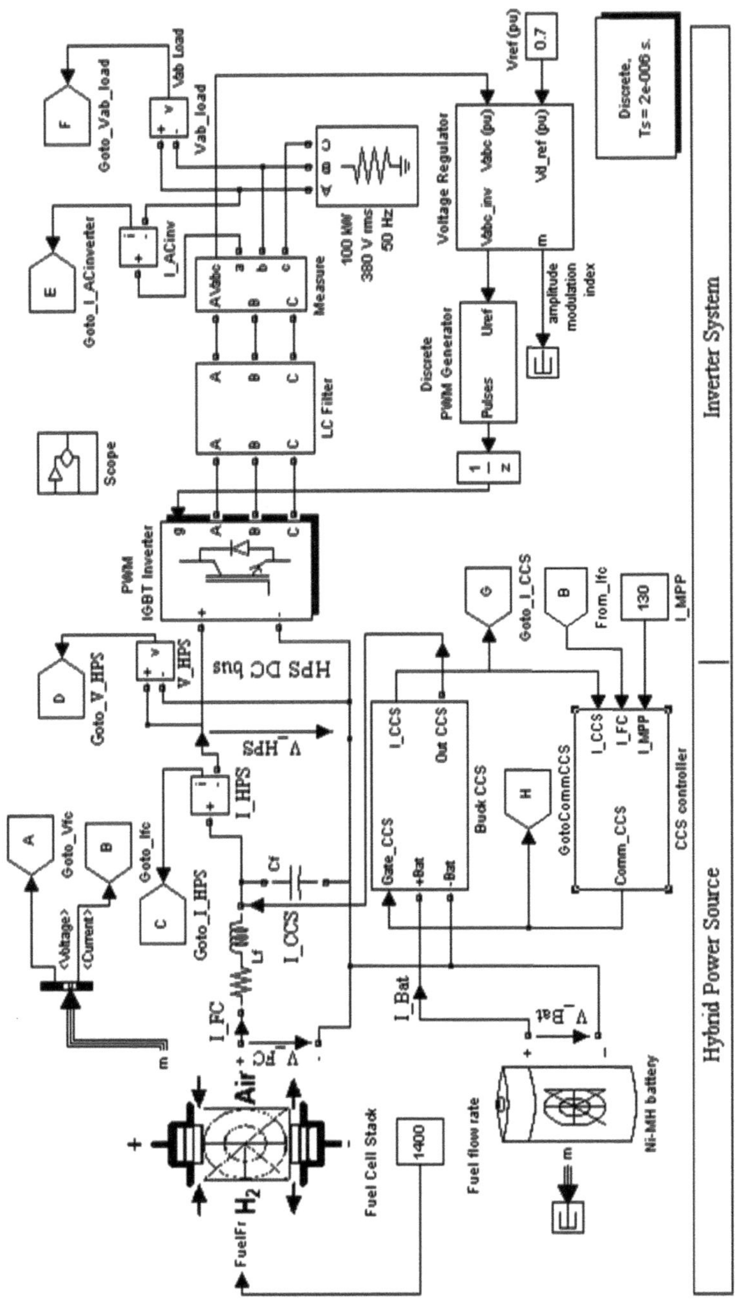

Fig. 6 The EGS architecture that use a FC/battery HPS topology having a control loop to mitigate the FC current ripple; adapted from [33, 35]

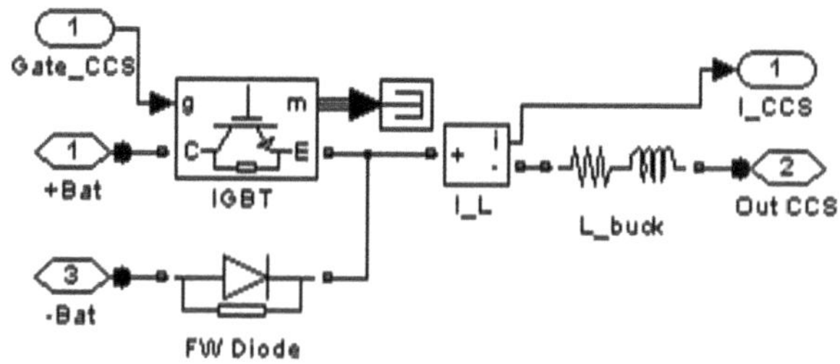

Fig. 7 The structure of the buck CCS

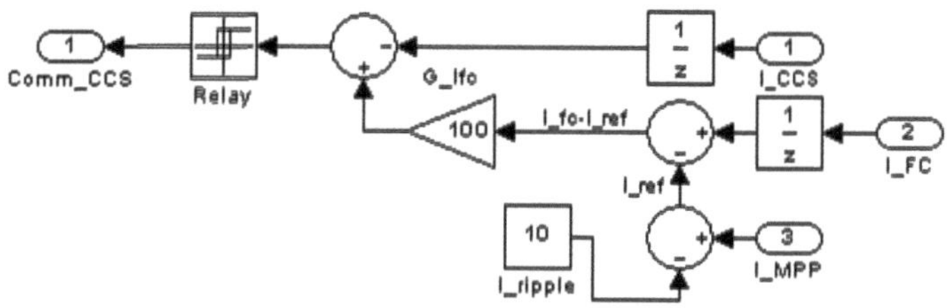

Fig. 8 The structure of the CCS controller [35]

the hysteresis value. The G_{Ifc} gain value set the mitigation performance, defining the tracking accuracy of the ripple shape for the inverter current that is propagated back to the PEMFC stack. Thus, the buck CCS will generate an I_{CCS} current, which is in fact an anti-ripple that will be injected in the HPS output node to mitigate the inverter current ripple. The anti-ripple will be generated based on the gained FC ripple (not based on the gained inverter ripple) to reduce the HF ripple in the FC ripple. Also, it is obvious that both FC and inverter ripples have the same LF shape in the FC EGS architecture without current ripple mitigation control (Fig. 9). This architecture without current ripple mitigation control is used to compare the mitigation performances.

The FC HPS topology that is shown in Fig. 10 has an equivalent load that replaces the inverter system.

As it was mentioned, the use of an equivalent load instead of the inverter system will speed-up simulation, without affecting the analysis of the mitigation performances that is performed in the LF range for different shapes of ripples.

3.2 Simulation Results

Some simulation results considering the FC EGS architecture, with and without control feature to mitigate the inverter ripple, are shown in Figs. 11 and 12. The shape of the buck CCS current will track the LF shape of the input inverter current (Fig. 12b). This is put in evidence by the magnitudes of the LF harmonics, which

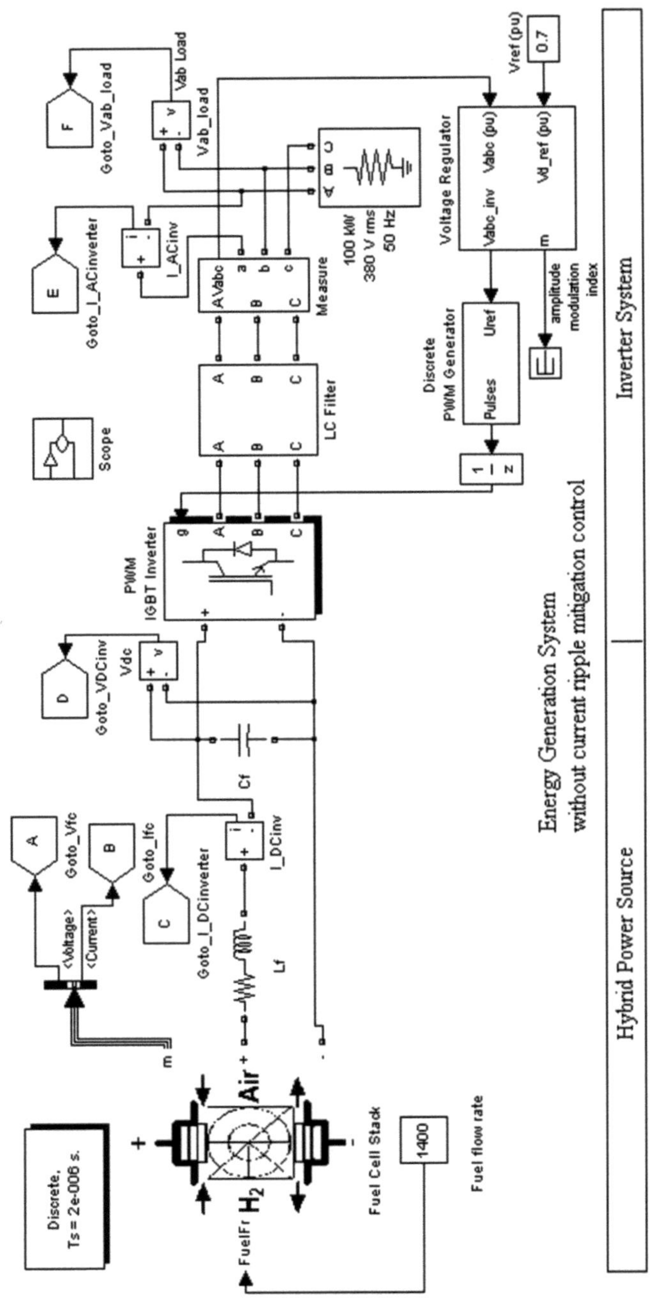

Fig. 9 The FC EGS architecture without current ripple mitigation control

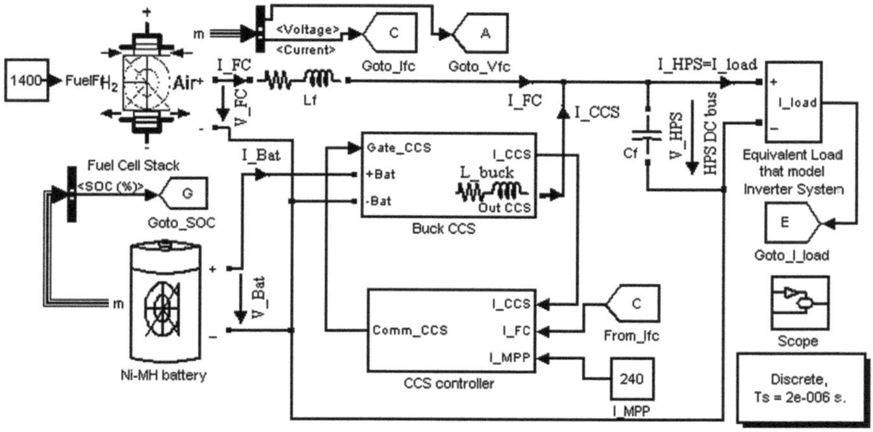

Fig. 10 The FC HPS topology with a current ripple mitigation control and equivalent load

are almost the same for the both currents [35]. The LF harmonics magnitudes of the FC current for two G_{Ifc} gains are shown in Fig. 11. Note that the 100 Hz harmonic magnitude decreases from 0.6661 A for $G_{\text{Ifc}} = 10$ (Fig. 11a) to 0.07755 A for $G_{\text{Ifc}} = 100$ (Fig. 11b), but not in a linear manner. Further simulations performed showed that the mitigation ratio is not linear versus the G_{Ifc} gain, which vary in range 1–100 [35]. The mitigation ratio is computed as ratio of the ripples in the HPS and the FC outputs. The mitigation ratio could be also computed as ratio of RF values, $\text{RF}_{\text{HPS}}/\text{RF}_{\text{FC}}$, considering that average values are almost equal for the FC and the HPS currents. In the same manner the mitigation ratios for different LF harmonics can be defined. For example, the load current for the equivalent load shown in Fig. 1 has the base value, $I_{\text{HPS(base)}}$, 240 A, the ripple peak-to-peak 30 $\text{A}_{\text{p-p}}$ and the 100 Hz harmonic magnitude 7.397 A. Thus, the mitigation ratio of the 100 Hz harmonic is about $7.397/0.6661 \cong 11.1$ and $7.397/0.07755 \cong 95.4$ for case of $G_{\text{Ifc}} = 10$ and $G_{\text{Ifc}} = 100$, respectively.

The effective mitigation ratio of the 100 Hz harmonic is about $0.1629/0.07755 \cong 2.1$ for $G_{\text{Ifc}} = 100$ (see Figs. 11b and 12a). Also, the effective mitigation ratios for the 300 Hz and 600 Hz harmonics are about 2.1 165/65 $\cong$ 5.3 and $2.1{\cdot}45/17 \cong 5.6$, respectively.

3.3 The Design of the Nonlinear Control Law Based on Simulation Results

The characteristic of the FC current ripple versus G_{Ifc} gain is shown in Fig. 13a. It is obvious that this is a nonlinear law. The ripple mitigation ratio, RM, is defined as a ratio of the load current ripple and FC current ripple, $\text{RM} = \Delta I_{\text{load}}/\Delta I_{\text{FC}}$.

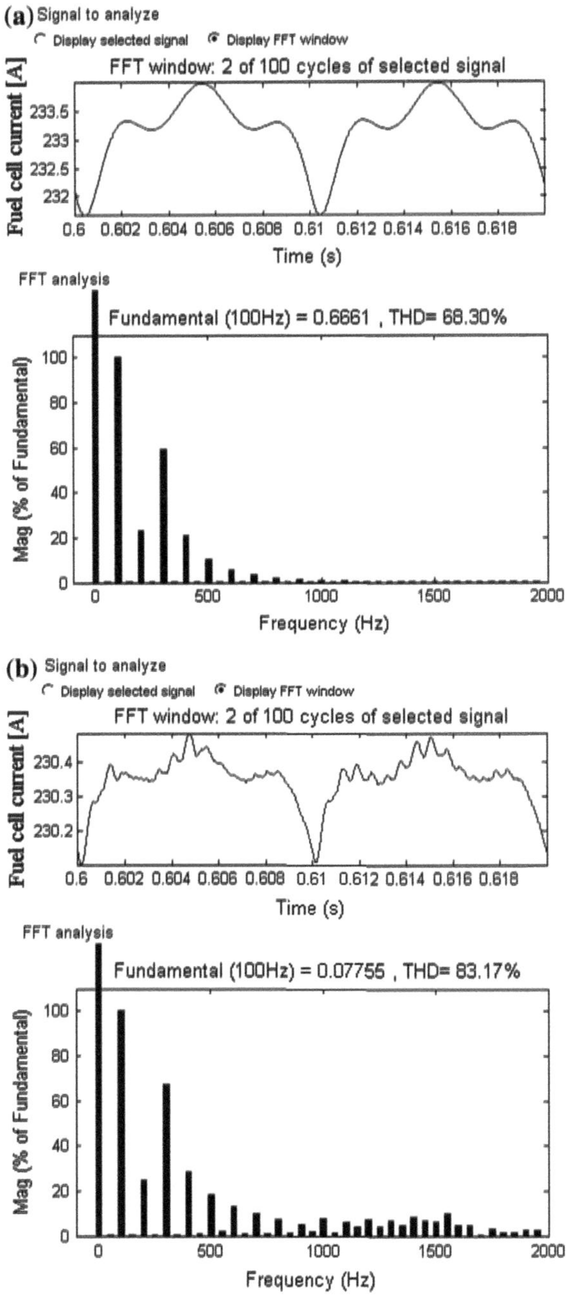

Fig. 11 The simulation results for the FC EGS architecture with current ripple mitigation control, **a** case $G_{Ifc} = 10$, **b** case $G_{Ifc} = 100$

Fig. 12 The simulation results for the FC EGS architecture without current ripple mitigation control, **a** The fuel cell current ripple (*top*) and its power spectrum (*bottom*), **b** The input inverter current (*top*) and its power spectrum (*bottom*)

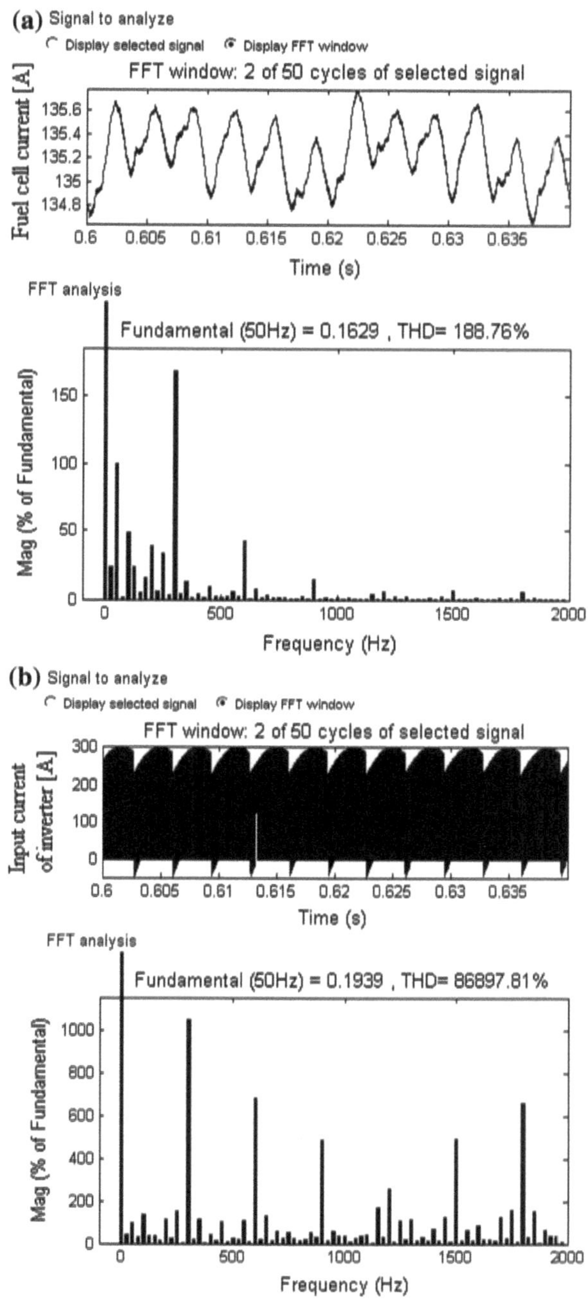

The characteristic of the RM versus the G_{Ifc} gain is shown in Fig. 13b.

Consequently, the characteristic of the Ripple mitigation ratio, RM, versus the FC current ripple can be computed, as it is shown in Fig. 13c (marker ■), considering different G_{Ifc} gain in range. If the control goal is to have ripple mitigation almost constant for different load current ripple, then the G_{Ifc} must have the shape of the nonlinear gain (marked with ● in Fig. 13c) that is symmetrically against vertical axis (the dashed line).

This nonlinear gain could be simply implemented by a piecewise linear (PWL) function, using for example a look-up table (see Fig. 14a):

$$\mathrm{X} = [0,\ 2.5,\ 3.5,\ 4,\ 4.49,\ 4.5] \quad \text{and} \quad \mathrm{Y} = [10,\ 20,\ 40,\ 100,\ 200,\ 200].$$

The PWL nonlinear gain is shown in Fig. 14. The nonlinear CCS controller structure is shown in Fig. 14, too. The control gain has a nonlinear part (the PWL nonlinear gain) and a linear part (G_{Ifc}), which increases the mitigation performance by choosing a gain value in range 1–10. Note that a higher value than 10 will increase the switching frequency over 50 kHz [35].

Simulation results for the FC HPS topology with the CCS nonlinear controller that uses a PWL nonlinear gain are extensively presented in [35]. For example, the ripple mitigation ratio of the 100 Hz harmonic value is about 7.397/0.032 $\cong$ 231, so its effective RM will be about 231/95.4 $\cong$ 5.1. The nonlinear control goal is validated based on the simulations performed for all the LF harmonics, resulting that the effective RM has almost the same value [35].

3.4 The Design of the Nonlinear Law Based on a Fuzzy Logic Controller

The nonlinear control law will be also designed through a FLC to validate the obtained RM characteristic versus the FC current ripple. If the G_{Ifc} constant gain is set to 10, then the X vector will be scaled with 10 and renamed as Xg. Consequently, the nonlinear gain that includes both constant and variable gains can be implemented by the following PWL function:

$$\mathrm{Xg} = [0,\ 0.25,\ 0.35,\ 0.4,\ 0.449,\ 0.45] \quad \text{and} \quad \mathrm{Y} = [10,\ 20,\ 40,\ 100,\ 200,\ 200].$$

The methodology to design the FLC is detailed in [34]. The shapes of the membership functions that result for the FC current ripple (fc), the ripple mitigation ratio (rm), which are the input variables and the output command signal (com) are shown in Fig. 15, plot a, b and c, respectively. Five membership functions are defined for both input variables in correlation with pair of vectors (Xg, Y). These are named as VS = Very Small, S = Small, M = Medium, B = Big and VB = Very Big. It can be observed that their peaks are located at

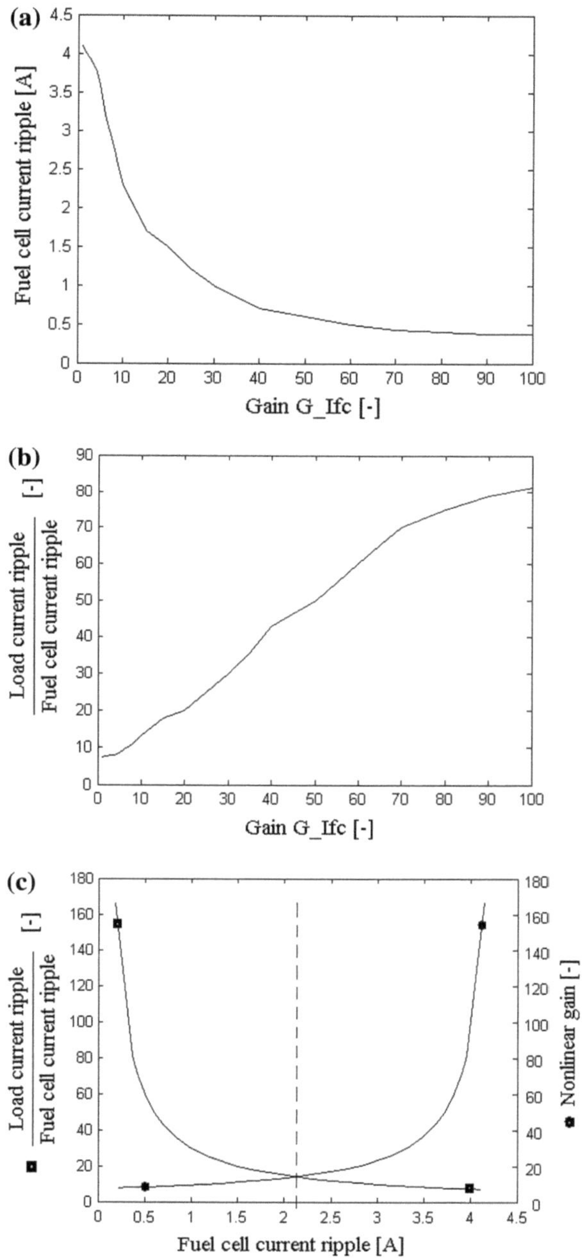

Fig. 13 The performance characteristics of the FC HPS topology with current ripple mitigation control [35], **a** FC current ripple versus G_{Ifc} gain, **b** RM versus G_{Ifc} gain, **c** RM versus FC current ripple

values of the Xg and Y vectors. Also, five membership functions are uniformly defined for the output variable in range 0 to 1. They are named as VS = Very Small, S = Small, M = Medium, B = Big and VB = Very Big.

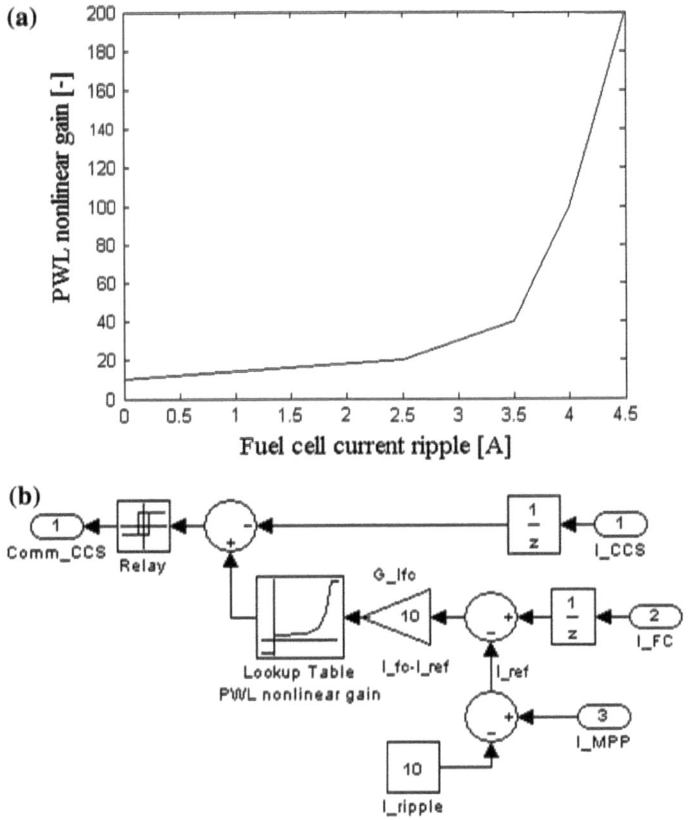

Fig. 14 The nonlinear CCS controller, **a** PWL nonlinear gain, **b** The structure of the nonlinear CCS controller [35]

The rules base is shown in Table 1. The proposed CCS controller that uses this FLC is shown in Fig. 15d. The Mamdani implication and centre of gravity defuzzification method are used. The resulting control surface and the contour projections for different levels of the command signal are shown in Fig. 15, plot e and f, respectively. It can be observed that the projection contour of the 0.7 level is so similar with the shape of the PWL nonlinear gain shown in Fig. 14a, considering the constant gain, $G_{\mathrm{Ifc}} = 10$, i.e. using the pair of vectors (Xg, Y). This result validates the previous design made through the trial and the error method using the simulation results. The 0.7 level will be set for the threshold of the relay used to convert the FLC output into a PWM command. A 0.2 hysteresis is set for the same relay.

As a conclusion, in this section a systematic design of a nonlinear controller is presented. Two ways to design the nonlinear control law are proposed. The first one is based on simulations to draw the characteristic of the ripple mitigation ratio versus the FC current ripple. The nonlinear control law is designed by symmetry.

Fig. 15 Design of the FLC for the CCS controller [35], **a** Fuel cell current ripple memberships, **b** Ripple mitigation ratio memberships, **c** Command signal memberships, **d** CCS controller using a FLC, **e** FLC control surface, **f** Contours of the FLC control surface

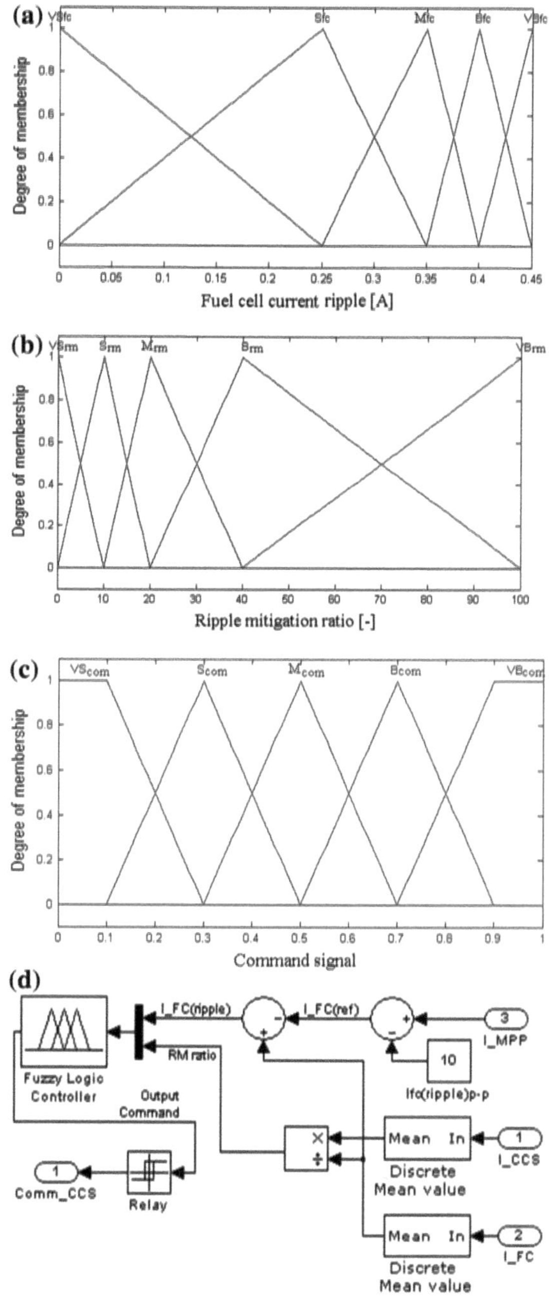

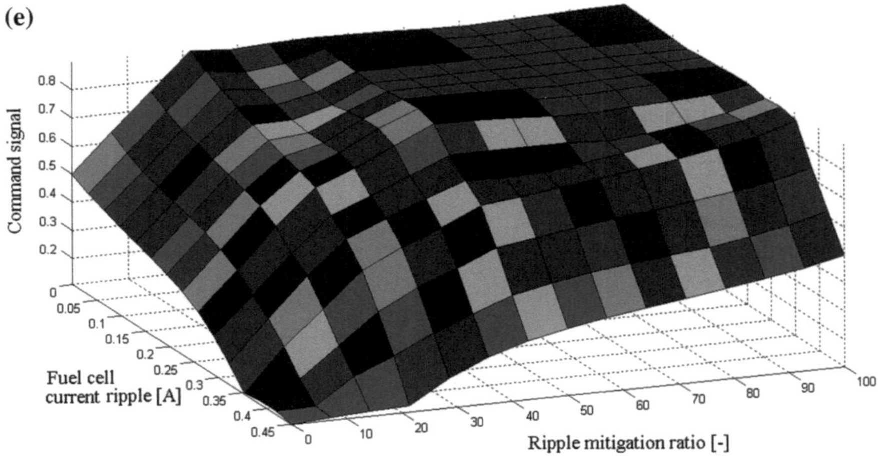

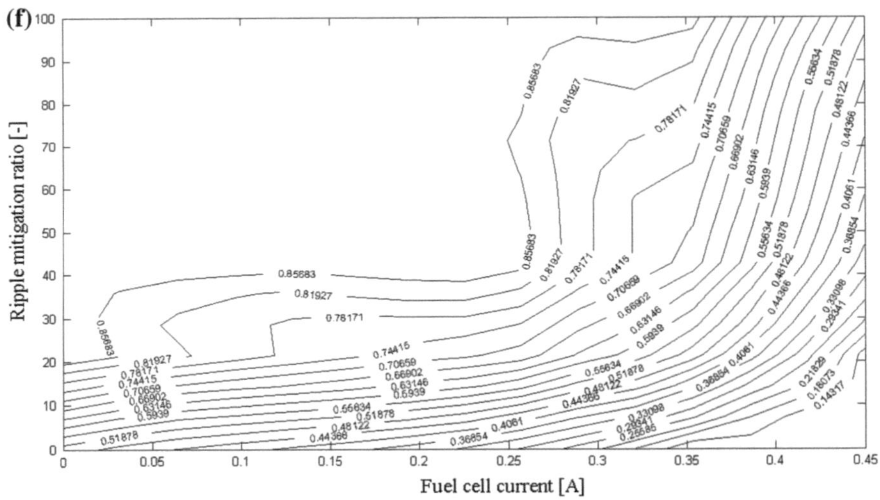

Fig. 15 continued

Table 1 FLC rules base [35]

Command signal		Fuel cell current ripple (A)				
		VSfc	Sfc	Mfc	Bfc	VBfc
Ripple mitigation	VSrm	Mcom	Scom	VScom	VScom	VScom
	Srm	Bcom	Mcom	Scom	VScom	VScom
	Mrm	VBcom	Bcom	Mcom	Scom	VScom
	Brm	VBcom	VBcom	Bcom	Mcom	Scom
	VBrm	VBcom	VBcom	VBcom	Bcom	Mcom

The second one is based on the FLC control surface. The 0.7—cut of this surface is projected in plane of the input variables, defining almost the same nonlinear control law.

4 Nonlinear Control of Medium Power FC HPS

Some interesting control solutions to mitigate the FC current ripple are presented in [7, 36, 37] for medium power FC HPS. The FC power is considered constant, near to MPP, and this operating regime will be set in the proposed HPS topology shown in Fig. 16a. The HPS topology proposed is based on a bi-buck structure. Using an appropriate control the FC current ripple is mitigated and spread in wide frequency band. This section is organised as follows. The Sect. 4.1 briefly presents the HPS proposed. Modelling and designing of the HPS based on bi-buck topology are shown in the second subsection. Designing of the proposed nonlinear law for current and voltage control is presented in Section 3 and 4, respectively. Some selected simulation results are shown, too. The Sect. 4.2 concludes this section.

4.1 The Medium-Power HPS Topology

As it is known, the boost or full-bridge converter topologies are suitable to boost the FC voltage and to mitigate the FC current ripple with appropriate control [7, 36, 37], but here a bi-buck topology will be used [38]. One of the buck converters will operate as CVS, while the other will operate as CCS. The buck CCS will generate an anti-ripple via the tracking control implemented in the CCS controller to mitigate the load current ripple. The buck CCS will operate as an active LF ripple filter, spreading the LF ripple in wide frequency band via the anti-control scheme implemented in the CVS controller. The CVS controller must assure a stabilised output voltage, V_{out}, too. So the voltage error will be used as an input. An anti-control scheme to chaotify the switching command of a buck converter is proposed in [39]. In this way, the remained LF ripple power spectrum is spread in the HF band, increasing the PEMFC life cycle. Consequently, the HPS power interface has two control loops: One for adjusting the output voltage at the imposed value by the reference voltage, V_{ref}, and other for mitigating the LF current ripple through compensation. A nonlinear control of the voltage-mode and the current-mode will be designed for this power interface. The HPS power interface based on a bi-buck topology is shown in Fig. 16b.

The current ripple mitigation technique proposed in [40] is tested here by simulation. The nonlinear controller law is designed by the trial and the error method. Also, this innovative bi-buck converter topology is proposed in [33, 34] as a multi-port power interface (see Fig. 17) and its appropriate voltage- and current-mode control will be further analysed in the Sects. 4.1 and 4.2.

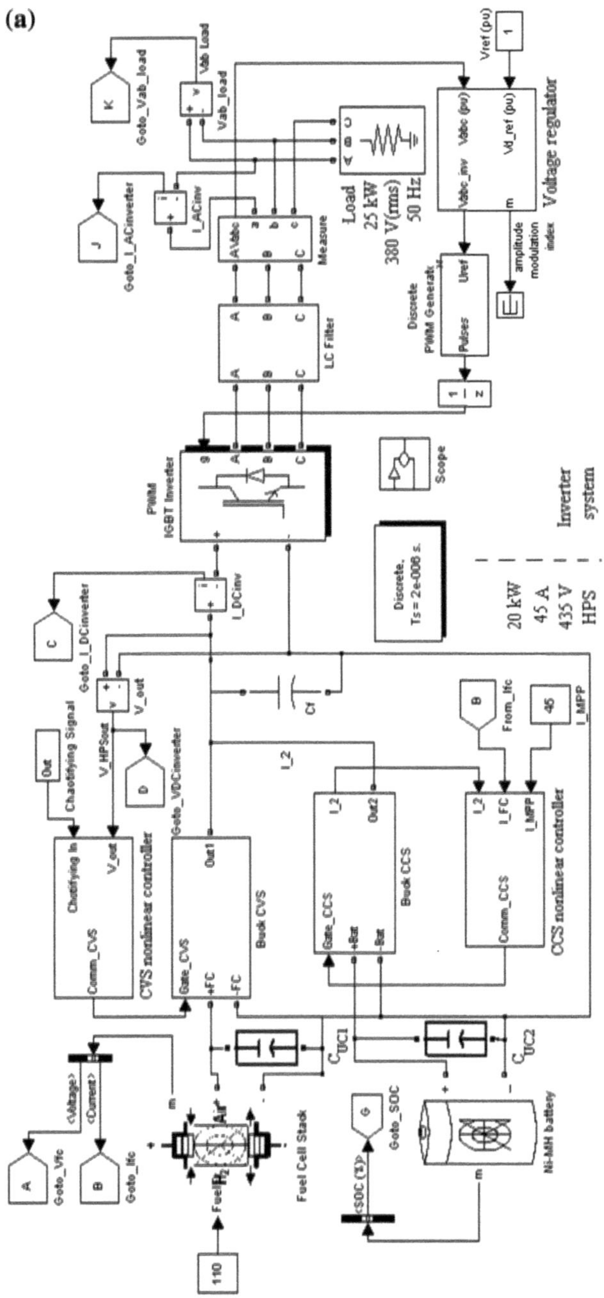

Fig. 16 **a** A three-phase inverter system (with pure sine PWM command) powered by the 20 kW (435 V, 44 A) HPS, **b** The bi-buck HPS topology [33]

(b)

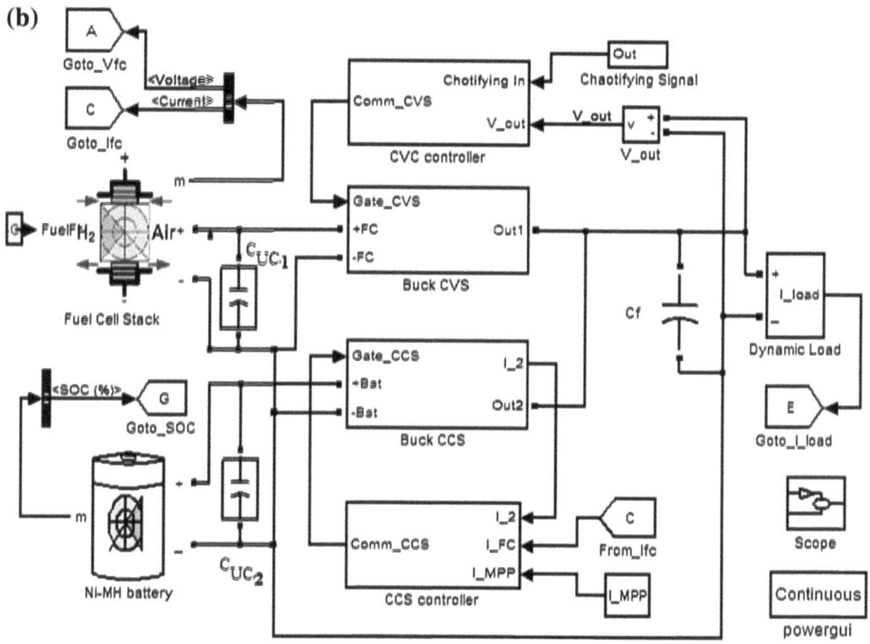

Fig. 16 continued

4.2 Modelling and Designing of the HPS Power Interface

For a 1.26 kW HPS (1.26 kW PEMFC/1 F ultracapacitor/20 Ah NiMH battery) the design parameters are set to $V_{out} \cong V_{ref} = 25$ V, $I_{FC(AV)} \cong I_{MPP} = 45$A (so $V_{FC(MPP)} \cong 27$ V based on FC characteristic), $V_{Bat} = 40$ V and the profile of the load current was shown in Fig. 5. These parameters will be considered below in the HPS design. The load current, i_{out}, is given by relationships written based on average (AV) and alternative (AC) components:

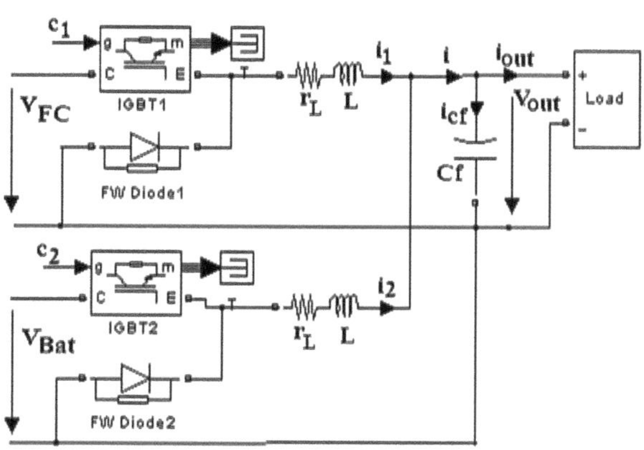

Fig. 17 The HPS power interface of bi-buck type; adapted from [33, 34]

$$i_{out} + i_{Cf} = i_1 + i_2 \Rightarrow \begin{cases} I_{out(AV)} = I_{1(AV)} + I_{2(AV)} \\ i_{out(AC)} + i_{Cf(AC)} = i_{1(AC)} + i_{2(AC)} \end{cases} \quad (1)$$

If the output voltage ripple is considered small, then the AC current through the filtering capacitor, $i_{Cf(AC)}$, will be much lower than the AC load current, $i_{out(AC)}$:

$$i_{1(AC)} + i_{2(AC)} = i_{out(AC)} + i_{Cf(AC)} \cong i_{out(AC)} \quad (2)$$

If the LF current ripple compensation control loop operates correctly, then $i_{1(AC)} \ll i_{2(AC)}$, too, resulting:

$$\begin{aligned} & i_{2(AC)} \cong i_{out(AC)} \\ & I_{2(AV)} = I_{out(AV)} - \frac{\eta_1 V_{FC(MPP)}}{V_{out}} \cdot I_{MPP} \end{aligned} \quad (3)$$

where η_1 is the energy efficiency of the buck CVS. This is calculated as the ratio between the output power (delivered to the load) and the input power (delivered by the PEMFC stack).

Because the unidirectional CCS topology is cheaper than a bidirectional CCS topology, the first topology was considered here. Consequently, the relationships for the AC components must be redefined in terms of a positive ripple. If a base current is defined as a value slightly smaller than the minimum value of the respective current (as it is shown in Fig. 5 for the load current, i_{out}), then the difference from the base current can be defined as a positive ripple. Consequently, the above relationships may be rewritten as:

$$\begin{cases} i_{out} = I_{out(base)} + i_{out(ripple)} \\ i_2 = I_{2(base)} + i_{2(ripple)} \end{cases} \Rightarrow \begin{cases} i_{2(ripple)} \cong i_{out(ripple)} \\ I_{2(base)} = I_{out(base)} - \frac{\eta_1 V_{FC(base)}}{V_{out}} \cdot I_{FC(base)} \end{cases} \quad (4)$$

where $I_{FC(base)} \cong I_{MPP} - I^*_{FC(ripple)p-p}$, and $I^*_{FC(ripple)p-p}$ is the FC current ripple (peak-to-peak) without the buck CCS.

Based on this definition the PEMFC current does not exceed I_{MMP}, even if the compensation loop is not yet in the operating phase or it is accidentally malfunctioned. If $I^*_{FC(ripple)p-p} \cong 10\ \%\ I_{MPP}$ is chosen, then the base operating point is $I_{FC(base)} \cong 40$ A (where $V_{FC(base)} \cong 33$ V). In this case all HPS operating phases are carried out normally (without exceeding the allowable limit parameters) and the PEMFC base operating point is still close enough to the MPP. If the compensation loop starts to operate, then the anti-ripple is generated in the range defined by the gap chosen towards the MPP. The energy delivered by the auxiliary power source (for example, a batteries stack) is minimised if the base current, $I_{2(base)}$, is set to be zero (or slightly greater than zero). In this case:

$$I_{out(base)} \cong \frac{\eta_1 V_{FC(base)}}{V_{out}} \cdot I_{FC(base)} = I_{1(base)} \quad (5)$$

Considering $\eta_1 \cong 0.9$, then $I_{out(base)} \cong 48$ A. Also (see Fig. 5):

$$I_{2(AV)} \cong I_{2(ripple)(AV)} \cong I_{out(ripple)(AV)} \approx 23\,\mathrm{A} \tag{6}$$

Considering $\eta_2 \cong 0.9$ (where η_2 is the power efficiency of the buck CCS), then:

$$I_{Bat(AV)} \cong \frac{V_{out}}{\eta_2 V_{Bat}} \cdot I_{2(AV)} \approx 16\,\mathrm{A} \tag{7}$$

As it was mentioned before, the current levels of the batteries stack could be temporally higher than this minimum level that is computed with Eq. 7, due to the relatively large response time of the FC stack. In this transitory regime of the FC stack, the supplementary power flow is supplied via the buck CCS from the batteries stack. If it is necessary, a mixed batteries and ultracapacitors stack may be used to assure high energy demands in a short time. Of course, it is necessary to have a small response time for the buck CCS, too. Generally, the basic PWM converters (like buck, boost and buck-boost topologies) are second-order systems, in which one state variable is the inductor current and the other state variable is the capacitor voltage (which is also equal to the output voltage). Of course, the bi-buck HPS topology (see Fig. 17) can be modelled by a third-order system that usually uses as state variables the filtering capacitor voltage (which is equal to the output voltage) and both inductor currents. It will be shown below that the HPS topology can be also modelled by second-order systems, if the load power profile is given by the load current modelled through a CCS. Modelling analysis is focused on the current-mode control of the buck CCS that can compensate the inverter current ripple via the anti-ripple generated by the buck CCS. The DC components and the LF components are of interest in designing the current-mode control [14]. Consequently, the LF components are used to characterize the buck CCS dynamic and the proposed model tries to emulate the behaviour of the bi-buck converter in the LF range.

If c1 and c2 are switching command for the IGBT transistors ($c_1 = 1/0 \Rightarrow \mathrm{IGBT}_1$ on/off and $c_2 = 1/0 \Rightarrow \mathrm{IGBT}_2$ on/off), then (neglecting the series resistance of the inductors and filter capacitor) the operating equations are:

$$\begin{aligned} c_1 v_{FC} &= L_1 \frac{di_1}{dt} + v_{out} \\ c_2 v_{Bat} &= L_2 \frac{di_2}{dt} + v_{out} \\ i_1 + i_2 &= i_{out} + i_{Cf} \\ i_{Cf} &= C_f \frac{dv_{out}}{dt} \end{aligned} \tag{8}$$

With a simple manipulation of the equations above, a second-order differential equation is obtained:

$$\frac{L_2}{L_1+L_2}c_1 v_{FC} + \frac{L_1}{L_1+L_2}c_2 v_{Bat} = \frac{L_1 \cdot L_2}{L_1+L_2}\frac{di_{out}}{dt} + v_{out} + C_f \cdot \frac{L_1 \cdot L_2}{L_1+L_2}\frac{d^2 v_{out}}{dt^2} \quad (9)$$

Considering identical inductors ($L_1 = L_2 = L$, having the same series resistance, r_L), the second-order differential Eq. (9) can be rewritten as:

$$\frac{1}{2}(c_1 V_{FC} + c_2 V_{Bat}) - \frac{r_L}{2} i_{out} - \frac{L}{2}\frac{di_{out}}{dt} = V_{out} + \frac{C_f r_L}{2}\frac{dv_{out}}{dt} + \frac{C_f L}{2}\frac{d^2 v_{out}}{dt^2} \quad (10)$$

where the second-order system parameters are the natural frequency, ω_n [rad/s], and the dimensionless damping ratio, ξ:

$$\omega_n = \sqrt{\frac{2}{C_f L}}, \quad \xi = \frac{C_f r_L}{4}\omega_n = \frac{r_L \sqrt{2}}{4}\sqrt{\frac{C_f}{L}} \quad (11)$$

As it was mentioned above, this model can show the output voltage dependence to the load current. If series resistance of the inductor, r_L, and of the filter capacitor, r_C, will be considered in modelling, then the operating equations are:

$$\begin{aligned} c_1 v_{FC} &= r_L i_1 + L\frac{di_1}{dt} + v_{out} \\ c_2 v_{Bat} &= r_L i_2 + L\frac{di_2}{dt} + v_{out} \\ i_1 + i_2 &= i = i_{out} + i_{Cf} \\ v_{out} &= r_C i_{Cf} + \frac{1}{C_f}\int i_{Cf} dt \end{aligned} \quad (12)$$

The model of the HPS system is obtained by summing the first two relations in (11). If the voltage of FC and batteries stacks will be considered almost constant during a LF period ($dv_{FC}/dt \cong 0$ and $dv_{Bat}/dt \cong 0$; see Fig. 18), then, by differentiating the last relationship, a second-order differential equation is obtained, too:

$$\frac{1}{2}(c_1 v_{FC} + c_2 v_{Bat}) - v_{out} = \frac{r_L}{2} i + \frac{L}{2}\frac{di}{dt} \Rightarrow 0 \cong \frac{dv_{out}}{dt} + \frac{r_L}{2}\frac{di}{dt} + \frac{L}{2}\frac{d^2 i}{dt^2} \quad (13)$$

The second-order differential equation for the filtering current, i_{Cf}, is obtained by differentiating the last two relations of (12) and then using them in (13):

$$\left.\begin{aligned} \frac{di}{dt} &= \frac{di_{out}}{dt} + \frac{di_{Cf}}{dt} \\ \frac{i_{Cf}}{C_f} + r_C\frac{di_{Cf}}{dt} &= \frac{dv_{out}}{dt} \end{aligned}\right\} \Rightarrow \begin{aligned} & -\frac{r_L}{2}C_f\frac{di_{out}}{dt} - \frac{L}{2}C_f\frac{d^2 i_{out}}{dt^2} \\ & \cong i_{Cf} + \left(\frac{r_L}{2} + r_C\right)\frac{di_{Cf}}{dt} + \frac{L}{2}C_f\frac{d^2 i_{Cf}}{dt^2} \end{aligned} \quad (14)$$

The switching frequency is chosen to be in a range of approximately 10 kHz value and the designed range for the filtering capacitor is from 10 μF to 100 μF.

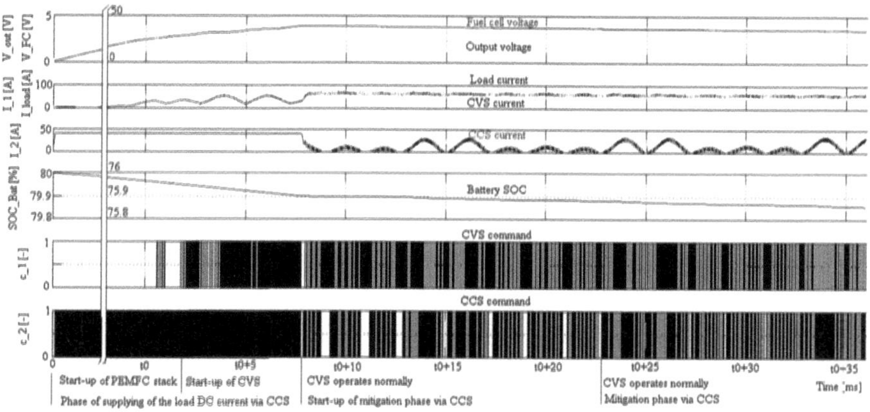

Fig. 18 The HPS behaviour powered by 6 kW PEMFC [34]

Neglecting the series resistor of the capacitor, the amplitude (peak-to-peak) of the filtering current can be estimated by using the last equation of (8):

$$I_{\mathrm{Cf(p-p)}} \cong C_f f_s V_{\mathrm{FC}} \cdot RF_V \tag{15}$$

where $\mathrm{RF_V}$ is the voltage ripple factor defined for the output voltage:

$$\mathrm{RF}_V = \frac{\Delta v_{\mathrm{out}}}{V_{\mathrm{out}}} \tag{16}$$

The design goal for the CVS controller is to obtain a $\mathrm{RF_V} < 4\,\%$ using a filtering capacitor in range from 10 to 100 μF.

The amplitude of the filtering current is about 0.6 and 2 A for the test loads considered (Figs. 1 and 5, respectively). Thus, the assumption regarding the level of the filtering current was correct (see last relation of (1)). This assumption mentions that this current is smaller than the ripple of the output current, so:

$$i = i_{\mathrm{out}} + i_{\mathrm{Cf}} \cong i_{\mathrm{out}} \tag{17}$$

If the LF range is considered to be from 50 to 1000 Hz, then the time of a ripple pulse, $2\Delta t_{\mathrm{p}}$, (named above as a LF period) will be in range from 1 to 20 ms (see Figs. 19 and 20). The voltages of both FC and batteries stacks and also output voltage can be considered constant during this short time:

$$v_{\mathrm{out}} \cong V_{\mathrm{out}},\; v_{\mathrm{FC}} \cong V_{\mathrm{FC}},\; v_{\mathrm{Bat}} \cong V_{\mathrm{Bat}} \tag{18}$$

Also, the voltage over the series resistor of the inductor is much smaller than the output voltage:

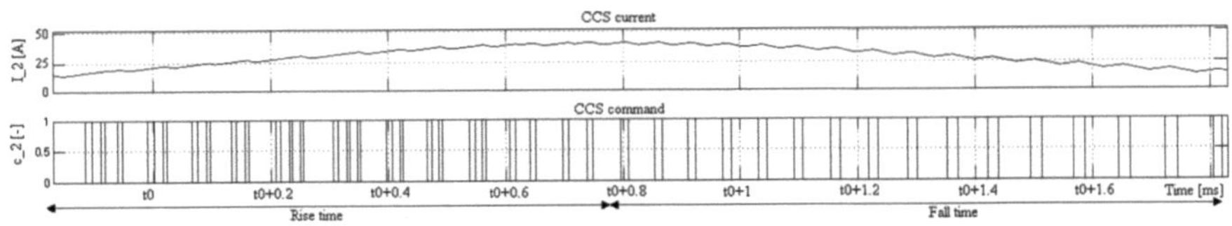

Fig. 19 Zoom of the buck CCS behaviour during the mitigation phase (6 kW PEMFC case) [34]

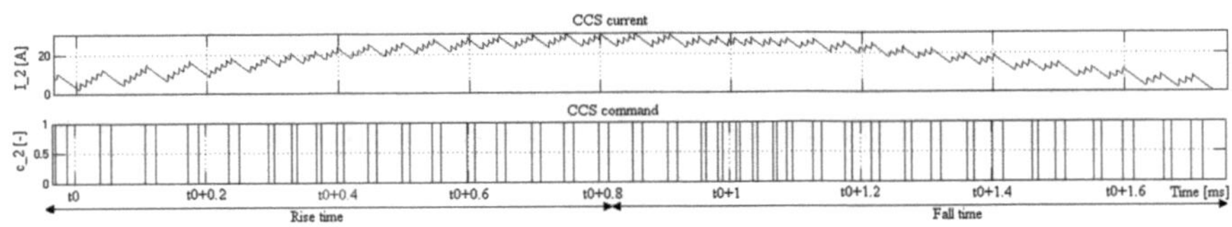

Fig. 20 Zoom of the buck CCS behaviour during the mitigation phase (1.26 kW PEMFC) [34]

$$\frac{r_L}{2} i_{\text{out}} << V_{\text{out}} \tag{19}$$

Taking into account these assumptions, the first two relations of (8) can be rewritten. Summing both relations, the relation that can model the HPS behaviour in all operation phases will be obtained (see Fig. 18):

$$\frac{(c_1 V_{FC} + c_2 V_{\text{Bat}})/2 - V_{\text{out}}}{L/2} = \frac{\mathrm{d}i_{\text{out}}}{\mathrm{dt}} = \frac{\mathrm{d}i_{\text{out(ripple)}}}{\mathrm{dt}} \tag{20}$$

The CVS starts to operate after the start-up phase of the PEMFC stack, when the PEMFC voltage became bigger than the output voltage reference, $V_{\text{ref(out)}}$. In all this time the load is powered via the CCS. If the CVS starts to operate, then the FC stack current appears. The mitigation phase starts when the FC stack current become bigger than $I_{\text{FC(base)}}$ value. From now on, the load is mainly powered via the CVS and the load ripple will be mitigated via the CCS. During this phase both converters operate and four states of the transistor's conduction appear (see Table 2).

Table 2 The conduction states of the HPS operation during the start-up of the mitigation phase [34]

State	C1	C2	The duration of state (s)	The current slope (A/s)	The average number of switching periods into a state
I	1	1	t_{on1on2}	s_1	n_1
II	1	0	t_{on1off2}	s_2	n_2
III	0	1	t_{off1on2}	s_3	n_3
IV	0	0	t_{off1off2}	s_4	n_4

For each conduction state, the current slope for the output current can be estimated using the Eq. (20), as shows:

$$s_1 = \frac{\Delta i_{\mathrm{out(I)}}}{t_{\mathrm{on1on2}}} \cong \frac{(V_{\mathrm{FC}} + V_{\mathrm{Bat}})/2 - V_{\mathrm{out}}}{L/2}; \quad s_2 = \frac{\Delta i_{\mathrm{out(II)}}}{t_{\mathrm{on1off2}}} \cong \frac{V_{\mathrm{FC}}/2 - V_{\mathrm{out}}}{L/2}$$
$$s_3 = \frac{\Delta i_{\mathrm{out(III)}}}{t_{\mathrm{off1on2}}} \cong \frac{V_{\mathrm{Bat}}/2 - V_{\mathrm{out}}}{L/2}; \quad s_4 = \frac{\Delta i_{\mathrm{out(IV)}}}{t_{\mathrm{off1off2}}} \cong \frac{-V_{\mathrm{out}}}{L/2} \tag{21}$$

The relationship that gives the LF ripple amplitude of the load pulse is:

$$\Delta I_{\mathrm{out(ripple)}} = n_1 s_1 t_{\mathrm{on1on2}} + n_2 s_2 t_{\mathrm{on1off2}} + n_3 s_3 t_{\mathrm{off1on2}} + n_4 s_4 t_{\mathrm{off1off2}} \tag{22}$$

If both converters normally operate, then the same Eq. (20) is operational, considering:

$$v_{\mathrm{out}} \cong V_{\mathrm{out}} \cong V_{\mathrm{ref(out)}},\ v_{\mathrm{FC}} \cong V_{\mathrm{FC(base)}},\ v_{\mathrm{Bat}} \cong V_{\mathrm{Bat(base)}} \tag{23}$$

During the mitigation phase, the buck CCS generates an anti-ripple current that makes an active compensation of the output current ripple. In this HPS operation regime, the buck CCS behaviour can be shown using the second relationship of (8):

$$\frac{c_2 V_{\mathrm{Bat}} - V_{\mathrm{out}}}{L} \cong \frac{\mathrm{d}i_2}{\mathrm{d}t} = \frac{\mathrm{d}i_{2\mathrm{(ripple)}}}{\mathrm{d}t} \cong \frac{\mathrm{d}i_{\mathrm{out(ripple)}}}{\mathrm{d}t} \tag{24}$$

A zoom of the buck CCS current shape is shown in Figs. 19 and 20 for the HPS powered by a 6 and 1.26 kW PEMFC stack, respectively. A pulse of the LF buck CCS current shape has a rise time (Δt_{r}) and a fall time (Δt_{f}). During the rise and fall time phase the buck CCS current rises to $\Delta I_{\mathrm{out(rise)}}$ value and falls from $\Delta I_{\mathrm{out(fall)}}$ value. The positive and negative slopes are (see also the Table 3):

$$s_5 = \frac{\Delta i_{2(r)}}{t_{\mathrm{on2}(r)}} = \frac{\Delta i_{2(f)}}{t_{\mathrm{on2}(f)}} \cong \frac{V_{\mathrm{Bat}} - V_{\mathrm{out}}}{L}; \quad s_6 = \frac{\Delta i_{2(r)}}{t_{\mathrm{off2}(r)}} = \frac{\Delta i_{2(f)}}{t_{\mathrm{off2}(f)}} \cong \frac{-V_{\mathrm{out}}}{L} \tag{25}$$

In this phase, when both converters operate normally, the behaviour of the buck CVS can be shown using relationships (8). The converters operate independently to satisfy the control goals under the switching command generated by each controller (see Fig. 21). The simulation results shown in Fig. 18 are obtained using a hysteretic current-mode control. In the next section the simulation results will be

Table 3 The CCS behaviour during the mitigation phase [34]

CCS current [phase]	C2	The average time of each phase (s)	The current slope (A/s)	The average number of switching periods into a phase
Rise	1	$t_{\mathrm{on2(r)}} = \mathrm{T_2 D_{2r}}$	s_5	n_{r}
Rise	0	$t_{\mathrm{off2(r)}} = \mathrm{T_2(1-D_{2r})}$	s_6	n_{r}
Fall	1	$t_{\mathrm{on2(f)}} = \mathrm{T_2 D_{2f}}$	s_5	n_{f}
Fall	0	$t_{\mathrm{off2(f)}} = \mathrm{T_2(1-D_{2f})}$	s_6	n_{f}

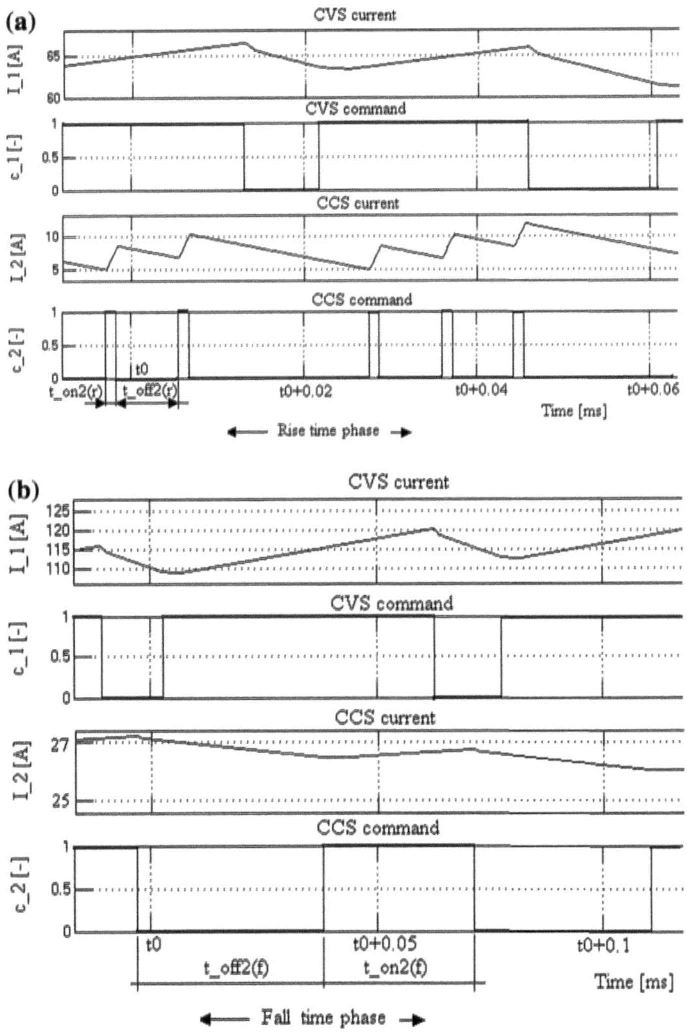

Fig. 21 Zoom of the bi-buck converter behaviour during the mitigation phase [34], **a** 1.26 kW PEMFC case, **b** 6 kW PEMFC case

also shown for a constant frequency method, the peak-current-mode control (PCC) method.

Some design relationships will be written below. For example, the ripple amplitude for the buck CCS current during the rise phase is about:

$$\Delta I_{\mathrm{out(rise)}} = n_r\left(s_5 t_{\mathrm{on2}(r)} + s_6 t_{\mathrm{off2}(r)}\right) = n_r T_2\left[s_5 D_{2r} + s_6(1 - D_{2r})\right] \qquad (26)$$

Consequently, the average number of the switching periods into the rise time phase is:

$$n_r \cong \frac{\Delta t_r}{T_2} \cong \frac{\Delta I_{out(rise)}}{T_2[s_5 D_{2r} + s_6(1 - D_{2r})]} \tag{27}$$

Thus, the average value for the duty cycle of the CCS command is:

$$D_{2r} = \frac{\Delta I_{out(rise)} - n_r T_2 s_6}{n_r T_2 (s_5 - s_6)} \tag{28}$$

In the same manner, during the fall time phase the average value for the duty cycle of the CCS command will be:

$$D_{2f} = \frac{-\Delta I_{out(fall)} - n_f T_2 s_6}{n_f T_2 (s_5 - s_6)} \tag{29}$$

The ratio of the duty cycles, r_D, can be compute as:

$$r_D \triangleq \frac{D_{2r}}{D_{2f}} \tag{30}$$

Obviously, if the rise time is almost equal to the fall time, then:

$$\Delta t_r \cong \Delta t_f \Rightarrow \begin{cases} \Delta I_{out(rise)} \cong \Delta I_{out(rise)} = \Delta I_{out(ripple)} \\ n_r \cong n_f = n \end{cases} \tag{31}$$

Considering the above data and a switching frequency, f_s, of 10 kHz for both PWM controllers (the buck CVS and the buck CCS controllers), the duty cycle of the switching commands ($c1$ and $c2$, respectively) will be:

$$D_1 = \frac{t_{on1}}{T} = \frac{V_{out}}{V_{FC}} \cong 0.75, \quad D_2 = \frac{t_{on2}}{T} = \frac{V_{out}}{V_{Bat}} \cong 0.625 \tag{32}$$

The LF to HF current ripple ratio, K_{ripple}, is defined as a ratio between the LF current ripple (peak-to-peak), $I_{(LF\ ripple)p-p}$ and the HF current ripple (peak-to-peak), $I_{(HF\ ripple)p-p}$:

$$K_{ripple} \triangleq \frac{I_{(LF\,ripple)p-p}}{I_{(HF\,ripple)p-p}} \cong \frac{f_{(HF\,ripple)}}{f_{(LF\,ripple)}} \tag{33}$$

where $f_{(HF\,ripple)} = f_s = 10$ kHz. Considering the LF power spectrum up to 500 Hz ($f_{(LF\,ripple)} = 500$ Hz), the HF current ripple is more than twenty times lower than the LF current ripple. Taking into account that the PEMFC stack is tolerant to the HF current ripple, it is obvious why the HF current ripple is not considered in the model of load, even if this HF ripple appears (see Fig. 12).

When IGBT1 is on (see Fig. 17) the second Eq. (8) can be written as:

$$V_{\text{Bat}} = L_2 \frac{\Delta i_2}{t_{\text{on}2}} + V_{\text{out}} \Rightarrow L_2 \cong \frac{V_{\text{out}}(1 - D_2)T}{I_{(\text{HF ripple})\text{p-p}}} \tag{34}$$

The value of the buck CCS inductance defines the response time in the tracking loop of the load current shape. Therefore, the mitigation performance for the LF current ripple is shown in Fig. 5 will be obtained for:

$$L_2 \cong \frac{V_{\text{out}}(1 - D_2)K_{\text{ripple}}}{f_s \cdot I_{(\text{LF ripple})\text{p-p}}} \cong 0.625 \cdot 10^{-3}\,\text{H} \tag{35}$$

As it is known, if the PWM voltage-mode control is used, then the output voltage ripple factor is given by:

$$\frac{\Delta v_{\text{out}}}{V_{\text{out}}} \cong \frac{\pi^2}{2} \cdot (1 - \text{D}_1) \cdot \left(\frac{f^*_{\text{LC}}}{f_s}\right)^2, \quad f^*_{\text{LC}} = \frac{1}{2\pi\sqrt{LC_f}} \tag{36}$$

If CVS and CCS use the same type of inductor, $L_1 = L_2 = L = 100\ \mu\text{H}$, having a series resistance about $r_L = 100\ \text{m}\Omega$, and the RF for the output voltage is set to 4 %, then the filter capacitance, C_f, must be greater that 10 μF. A value of 47 μF is chosen to assure a ripple factor less than 4 %. If a variable-frequency control method will be used, then the switching frequency will be in a range of around 10 kHz. The current ripple amplitude of the filtering capacitors, $i_{\text{Cf(AC)}}$, can be estimated using the Eq. (15):

$$I_{\text{Cf(p-p)}} \cong C_f f_s \frac{\Delta v_{\text{out}}}{D_1} = C_f f_s V_{FC} \frac{\Delta v_{\text{out}}}{V_{\text{out}}} \cong 0.6\,\text{A} \tag{37}$$

Consequently, the assumptions that were used to write the above relations have been checked.

If the load current is constant, then an equivalent load resistance can be defined as $R_{\text{out}} = V_{\text{out}}/I_{\text{out}}$. Thus, the associate frequencies are:

$$f_{\text{RC}} = \frac{1}{2\pi} \cdot \frac{1}{C_f R_{\text{out}}}, \quad f_{\text{RL}} = \frac{1}{2\pi} \cdot \frac{R_{\text{out}}}{L/2}, \quad f_{\text{LC}} = \frac{\omega_n}{2\pi} \tag{38}$$

If the load current is set in the range of the continuous current mode (CCM) operation for the buck CVS, which means $I_{\text{out}} < I_{\text{out(base)}} \cong 48$ A and $R_{\text{out}} > V_{\text{out}}/I_{\text{out}} \cong 0.5\ \Omega$, then the associated frequencies will have the same order of magnitude. For example, if $I_{\text{out}} \cong 25$ A, then $R_{\text{out}} = 1\ \Omega$, and the associated frequencies are $f_{\text{RC}} = 3.3863$ kHz, $f_{\text{RL}} = 1.5915$ kHz and $f_{\text{LC}} = 2.3215$ kHz. The same order of magnitude for the associated frequencies of the second-order system means that the system response time is about of $2\pi/\omega_n$, i.e. 500 μs. The switching period must be lower than the system response time, so the decision to use a 10 kHz switching frequency (or up to this value) represents a good designing option towards applying of the stiff system theory.

5 Voltage and Current-Mode Control Techniques Applied to FC HPS

5.1 Current-Mode Control

In this section, the simulation results for three types of current-mode control (hysteretic control, PCC and nonlinear control) for the buck CCS will be presented. Because the first two control methods are classical and easy to be designed, then only the controller structure and the used parameters are shown in the subsections below. Besides those, the systematic design of the nonlinear control law is presented.

5.1.1 Hysteretic Current-Mode Control

The structure of the CCS hysteretic controller is shown in Fig. 22. The fuel cell base current, $I_{\mathrm{FC(base)}}$, is chosen to be close to the MPP value of the FC current, I_{MPP}. The positive ripple of the FC current, $I_{\mathrm{fc(ripple)}}$, is quite a lot amplified in the control loop in order to be easily tracked by the CCS current, I_2.

The Ripple Factor for a current type signal, RF_I, is defined as a ratio between the peak-to-peak ripple and its base value:

$$\mathrm{RF}_I = \frac{\Delta I_{\mathrm{(ripple)}}}{I_{\mathrm{(base)}}} \tag{39}$$

Consequently:

$$\mathrm{RF}_{\mathrm{out}} = \frac{\Delta I_{\mathrm{out(ripple)}}}{I_{\mathrm{out(base)}}} \tag{40}$$

$$\mathrm{RF}_{\mathrm{FC}} = \frac{\Delta I_{\mathrm{FC(ripple)}}}{I_{\mathrm{FC(base)}}} \tag{41}$$

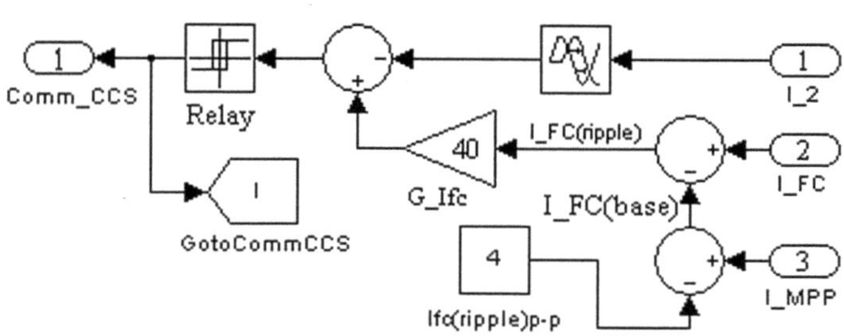

Fig. 22 The structure of the CCS hysteretic controller [34]

So, considering the relationships (41) and (42), the mitigation loop gain, G_{Ifc}, must be established in accordance:

$$G_{Ifc} = \frac{\Delta I_{CCS(ripple)}}{\Delta I_{FC(ripple)}} \cong \frac{\Delta I_{out(ripple)}}{\Delta I_{FC(ripple)}} \cong \frac{RF_{out}}{RF_{FC}} \cdot \frac{\eta_1 V_{FC(base)}}{V_{out}} \Rightarrow G_{Ifc} \cong \frac{RF_{out}}{RF_{FC}} \cdot G_{1(base)} \tag{42}$$

Thus, considering the imposed value of RF_{FC}, the mitigation loop gain, G_{Ifc}, can be designed based on the known value of the RF_{out}. For example, if $I_{fc(ripple)p-p} \cong 10\,\%\, I_{MPP}$ is chose, then the base operating point ($I_{FC(base)}$, $V_{FC(base)}$) is about (40 A, 33 V) and (108 A, 55 V) for the 1,26 kW and 6 kW HPS, respectively. Using the simulation results performed and shown above for $G_{Ifc} = 100$, the RF and G^*_{Ifc} values can be computed for the 1,26 kW PEMFC stack, as below:

$$\left.\begin{aligned} RF_{out} &\cong \frac{95-50}{50} = 90\% \\ RF_{FC} &\cong \frac{43.38-42.9}{42.9} \cong 1.1\% \end{aligned}\right\} \Rightarrow G^*_{Ifc} \cong \frac{90}{1.1} \cdot \frac{0.9 \cdot 33}{25} \cong 95.5 \tag{43}$$

The RF and G^*_{Ifc} values for the 6 kW PEMFC stack and $G_{Ifc} = 100$ are:

$$\left.\begin{aligned} RF_{out} &\cong \frac{240-120}{120} \cong 100\,\% \\ RF_{FC} &\cong \frac{95.5-91.5}{91.5} \cong 4.4\,\% \end{aligned}\right\} \Rightarrow G^*_{Ifc} \cong \frac{92}{4.4} \cdot \frac{0.9 \cdot 55}{40} \cong 28.1 \tag{44}$$

Also, the simulation results for the 1.26 kW PEMFC stack and $G_{Ifc} = 80$ are shown in Fig. 23. In this case, the G^*_{Ifc} gain is computed:

$$\left.\begin{aligned} RF_{out} &\cong \frac{95-50}{50} = 90\,\% \\ RF_{FC} &\cong \frac{43.85-43.1}{43.1} \cong 1.74\,\% \end{aligned}\right\} \Rightarrow G^*_{Ifc} \cong \frac{90}{1.74} \cdot \frac{0.9 \cdot 33}{25} \cong 61 \tag{45}$$

From these results it can be concluded that between the set gain (G_{Ifc}) and the computed gain (G^*_{Ifc}) there is a nonlinear relation of dependence. This nonlinear law will be analysed in the subsection dedicated to the nonlinear control.

5.1.2 Peak-Current-Mode Control

The structure of the CCS PCC controller is shown in Fig. 24. The buck converter will operate at the switching frequency of 10 kHz set by the pulse generator.

The simulation results are shown in Fig. 25 for the 1.26 kW HPS using $G_{Ifc} = 80$. It can be observed that the mitigation performances are almost the same for both hysteretic and PCC controllers (see Figs. 23 and 25).

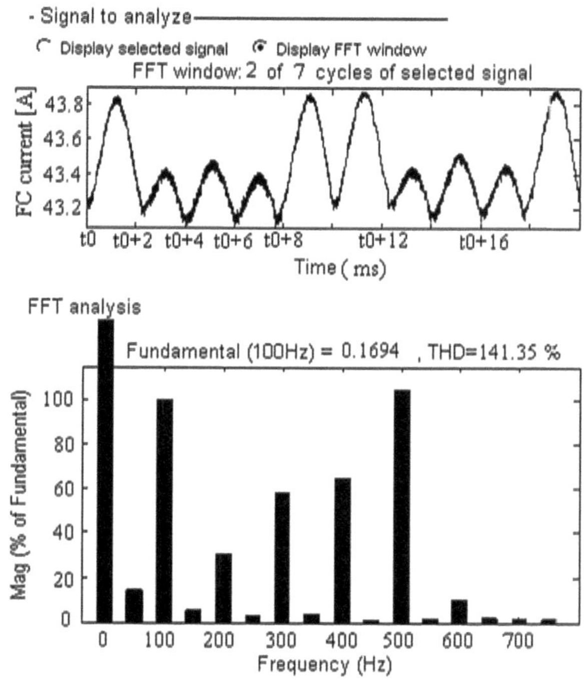

Fig. 23 The FC current (*top*) and its power spectrum (*bottom*): 1.26 kW HPS case, having the CCS Hysteretic controller with $G_{Ifc} = 80$ [34]

The output voltage is shown in Fig. 26. The power spectrum is spread in the LF band based on the anti-control technique implemented in the CVS controller, which will be explained later in this chapter. Note that the switching frequency harmonics can still be seen, thus this kind of control could be further improved.

Taking into account that the spectrum of the FC current is almost the same for the both CCS controllers (see Figs. 23 and 25), it is obvious that the buck CCS current spectrum will be almost the same, too (see Fig. 27). Because the FC ripple

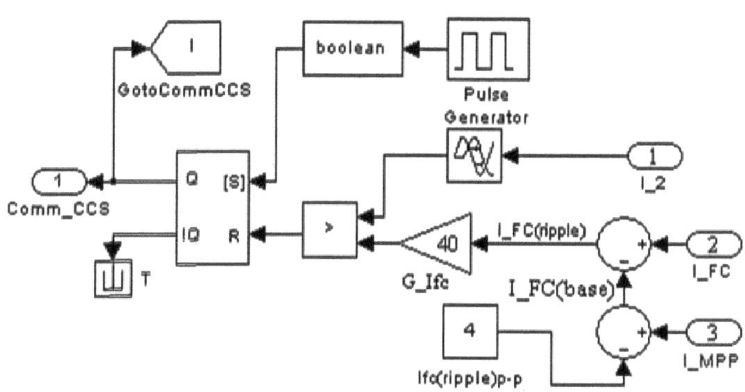

Fig. 24 The structure of the CCS PCC controller [34]

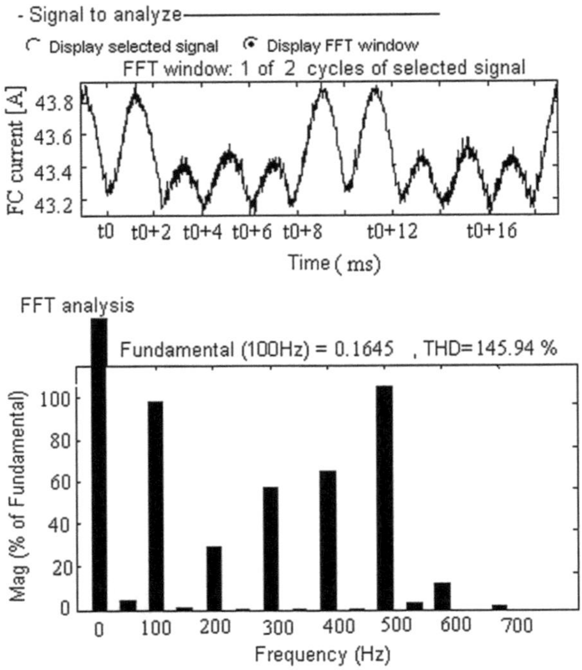

Fig. 25 The FC current (*top*) and its power spectrum (*bottom*): 1.26 kW HPS case, having the CCS PCC controller with $G_{\mathrm{Ifc}} = 80$ [34]

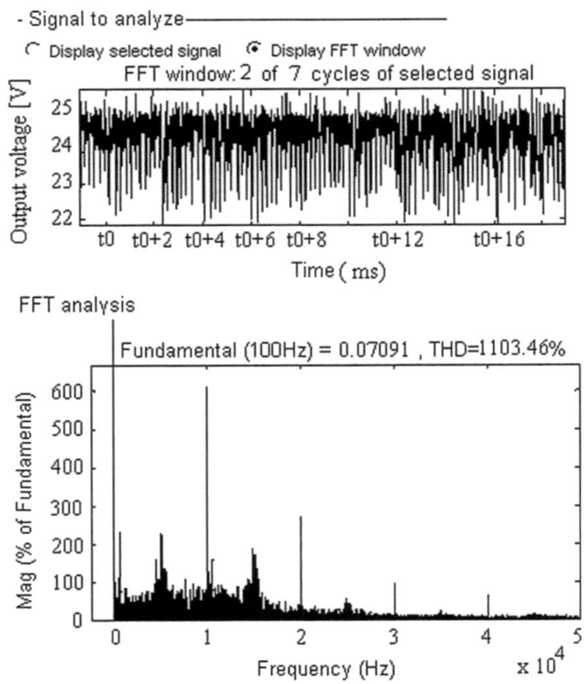

Fig. 26 The output voltage (*top*) and its power spectrum (*bottom*): 1.26 kW HPS case, having the CCS PCC controller with $G_{\mathrm{Ifc}} = 80$ [34]

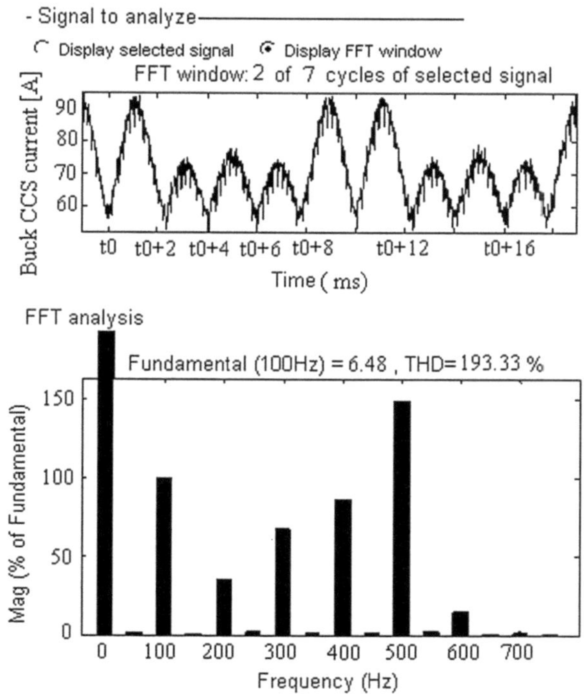

Fig. 27 The buck CCS current (*top*) and its power spectrum (*bottom*): 1.26 kW HPS case, having the CCS PCC controller with $G_{\mathrm{Ifc}} = 80$ [34]

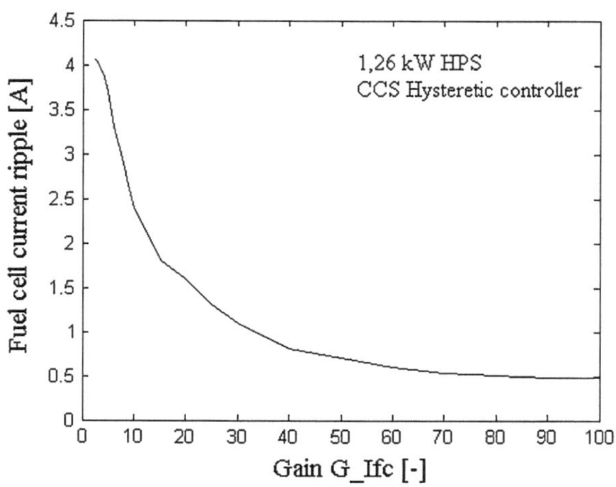

Fig. 28 The FC current ripple versus G_{Ifc} gain set [34]

is much smaller than the load ripple, it is observed that the magnitudes of the fundamental frequency for the buck CCS current and the load current are almost equal (see Figs. 5 and 27). Obviously, the magnitudes difference will be propagated back to the PEMFC stack (see Fig. 25).

The same results are obtained for the 6 kW HPS using both CCS controllers.

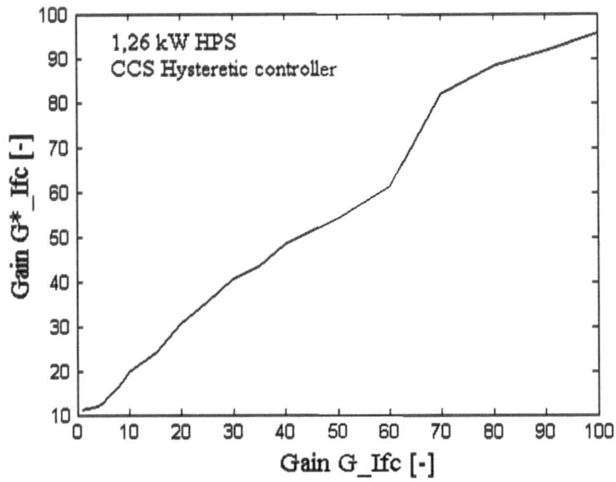

Fig. 29 The computed gain G^*_{Ifc} versus G_{Ifc} gain set [34]

5.1.3 Nonlinear Current-Mode Control

In the previous two sections it was shown that the mitigation performance of the FC current ripple depends on G_{Ifc} gain. Using simulation for a 1.26 kW HPS that uses a CCS hysteretic controller, the characteristics of the FC current ripple versus the set G_{Ifc} gain and the FC harmonic magnitude versus the set G_{Ifc} gain can be drawn (see Fig. 28). It can be observed that almost the same shape as in Fig. 13a is obtained. Thus, the mitigation ripple based on CCS gives almost the same FC ripple whatever level power is. The characteristic of the computed gain (G^*_{Ifc}) versus the set gain (G_{Ifc}) is shown in Fig. 29. These characteristics were drawn in the same manner used for the 6 kW HPS case study. These characteristics also show that the G_{Ifc} gain must be nonlinear in order to obtain the same FC ripple factor for different load ripple. So the CCS controller must have both ripples as input variable. The design of the CSS controller will be shown in the next subsection based on the FLC controller.

5.1.4 Designing of the Nonlinear Control Law

An accurate model of the HPS system is too complex due to the nonlinearities that are included in all models of the HPS subsystems. Consequently, the nonlinear control law will be designed based on FLC.

First of all, the input variables are defined as it was mentioned above. Obviously, the first input is the FC current ripple. The second variable was chosen to be the CCS current ripple. This signal tries to track the load ripple, so it is a measure of it, too. The CCS current is generated via the buck CCS as an anti-ripple of the inverter ripple, which is the HPS output ripple ($I_{\mathrm{inverter(ripple)}} \cong I_{\mathrm{load(ripple)}}$). Thus, this is also a measure of it ($I_{2\mathrm{(ripple)}} \cong I_{\mathrm{load(ripple)}}$). Compared to the CCS current

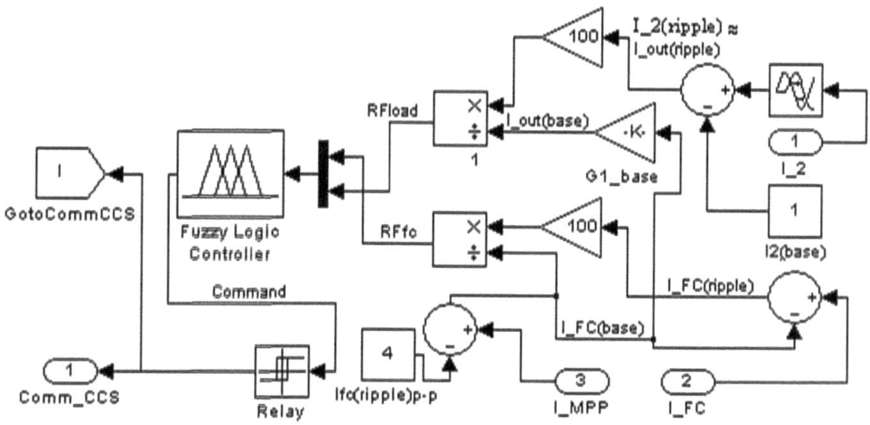

Fig. 30 The structure of the CCS FLC controller [34]

ripple, the inverter ripple has HF components with high magnitudes (see Fig. 12) that could perturb the tracking loop. If a low-pass filter will be used to obtain the LF profile of the inverter ripple, then a variable phase shift will appear between the input variables. Consequently, the CCS current ripple is better to be considered as a second variable, to overcome these problems (see Fig. 30).

Also, the input variables were normalised, considering their ripple factors (noted as RF_{fc} and RF_{load}). The membership functions for both input variables, RF_{fc} and RF_{load}, are shown in Fig. 31. Four membership functions are defined for both input variables in correlation with the design goal. They are named as VS = Very Small, S = Small, B = Big and VB = Very Big (see Fig. 31). Also, five membership functions are uniformly defined for the output variable, named the CCS command signal (com). They are named as VS = Very Small, S = Small, N = Nominal, B = Big and VB = Very Big.

The rules base is shown in Table 4.

The Mamdani implication and the centre of gravity defuzzification method are used. The desired control surface is obtained through the position of the membership functions for the input variables (see Fig. 32). The contours projected for different levels of the CCS command signal are shown in Fig. 33.

The base reference for different currents is defined to have a positive ripple. The base reference for the CCS current, $I_{2(base)}$, was set close to zero (1 A) to reduce the size of the battery that supplies the buck CCS converter. The battery will be designed to meet the load ripple and some transient peaks of power. The base reference for the FC current, $I_{FC(base)}$, was chosen close to the MPP ($I_{FC(base)} = I_{FC(base)} - I_{fc(ripple)p-p}$) to increase the energy efficiency of the FC stack. The FC ripple, $I_{fc(ripple)p-p}$, was set to 4 A, up to the value that it is obtained without the use of the buck CCS converter (setting $G_{Ifc} = 0$ in the tracking loop; see Fig. 28). The base reference for the load current, $I_{out(base)}$, is computed using the $G_{1(base)}$ gain:

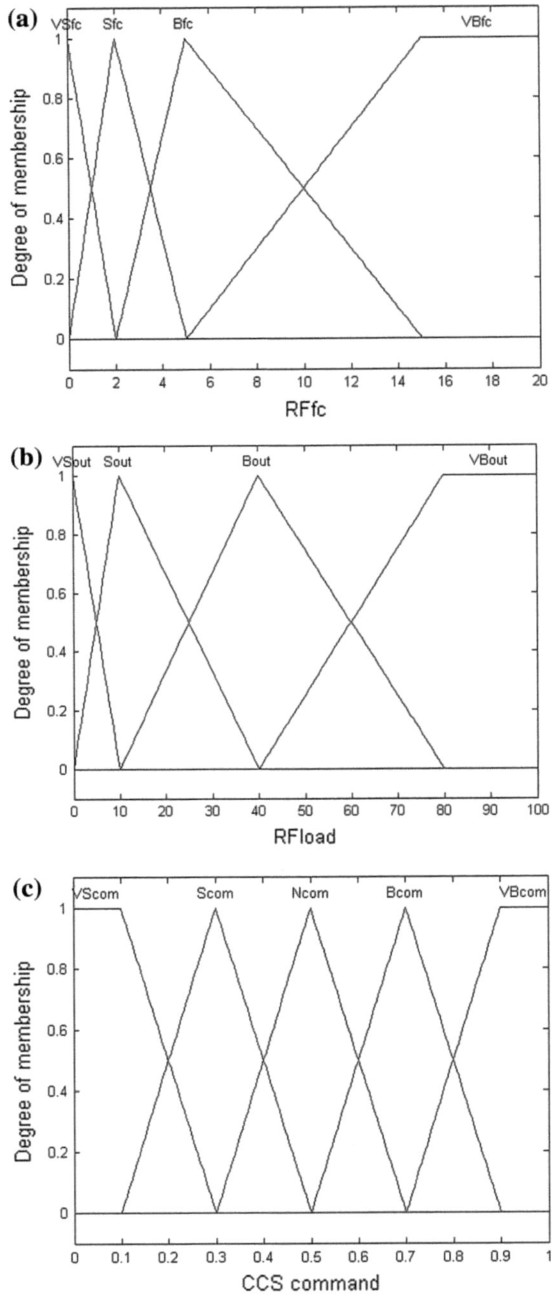

Fig. 31 The membership functions for the FLC variables [34], **a** Membership functions for the FC ripple factor, **b** Membership functions for the load ripple factor, **c** Membership functions for the CCS command signal

Table 4 The FLC rules base [34]

Command signal		RFfc			
		VSfc	Sfc	Bfc	VBfc
RFload	VSrm	Ncom	Scom	VScom	VScom
	Srm	Bcom	Ncom	Scom	VScom
	Brm	VBcom	Bcom	Ncom	Scom
	VBrm	VBcom	VBcom	Bcom	Ncom

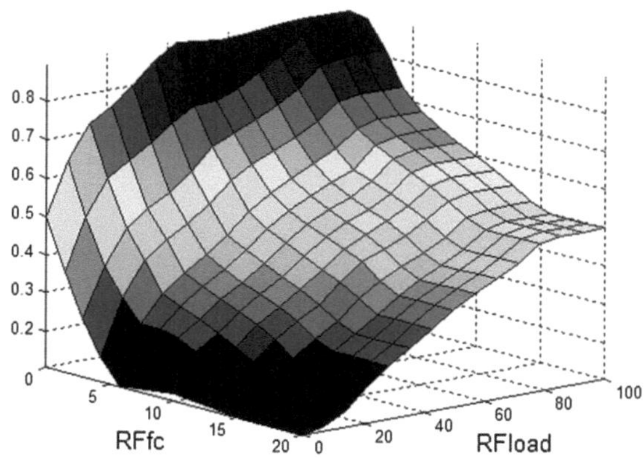

Fig. 32 The FLC control surface [34]

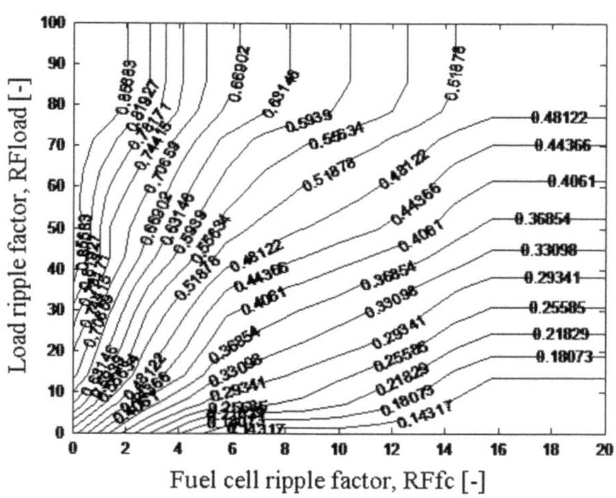

Fig. 33 The projected contours for the FLC control surface [34]

$$G_{1(\text{base})} \stackrel{\Delta}{=} \frac{\eta_1 V_{\text{FC(base)}}}{V_{\text{out}}} \cong \frac{I_{\text{out(base)}}}{I_{\text{FC(base)}}} \quad (46)$$

If the relay block has the on/off switching levels at 0.45 and 0.5 (0.475 ± 0.025), then the nonlinear characteristic of the RF_{fc} versus RF_{load}

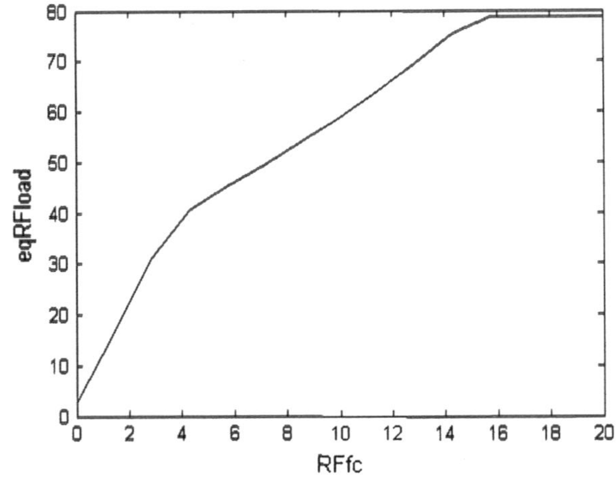

Fig. 34 The nonlinear control law of the EqRFload versus RFfc [34]

Table 5 The SISO FLC rules base [34]

	RFfc			
	VSfc	Sfc	Bfc	VBfc
EqRFload	VSload	Sload	Bload	VBload

obtained for the 0.475 command level (see Fig. 33) can be considered in the design of the transfer characteristic; the equivalent load RF ($EqRF_{load}$) versus RF_{fc} (see Fig. 34). This nonlinear control law of $EqRF_{load}$ versus RF_{fc} can be also obtained directly using a single input—single output (SISO) FLC. The designing of the SISO FLC is given below.

The input variable, RFfc, and the output variable, EqRFload, have the same membership functions defined as in Fig. 31. The rules base is shown in Table 5. The Mamdani implication and the centre of gravity defuzzification method are used. The FLC control surface obtained is shown in Fig. 27.

Obviously, this characteristic is close to one of the projected contours shown in Fig. 33, namely the one that is obtained for the level of command signal set to approximately 0.48. Thus, the structure of the CCS FLC controller is redesigned as in Fig. 35 based on this nonlinear control law shown in Fig. 34. The nonlinear control law can be simply implemented based on a PWL nonlinear gain. For example, the PWL nonlinear gain having the input vector [0, 4, 16, 20] and the output vector [0, 40, 80, 80]) is easy to be implemented.

5.1.5 Simulation Results

The start-up of the mitigation process is almost the same as in the case of the FC HPS that uses the CCS hysteretic controller (see Fig. 18). The simulation results

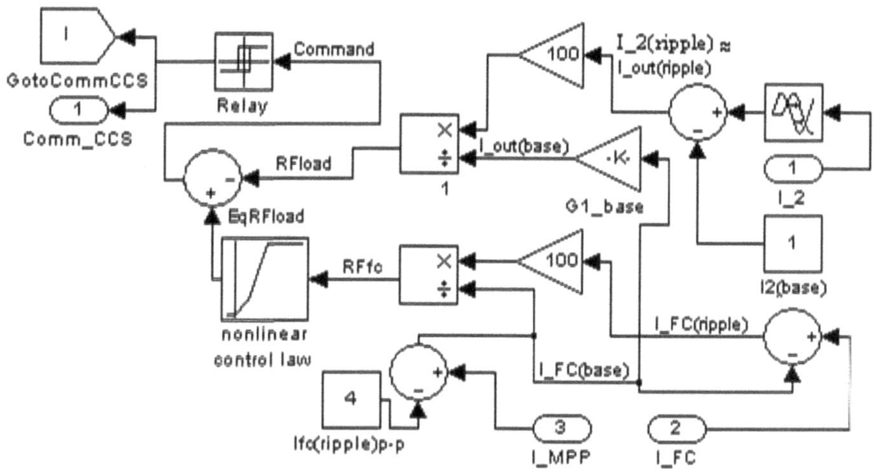

Fig. 35 The structure of the CCS nonlinear controller [34]

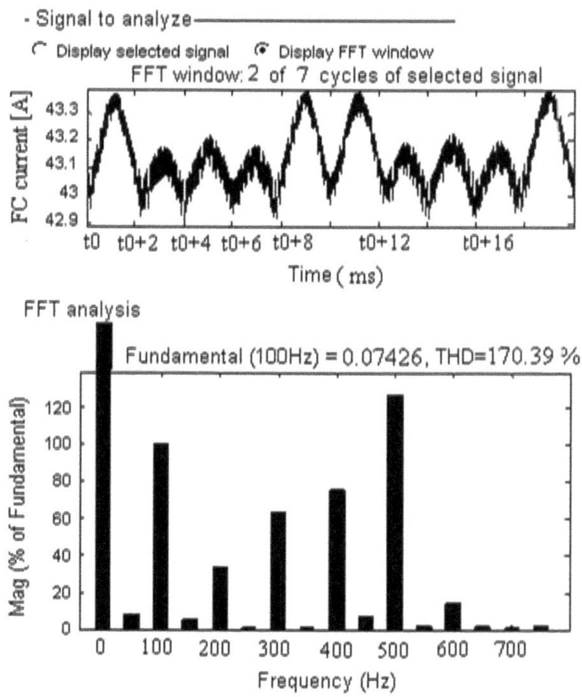

Fig. 36 The FC current (*top*) and its LF power spectrum (*bottom*): 1.26 kW HPS case with the CCS PWL nonlinear controller [34]

for the 1.26 kW HPS, which uses the PWL nonlinear controller proposed, are shown in Fig. 36. It can be observed that the mitigation performances are better compared to the use of the hysteretic or PCC controller (to compare, see Figs. 23

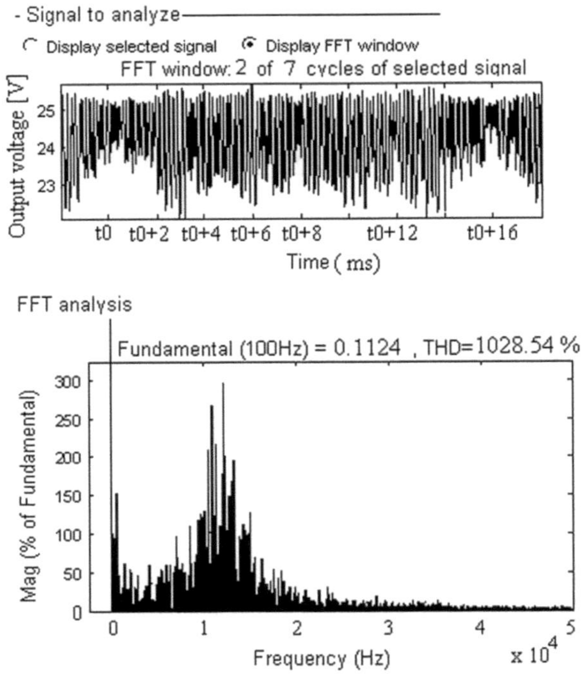

Fig. 37 The output voltage (*top*) and its power spectrum (*bottom*): 1.26 kW HPS case with CCS PWL nonlinear controller [34]

and 25). For example, the 100 Hz harmonics have a magnitude of approximately 0.074 (using PWL nonlinear controller) instead of approximately 0.17 (using hysteretic controller). Thus, the magnitude of 100 Hz harmonic is about two times smaller (0.074/0.17 $\cong$ 1/2).

The output voltage and its power spectrum are shown in Fig. 37 for a 1.26 kW HPS that uses the PWL nonlinear controller. The power spectrum was spread in the HF band up to 20 kHz. This HF ripple also appears in the power spectrum of the FC current (see Fig. 38), but it is much smaller than the LF ripple, being up to the allowable limits. Consequently, it is well tolerated by the PEMFC stack. The main advantage of the PWL nonlinear controller is its design that is not dependent to the level of HPS power. The simulation results for a 6 kW HPS are shown in Figs. 39 and 40 to prove this advantages. So using the simulation, results are shown in Figs. 39 and 40, the RF and GIfc values can be computed as below:

$$\left.\begin{array}{l} \mathrm{RF_{out}} \cong \dfrac{188-110}{110} = 71\,\% \\ \mathrm{RF_{FC}} \cong \dfrac{91-89.7}{89.7} \cong 1.4\,\% \end{array}\right\} \Rightarrow G^{*}_{\mathrm{Ifc}} = \left(\frac{90}{1.4}\right)\frac{(0.9*55)}{40} = 79 \qquad (47)$$

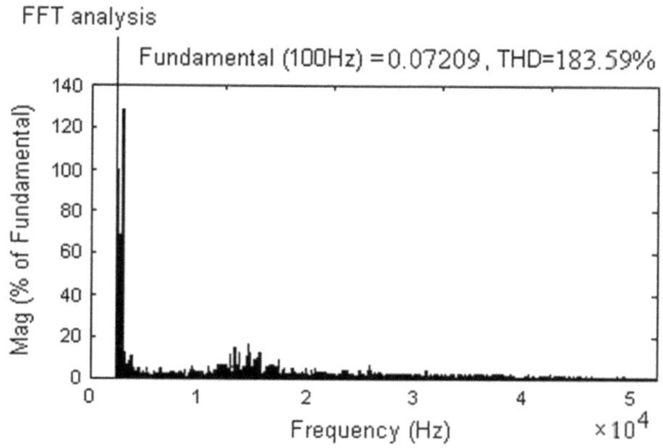

Fig. 38 The HF power spectrum of the FC current: 1.26 kW HPS case with CCS PWL nonlinear controller [34]

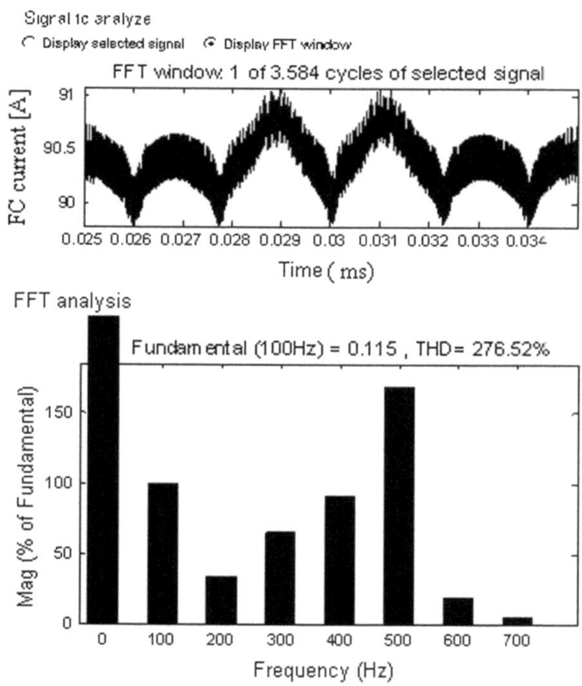

Fig. 39 The FC current (*top*) and its LF power spectrum (*bottom*): 6 kW HPS case with CCS PWL nonlinear controller [34]

It can be observed that the RF of the FC current, RF_{FC}, has now almost the same value (1.1 % and 1.4 %) for the both HPS power levels, not four times higher (see relationships 43, 44 and 47).

The case of a three-phase inverter system powered by a 20 kW HPS (see Fig. 16a) is analysed in order to show that the proposed PWL nonlinear controller will operate almost as well if an inverter system is used instead of the equivalent load. The current and the voltage of the 20 kW PEMFC stack are approximately

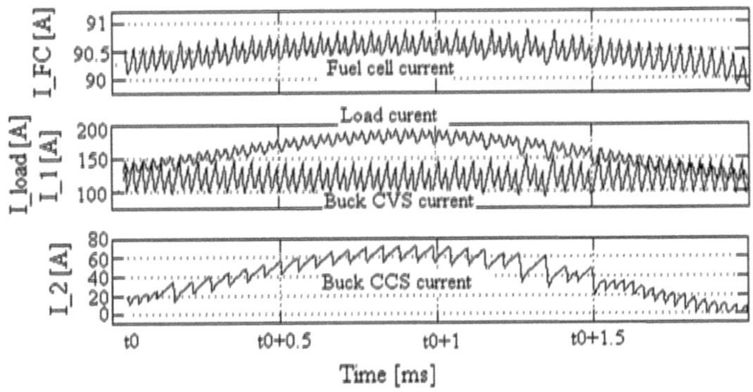

Fig. 40 The behaviour of the 6 kW HPS case with CCS PWL nonlinear controller [34]

45 A and 435 V, close to the MPP. The reference voltage of the buck CVS is set to 400 V.

The simulation results for the FC current and voltage are shown in Figs. 41 and 42 considering the bi-buck topology for the FC HPS (Fig. 17). The RF of the FC current is about of 0.8 %. The RF of the HPS output voltage is approximately 3.5 and 0.3 % using a C_f filter capacitance of 470 and 4700 μF, respectively (see Fig. 43).

Fig. 41 The FC current (*top*) and its power spectrum (*bottom*) for the 20 kW HPS case [34]

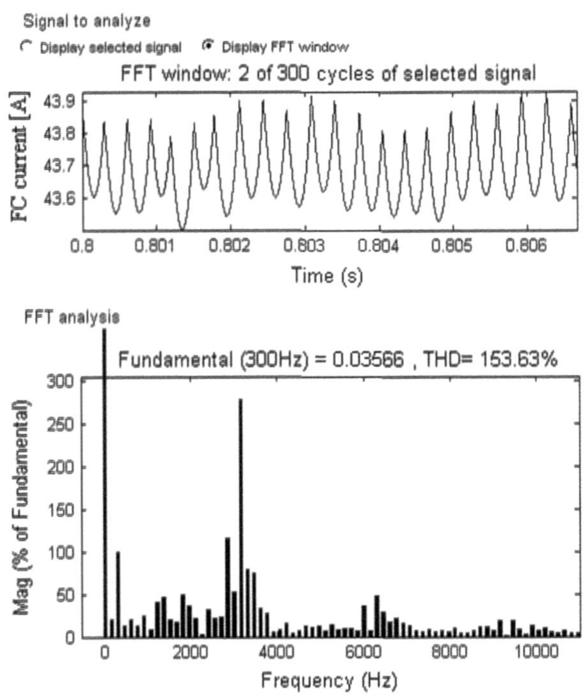

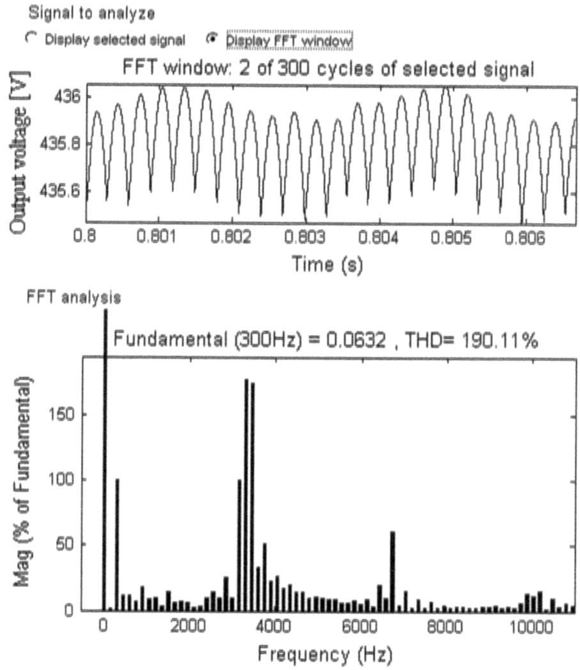

Fig. 42 The FC voltage (*top*) and its power spectrum (*bottom*) for the 20 kW HPS case [34]

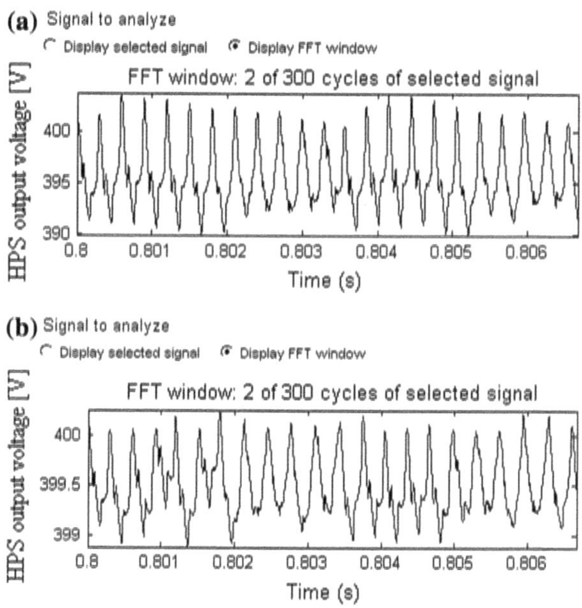

Fig. 43 The HPS output voltage for the 20 kW HPS case [34], **a** $C_f = 470\ \mu F$, **b** $C_f = 4{,}700\ \mu F$

The three-phase inverter system structure is a full-bridge that uses a switching command of pure sine PWM type (having the carrier frequency set to 10 kHz). A classic voltage control of the AC output voltage is used.

The input current and its power spectrum are shown in Fig. 44. This current has LF harmonics (see middle plot of Fig. 44) and, for example, the magnitude of the 300 Hz fundamental harmonic is 1.886 A. This LF current ripple will be mitigated by the injection of an anti-ripple current via the buck CCS. The buck CCS current and its power spectrum are shown in Fig. 45.

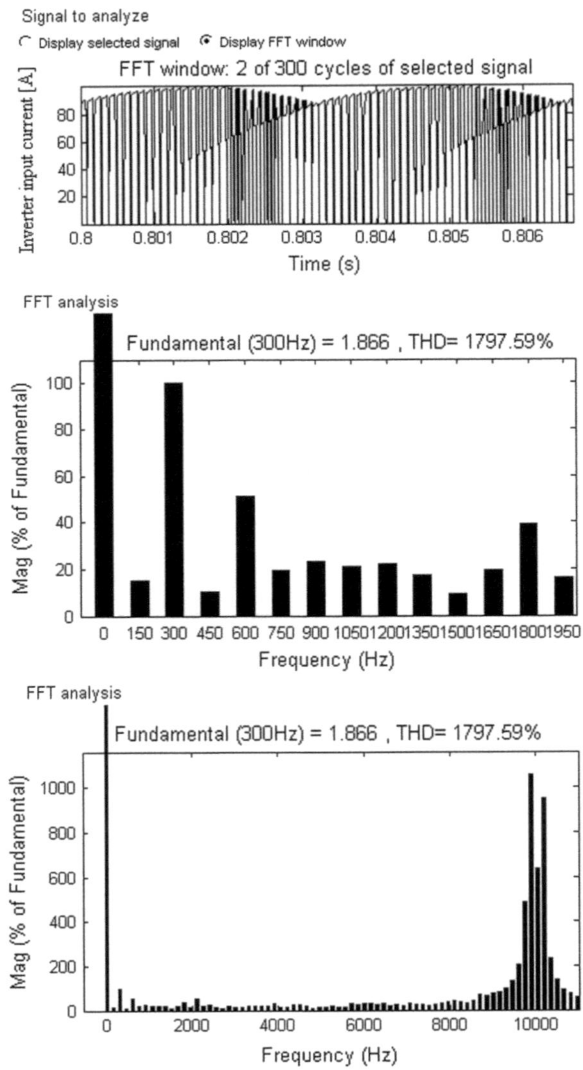

Fig. 44 The input current of the three-phase inverter system [34]

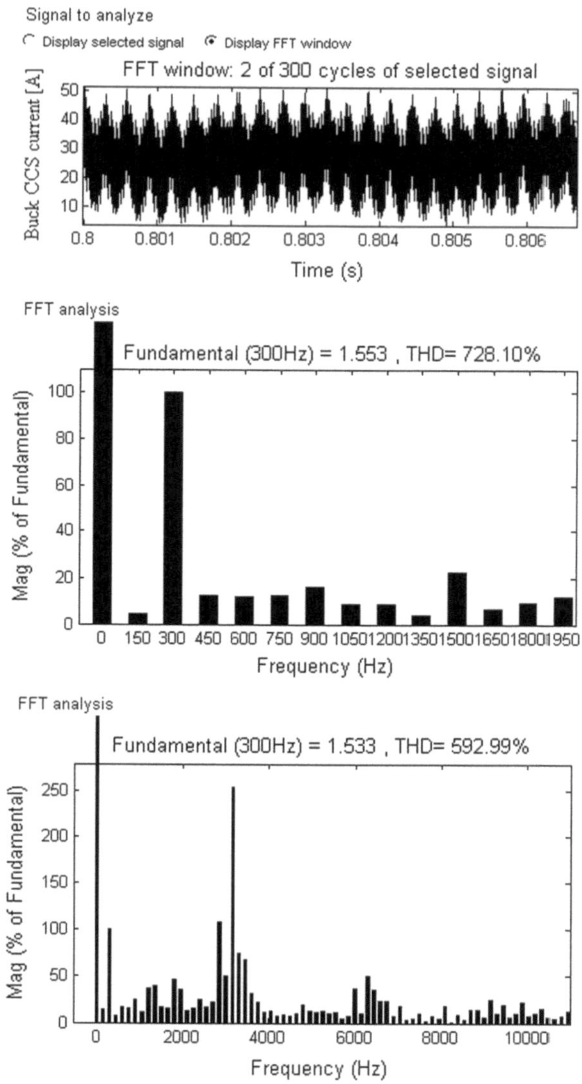

Fig. 45 The buck CCS current [34]

It can be observed that the magnitude of the 300 Hz fundamental harmonic is 1.553 A. The difference between those currents (0.333 A) is the output current of the CVS (see Fig. 40) and this current ripple is back propagated to the PEMFC stack. The 300 Hz fundamental magnitude of the PEMFC current is 0.03566 A (see Fig. 41). The mitigation ratio for the fundamental harmonic of the inverter ripple is approximately 1.8 %. Also, the mitigation of the fundamental harmonic based on the use of the buck CVS is about nine times ($0.333/0.03566 \cong 9.34$)

It can be observed that the LF shape of the buck CCS current (top plot of Fig. 45) tracks the LF shape of the input inverter current (top plot of Fig. 44). This is because the buck CVS mitigates the LF harmonics in the same ratio, which is about 9. The power spectrum of the HPS signals, which is shown in Figs. 41, 42 and 45, reveals that this is spread in a large band.

This section analyses the bi-buck topology as a solution to mitigate the inverter ripple. The simulation results have shown that the mitigation performance depends on the magnitude of the ripple and on the level of the load power. Consequently, a nonlinear controller for the buck CCS is necessary to be designed in order to overcome these issues. The RF for the inverter ripple is up to 3 % that is reported in [43].

The nonlinear controller for the buck CVS will be presented in next section.

5.2 Voltage-Mode Control

The nonlinear control law of the CVS controller will be designed based on a SISO FLC.

5.2.1 Designing of the Nonlinear Voltage Controller

Note that the buck CVS converter is the power interface of the FC stack (see Fig. 16) and the CVS controller structure is shown in Fig. 46.

The rules base used to obtain a nonlinear characteristic to mitigate the output voltage ripple is very simple: (1) if the output voltage ripple is small, then the loop gain must be small; (2) if the output voltage ripple increases, then the loop gain must rise quickly; (3) if the output voltage ripple is high, then the loop gain must be limited to a value that ensures the stability of the overall feedback loop. The main design questions are related to how small must be, how quickly must increase and how big should be the limit of the loop gain. The fuzzy logic reasoning based on systematic approach will be used to design the CVS nonlinear control law. The CVS controller is a single input—single output system. The input is the output voltage error, v_{error}, and the output is the v_2 voltage, having values in range of [−1 V, 1 V] and [−10 V, 10 V], respectively.

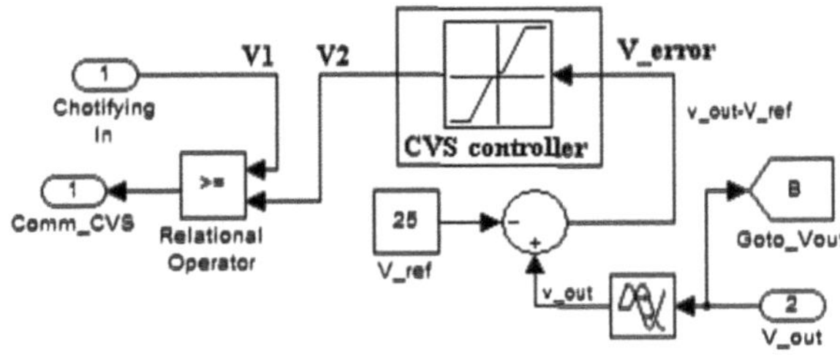

Fig. 46 The CVS controller that implement the voltage-mode nonlinear control proposed [33]

Fig. 47 The membership functions for the output voltage error [33]

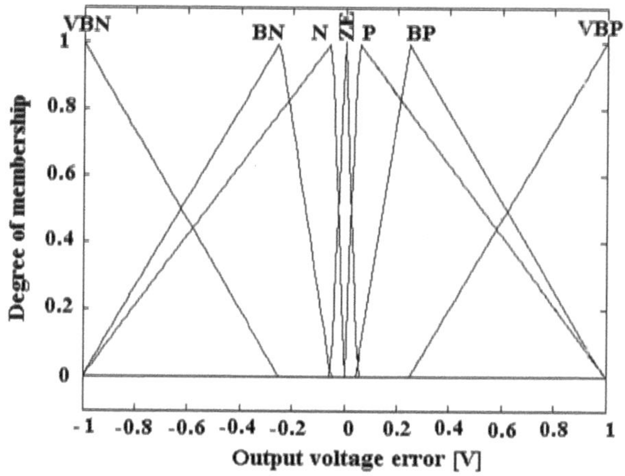

The v_2 voltage range must be correlated with the range of the chaotifying signal, which can be any periodic signal. The saw-tooth signal, $v_1(t)$, will be used in this section as a chaotifying signal:

$$v_1(t) = V_L + (V_H - V_L)\frac{t(\text{mod}\, T_{sw})}{T_{sw}} \tag{48}$$

This is a ramp voltage that decreases in a time period, $T_{sw} = 1/f_{sw}$, from a higher voltage, V_H, to a lower voltage, V_L. The values used in all simulations are $V_H = 9$ V, $V_L = 1$ V and $f_{sw} = 10$ kHz. These were chosen as in [41] in order to compare the obtained results.

The output voltage is set to 25 V, so a RF of 4 % means an output voltage error of 1 V. Consequently, the range of the output voltage error was chosen to be [−1 V, 1 V]. The CVS controller was replaced with a proportional controller. Through the trial and the error method a gain of 10 for which the RF is of 4 % was

Fig. 48 The membership functions for the output voltage the of nonlinear voltage-mode controller [33]

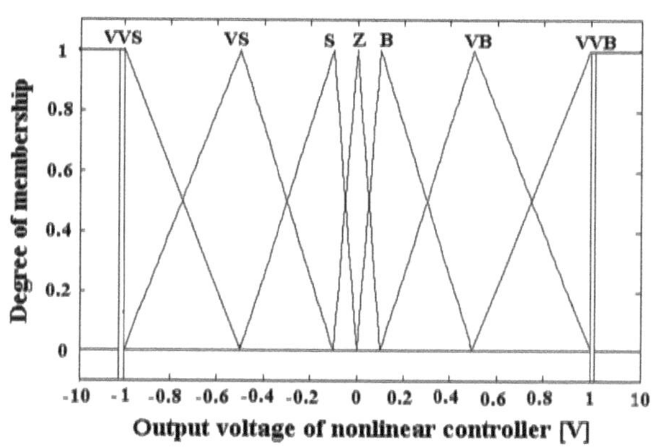

Table 6 The rules base for CVS fuzzy controller [33]

Rule	Output voltage error	Output voltage of the CVS controller
1	VBN	VVS
2	BN	VS
3	N	S
4	ZE	Z
5	P	B
6	VP	VB
7	VBP	VVB

found. So, the range [−10 V, 10 V] was chosen for the V_2 voltage. Thus, for the output voltage error, seven membership functions are symmetrically defined in these ranges (see Fig. 47: Very_Big_Negative (VBN), Big_Negative (VBN), Negative (N), Zero_Equal (ZE), Positive (P), Big_Positive (BP), and Very_Big_Positive (VBP)) and V_2 voltage (see Fig. 48: Very_Very_Small (VVS), Very_Small (VS), Small (S), Zero (Z), Big (B), Very_Big (VB) and Very_Very_Big (VVB)). Taking in account the basic idea to obtain the nonlinear characteristic of the CVS controller, the base of rules is shown in Table 6:

The control law is obtained by defining the position of the peaks for the membership functions. The Z and ZE triangular membership functions have the peak set to zero. The rest of the triangular membership functions for the output voltage error have the peaks set to ±50 mV, ±250 mV and ±1000 mV, which means a RF of 0.2, 1 and 4 %, respectively.

The triangular membership functions for the V_2 voltage (namely the B and VB, and their symmetrical membership functions, S and VS, respectively) have the peaks to ±100 mV and ±500 mV in order to result the same gain (100 mV/50 mV = 500 mV/250 mV = 2).

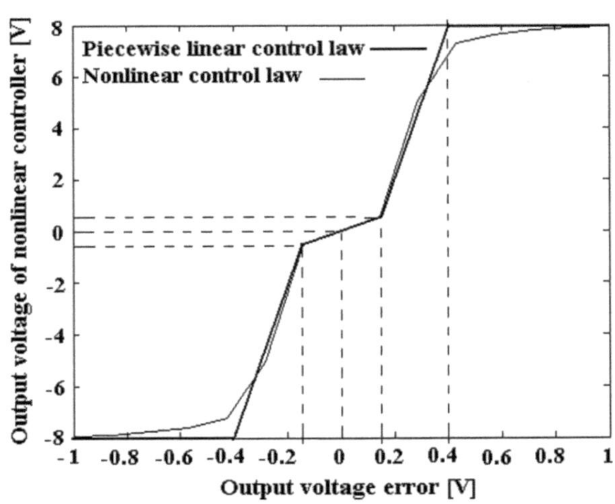

Fig. 49 The nonlinear voltage control characteristic (*thin line*) and PWL voltage control characteristic (*thick line*) which is fitted on it [33]

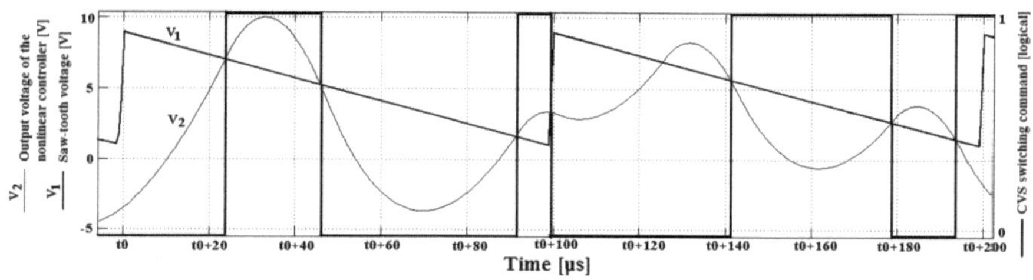

Fig. 50 Signals related to CVS controller operation with saw-tooth as chaotifying signal [33]

The VVS and VVB trapezoidal memberships are defined (as ±0.5, ±1, ±10, ±10) to result an output voltage ripple factor up to 4 %. The peaks of the VBN and VBP triangular membership functions were set to ±1 V in order to have a small gain at ±1 V output voltage error (equal with 1 V/1 V = 1) and a high gain at ±10 V output voltage error (equal with 10 V/1 V = 10).

The obtained nonlinear voltage-mode control law is shown in Fig. 49. Note that the Mamdani implication, max–min fuzzy connectives and the centre of the area defuzzification strategy were used. The PWL voltage-mode control law that fit the nonlinear voltage-mode control law is also shown in Fig. 49.

The PWL voltage-mode control law can be easily implemented by the following look-up table [42]:

- input vector: [−1, −0.5,−0.15, 0.15, 0.5, 1];
- output vector: [−10, −10, −0.4, 0.4, 10, 10].

A mathematical approach of the buck CVS in a closed control loop can be performed based on this PWL voltage-mode control law [32]. The saw-tooth signal operates as a chaotifying signal (see Fig. 50). It is observed that output voltage of the nonlinear controller is a distorted signal, obtained from the output voltage error via the PWL voltage-mode control law.

The performance of the output voltage regulation for the CVS controller is evaluated for different load currents (see Fig. 51). It can be seen that the output voltage is well regulated to the reference voltage of 25 V. The PEMFC time constant was set to 0.2 s.

5.2.2 Simulation Results

Two performance indicators, PI_1 and PI_2, are used to quantify the spreading level of the output power [32, 42]:

$$PI_1 = \frac{\Delta f_{10\,\%}}{f_{\mathrm{COG}}},\ PI_2 = \frac{S_{\mathrm{peak}}}{\mathrm{THD}} \tag{49}$$

where:

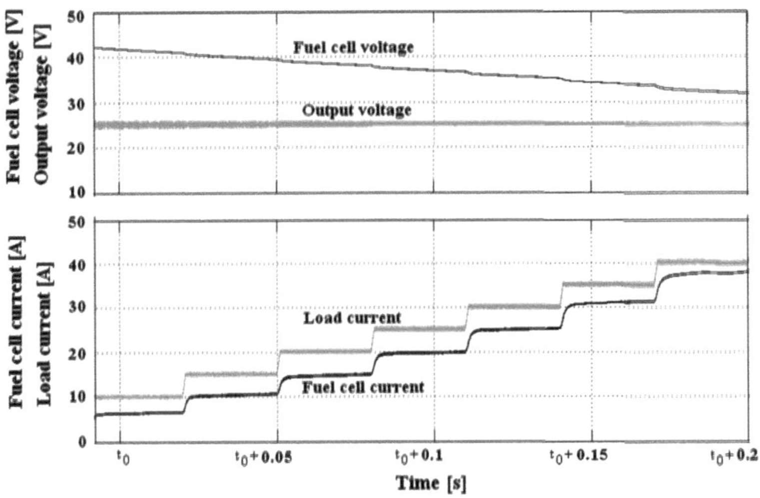

Fig. 51 The behaviour of the CVS controller tested with step-up load current [33]

S_{peak} is the maximum spectral magnitude (as % of DC component), excluding the harmonics of the chaotifying signal that can possibly occur (see Fig. 52);

$\Delta f_{10\ \%Sp}$—the frequencies band where the magnitude of the power spectrum is 10 % over the S_{peak};

f_{COG}—the frequency that is the centre-of-gravity of the power spectrum;

THD—the total harmonic distortion factor of the output voltage.

For example, the performance indicators could be interpreted as below:

- If $PI_1 > 50\,\%$ then $\Delta f_{10\%\ Sp} > f_{COG}/2$, and this means a large frequencies band where the most part of the spread power spectrum is situated;
- If $PI_2 < 50\ \%$, then $S_{peak} < THD/2$, and this means no high peak in the power spectrum.

The performance indicators are estimated for different load currents (see Fig. 52). Thus, the performance indicators are about $PI_1 \cong 27$ kHz/18 kHz $\cong$ 150 % and $PI_2 \cong 1400/6183 \cong 23$ %, and $PI_1 = 11$ kHz/17 kHz = 65 % and $PI_2 \cong 2200/8000 \cong 28$ %, for simulation results shown in Figs. 53 and 54, respectively. So, a well spreading performance is obtained in both cases. Note that the harmonics of the chaotifying signal occur in last case and all simulations shown in Fig. 51. This situation can be avoided by pseudo-randomise of the saw-tooth period. The output voltage ripple factor, $RF_{Vout} = \Delta v_{out}/V_{out}$, is approximately 0.24 % and 2.4 %, respectively.

The performance indicators PI_2, PI_1 and RF_{Vout} are estimated in Table 7, for the case study shown in Fig. 52. Note that the level of the load current influences the performance indicators by changing the associated frequencies of the buck CVS converter (f_{RL} and f_{RC}).

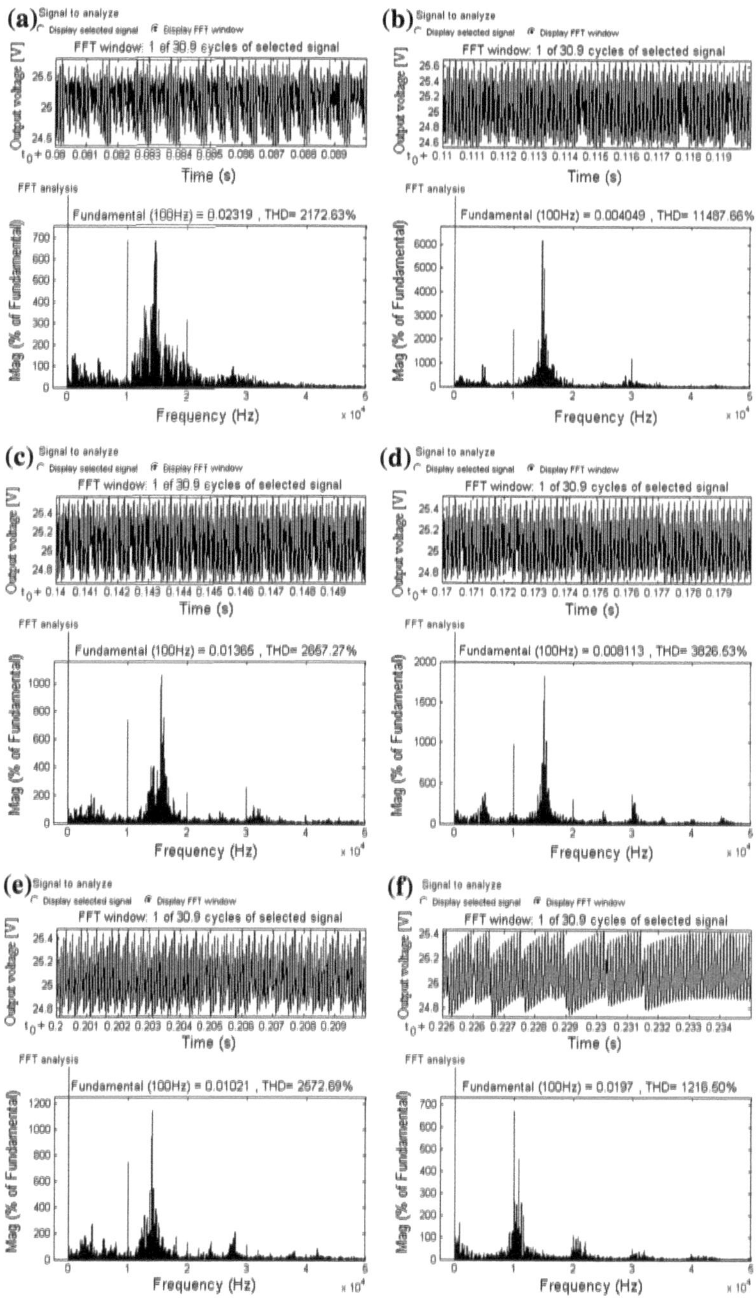

Fig. 52 The output voltage (*top*) and its power spectrum (*bottom*) for different load current, using the simulation parameters: $f_{sw} = 10$ kHz, $L = 100$ μH, and C = 47μF [33], **a** $I_{out} = 15$ A, **b** $I_{out} = 20$ A, **c** $I_{out} = 25$ A, **d** $I_{out} = 30$ A, **e** $I_{out} = 35$ A, **f** $I_{out} = 40$ A

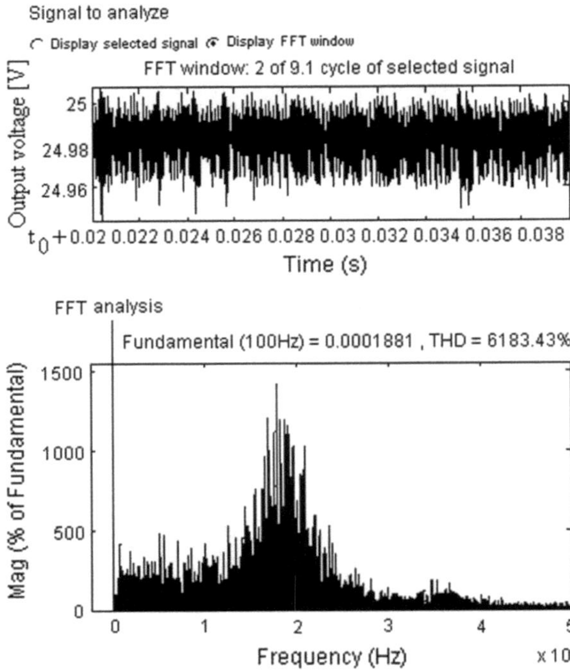

Fig. 53 The output voltage (*top*) and its power spectrum (*bottom*) with the parameters [33]: $f_{sw} = 15$ kHz, $L = 100$ μH, $C = 47$ μF, $R_{out} = 1\ \Omega$

If the buck parameters are $L = 100$ μH and $C = 47$ μF, then the natural frequency, f_{LC}, will be 2.3215 kHz. So, looking into Table 7 and also taking in account other simulation results, it can be concluded that a well spreading level of the power spectrum could be obtained if the associated frequencies have the same order of magnitude. The output voltage ripple factor is up to 4 % for the load current in the rated range. The response in the output voltage is lower than the voltage ripple if a 5 A step-up in the load current is used (Fig. 51). These results have shown the control robustness of the CVS controller. If a load current pulse over 10 A is used, then small voltage spikes appear on the output voltage during the rise and fall time of the load current pulse (Fig. 55). These spikes can be better compensated using a higher value for the filtering capacitor on the DC bus or via the buck CCS.

The resistive parameter of the load pulse, Rout(pulse), is 2/3Ω from 40 to 80 ms and 1 Ω in rest. Thus, the load current pulse is 12.5 A, and the FC current pulse can be estimated by (47):

$$I_{FC(pulse)} \cong \frac{V_{out}}{\eta_1 V_{FC(pulse)}} \cdot I_{out(pulse)} = \frac{V_{out}^2}{\eta_1 V_{FC(pulse)} R_{out(pulse)}} \tag{50}$$

The dynamic of the buck CVS converter is shown in phases plane (Fig. 56). The chaotifying effect is shown in a zoom, where the limit cycles are shown

Fig. 54 The output voltage (*top*) and its power spectrum (*bottom*) with the parameters [33]: $f_{sw} = 15$ kHz, $L = 100$ μH, $C = 47$ μF, $R_{out} = 2/3$ Ω

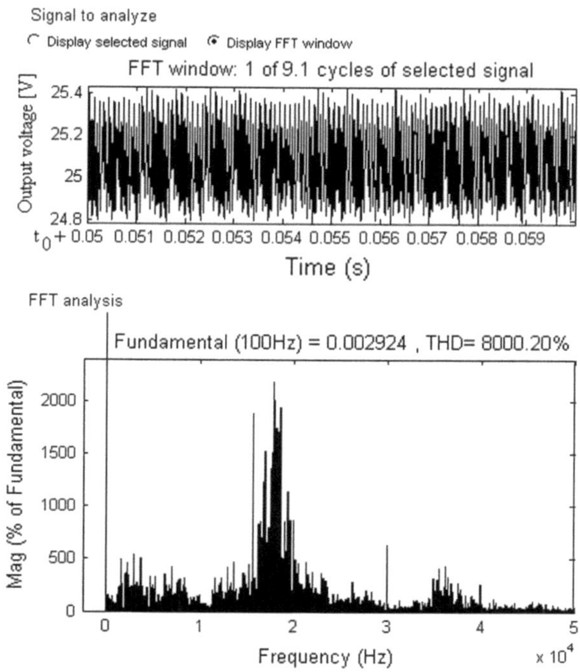

Table 7 The performance indicators for the case study shown in Fig. 52 [33]

I_{out} [A]	f_{RL} [kHz]	f_{RC} [kHz]	PI_1 [%]	PI_2 [%]	RF_{Vout} [%]
15	2.6526	2.0318	79	32	5.6
20	1.9894	2.7090	42	52	4
25	1.5915	3.3863	35	41	3.2
30	1.3263	4.0635	25	47	3.2
35	1.1368	4.7408	34	46	3.2
40	0.9947	5.4180	27	55	3.2

around the steady-state point of (25 V, 25 A). The simulations are performed for the 6 kW PEMFC with a fuel flow rate of 47 lpm, which set the MPP at approximately 120 A and 50 V. Considering the RF of the output voltage as before, i.e. 4 %, the CVS controller can be designed in the same manner. Thus, considering a switching frequency of 10 kHz, the HPS parameters are $L_1 = L_2 = L = 200$ μH and $C_f = 100$ μF, so $f_{LC} = 1.1254$ kHz. The performance indicators are shown in Table 8 for different load currents. These results validate the following conclusions: (1) the design of the CVS controller is less dependent to the load power level and (2) the performances regarding the voltage ripple and the spreading of the power spectrum are maintained for different load power levels in range. The restriction to power level means that the associated

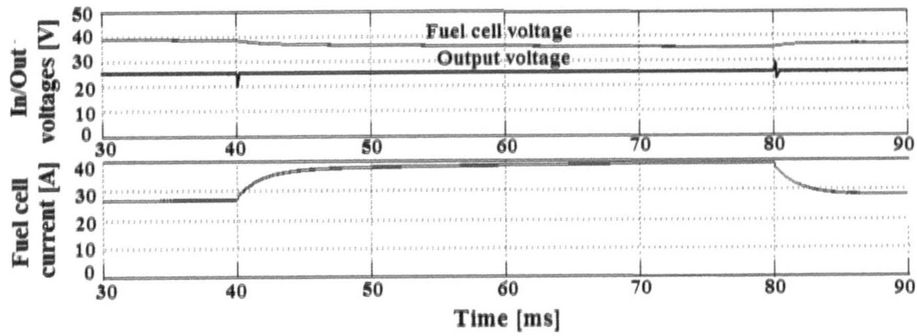

Fig. 55 The buck CVS tested with a pulse load current [33]

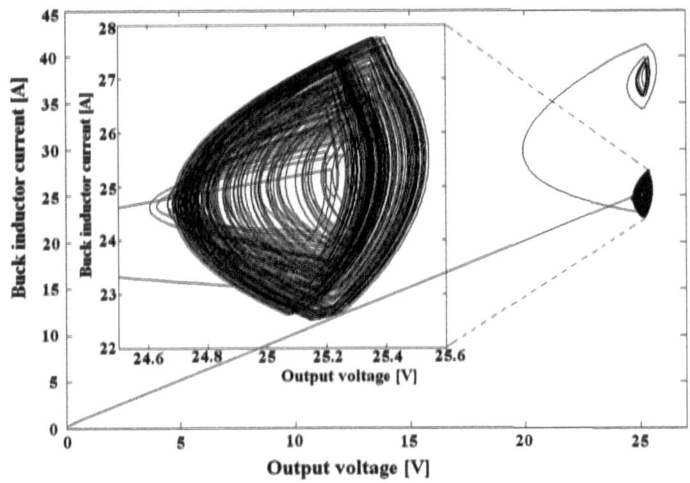

Fig. 56 The buck CVS dynamic in the phases plane [33]

frequencies of the buck CVS converter (f_{RL} f_{LC} and f_{RC}), the saw-tooth frequency (f_{sw}), and the closed loop frequency, f_0, must verify the Eq. (51):

$$\max(f_{RC}, f_{RL}, f_{LC}) < f_{sw} < f_0 \tag{51}$$

In this chapter the MPP was considered as an operating point for the PEMFC stack. If the load is dynamic, having an unknown power profile, then the load following control loop must be used to set the fuel flow in correlation with the requested power. As it is known the MPP tracking process is relatively slow, the searching time being dependent to the MPP algorithm chosen and to the PEMFC time constant [44, 45]. A new HPS topology was proposed and analysed. The multi-port topologies are proposed to overcome the issues that usually appear in

Table 8 The performance indicators for a 6 kW PEMFC/200 Ah battery HPS [33]

I_{out} (A)	f_{RL} (kHz)	f_{RC} (kHz)	PI_1 (%)	PI_2 (%)	RF_{Vout} (%)
20	1.9894	0.636	47	50	4.5
40	0.994	1.2732	37	44	3.7
60	0.663	1.9099	28	45	3.5
80	0.497	2.5465	30	41	3.8
100	0.397	3.1831	32	49	4

hybrid vehicle applications [46, 47]. In this chapter a multi-port topology of bi-buck type is proposed to mitigate the inverter ripple based on a buck CCS that generates an anti-ripple.

In this last section, the buck CVS is presented. The design goal for the CVS controller is to regulate the output voltage and reduce the electromagnetic interferences. The RF of the output voltage is up to 4 % even if the anti-control technique is used to spread the power spectrum. The spreading in the HF band of the LF ripple, which remains after active compensation via the buck CCS, leads to an increase of the PEMFC life cycle as well.

Two performance indicators are used to quantify the spreading level of the power spectrum. The first one is a measure of the width of the frequencies band where the most part of the power spectrum is situated. The second one is a measure of the peaks level in the power spectrum.

The proposed nonlinear CVS controller is designed to assure the best performances in both frequency and time domain. All the reported results have been validated in several simulations. The following performances are obtained: a RF of the output voltage up to 4 % for the load current in the rated range, about 3–37 kHz width for the frequencies band of the spread power spectrum, and power peaks in range of 30–55 % from THD.

6 Conclusion

The LF current ripple appears in the normal operation of the FC inverter system and this is propagated back via the power converters to the PEMFC stack. Also, in FC vehicle applications, high energy demands appear in a short time. That will cause high current pulses with high slopes, which are propagated back as a LF current ripple, too.

In this chapter, the bi-buck topology is analysed as a multi-port topology for the FC HPS. The bi-buck converter was designed to mitigate this LF current ripple via a buck CCS that will inject an anti-ripple into the point of the common coupling. In this node the dynamic load, the buck CVS and the buck CCS are connected. The anti-ripple is generated to track the shape of the LF inverter current ripple based on an active control implemented in the CCS controller. The use of a nonlinear control law improves the mitigation performances.

The buck CVS is used to spread the LF ripple that remains after active compensation. For high power HPS applications, the buck CVS can be removed to increase the HPS efficiency. The FC stack can operate under a dynamic load near the MPP by setting the fuel flow rate via a load following control and a MPP tracking control. For a dynamic load it is necessary to assure the power balance by adding an ESD (mixed stack of batteries and ultracapacitors) on a low or a high voltage bus. In the last case a bidirectional converter is necessary.

Some state-of-art architectures for the FC HPS are showed in this chapter. For the analysed HPS architectures nonlinear control laws that can effectively mitigate the current ripple to a RF of approximately 3 % (better than the RF reported in the literature), without increasing the voltage ripple factor over 4 %, were proposed.

Acknowledgments Some figures, tables and text are reproduced from [33–35] here with kind permission from Elsevier Limited, UK [February 16, 2013].

References

1. Kirubakaran A, Jain S, Nema RK (2009) A review on fuel cell technologies and power electronic interface. Renew Sustain Energy Rev 13:2430–2440
2. Wajiha S, Rahul AK, Arefeen M (2006) Analysis and minimization of input ripple current in PWM inverters for designing reliable fuel cell power systems. J Power Sources 156:448–454
3. Fontes G (2007) Interactions between fuel cells and power converters: Influence of current harmonics on a fuel cell stack. IEEE T Power Electron 22(2):670–678
4. Gemmen RS (2001) Analysis for the effect of the ripple current on fuel cell operating condition. ASME 369(4):279–289
5. Woojin C, Gyubum J, Prasad NE, Jo WH (2004) An experimental evaluation of the effects of ripple current generated by the power conditioning stage on a proton exchange membrane fuel cell stack. JMEPEG 13:257–264
6. Schmittinger W, Vahidi A (2008) A review of the main parameters influencing long-term performance and durability of PEM fuel cells. J Power Sources 180:1–4
7. Auld AE, Smedley KM, Mueller F, Brouwer J, Samuelsen GS (2010) Load-following active power filter for a solid oxide fuel cell supported load. J Power Sources 195:1905–1913
8. Thounthong P, Raël S, Davat B (2009) Energy management of fuel cell/battery/supercapacitor hybrid power source for vehicle applications. J Power Sources 193(1):376–385
9. NETL (2001) NETL published fuel cell specifications for future energy challenge 2001 competition (accessed in May 2, 2012, on http://netl.doe.gov)
10. Thounthong P, Davat B (2010) Study of a multiphase interleaved step-up converter for fuel cell high power applications. Energy Convers Manage 51:826–832
11. Hwang JC, Chen LH, Yeh SN (2007) Comprehensive analysis and design of multi-leg fuel cell boost converter. Appl Energy 84:1274–1288
12. Methekar RN, Patwardhan SC, Gudi RD, Prasad V (2010) Adaptive peak seeking control of a proton exchange membrane fuel cell. J Process Control 20:73–82
13. Zhong ZD, Huo HB, Zhu XJ, Cao GY, Ren Y (2008) Adaptive maximum power point tracking control of fuel cell power plants. J Power Sources 176:259–269
14. Bizon N (2010) On tracking robustness in adaptive extremum seeking control of the fuel cell power plants. Appl Energy 87(10):3115–3130

15. Thounthong P, Davat B (2007) A PEM fuel cell power source for electric vehicle applications with supercapacitor or battery as auxiliary. In: Alemo PV, Commack ED (eds.) Progress in fuel CELL research. Nova press, New York (chapter 8)
16. Lukic SM, Cao J, Bansal RC, Rodriguez F, Emadi A (2008) Energy storage systems for automotive applications. IEEE Trans Ind Electron 55:2258–2267
17. Kim MJ, Peng H (2007) Power management and design optimization of fuel cell/battery hybrid vehicles. J Power Sources 165:819–832
18. Tang Y, Yuan W, Pan M, Li Z, Chen G, Li Y (2010) Experimental investigation of dynamic performance and transient responses of a kW-class PEM fuel cell stack under various load changes. Appl Energy 87:1410–1417
19. Rodatz P, Paganelli G, Sciarretta A, Guzzella L (2005) Optimal power management of an experimental fuel cell/supercapacitor powered hybrid vehicle. Control Eng Practice 13(1):41–53
20. Thounthong P, Davat B, Raël S, Sethakul P (2009) An overview of power converters for a clean energy conversion technology. IEEE Electronic Mag 3(1):32–46
21. Thounthong P, Chunkag V, Sethakul P, Sikkabut S, Pierfederici S, Davat B (2010) Energy management of a fuel cell/solar cell/supercapacitor hybrid power source. J Power Sources 196(1):313–324
22. Corbo P, Corcione FE, Migliardini F, Veneri O (2009) Experimental study of a fuel cell power train for road transport application. J Power Sources 145:610–619
23. Uzunoglu M, Alam MS (2006) Dynamic modeling, design, and simulation of a combined PEM fuel cell and ultracapacitor system for standalone residential applications. IEEE T Energy Conver 21:767–775
24. Xu L, Li J, Hua J, Li X, Ouyang M (2009) Adaptive supervisory control strategy of a fuel cell/battery-powered city bus. J Power Sources 194:360–368
25. Bizon N, Lefter E, Oproescu M (2007) Modeling and Control of the energy sources power interface for automotive hybrid electrical system. In: 21st JUMV international automotive Conference on science and motor vehicles, paper NMV0752—Pages 12 (accessed in May 2, 2012, on http://riedal-dz.com/resources/52_bizon.pdf)
26. Bizon N, Oproescu M (2010) Control performances of the bi-buck interface used for inverter current ripple compensation. Int J Tech Phys Prob Eng 1(1):5–10
27. Bizon N (2010) Development of a fuel cell stack macro-model for inverter current ripple evaluation. Rev Roum Sci Techn–Électrotechn et Énerg 55(4):405–415
28. Woojin C, Jo HW, Prasad E (2006) Development of an equivalent circuit model of a fuel cell to evaluate the effects of inverter ripple current. J Power Sources 158:1324–1332
29. Kim JS, Choe GY, Kang HS, Lee BK (2010) Effect of load modeling on low frequency current ripple in fuel cell generation systems. J Electr Eng Technol 5:307–318
30. Spiegel C (2008) PEM fuel cell modeling and simulation using MATLAB, Elsevier Press, Amsterdam
31. Gou B, Na WK, Diong B (2010) Fuel cells: modeling, control, and applications. CRC Press, (chapter 3):13–52
32. Payman A, Pierfederici S, Tabar FM (2008) Energy control of supercapacitor/fuel cell hybrid power source. Energy Convers Manage 49:1637–1644
33. Bizon N (2011) Nonlinear control of fuel cell hybrid power sources: Part I–Voltage control. Appl Energy 88(7):2559–2573
34. Bizon N (2011) Nonlinear control of fuel cell hybrid power sources: Part II –current control. Appl Energy 88(7):2574–2591
35. Bizon N (2011) A new topology of fuel cell hybrid power source for efficient operation and high reliability. J Power Sources 196(6):3260–3270
36. Wang Y, Choi Y, Lee E (2009) Efficient and ripple-mitigating dc–dc converter for residential fuel cell system. Electr Power Energy Syst 31:43–49
37. Kim JS, Choe GY, Kang HS, Lee BK (2011) Robust low frequency current ripple elimination algorithm for grid-connected fuel cell systems with power balancing technique. Renewable Energy 36:1392–1400

38. Bizon N (2008) Control of the Bi-buck power interface used for inverter current ripple minimization. Sci J of the Electrical Eng Facult Valahia Univ 1:7–12
39. Bizon N (2009) DC–DC converter operation with power spectrum spreading using a nonlinear control law. Mediterr J Meas Control 5(2):78–90
40. Bizon N, Raducu M, Oproescu M (2008) Fuel cell current ripple minimization using a bi-buck power interface. In: 13th IEEE international power electronics and motion control conference 2008, IEEE Catalog CFP0834A-CDR 1:621–628
41. Morel C, Bourcerie M, Chapeau-Blondeau F (2005) Improvement of power supply electromagnetic compatibility by extension of chaos anticontrol. J Circuits, Syst, Comput 14(4):757–770
42. Bizon N (2009) DC–DC converter operation with power spectrum spreading using a nonlinear control law. Mediterr J Meas Control 5(2):78–90
43. Liu C, Lai JS (2007) Low frequency current ripple reduction technique with active control in a fuel cell power system with inverter load. IEEE Trans Power Elect 22(4):1453–1463
44. Bizon N (2013) Energy harvesting from the FC stack that operates using the MPP tracking based on modified extremum seeking control. Appl Energy 104:326–336
45. Bizon N (2013) FC energy harvesting using the MPP tracking based on advanced extremum seeking control. Int J Hydrogen Energy 38(14):1952–1966
46. Bizon N (2012) Energy efficiency for the multiport power converters architectures of series and parallel hybrid power source type used in plug-in/V2G fuel cell vehicles. Appl Energy 102:726–734
47. Bizon N (2012) Energy efficiency of multiport power converters used in plug-in/v2g fuel cell vehicles. Appl Energy 96:431–443

Erratum to: Analysis, Control and Optimal Operations in Hybrid Power Systems

Nicu Bizon, Naser Mahdavi Tabatabaei and Hossein Shayeghi

Erratum to:
N. Bizon et al. (eds.), *Analysis, Control and Optimal Operations in Hybrid Power Systems*, DOI 10.1007/978-1-4471-5538-6

The sequence of the editor names was incorrect. The correct sequence is "Nicu Bizon, Naser Mahdavi Tabatabaei and Hossein Shayeghi" instead of "Nicu Bizon, Hossein Shayeghi and Naser Mahdavi Tabatabaei".

The online version of the original book can be found under DOI 10.1007/978-1-4471-5538-6

N. Bizon (✉)
Faculty of Electronics, Communication and Computers, University of Pitesti, Pitesti, Romania
e-mail: nicu.bizon@upit.ro

N. Mahdavi Tabatabaei
Electrical Engineering Department, Seraj Higher Education Institute, Tabriz, Iran
e-mail: n.m.tabatabaei@gmail.com

H. Shayeghi
Departement of Technical Engineering, University of Mohaghegh, Ardabil, Iran
e-mail: hshayeghi@gmail.com

N. Bizon et al. (eds.), *Analysis, Control and Optimal Operations in Hybrid Power Systems*, Green Energy and Technology, DOI: 10.1007/978-1-4471-5538-6_8,

Index

N. Bizon et al. (eds.), *Analysis, Control and Optimal Operations in Hybrid Power Systems*, Green Energy and Technology, DOI: 10.1007/978-1-4471-5538-6,

MIX
Papier aus verantwortungsvollen Quellen
Paper from responsible sources
FSC® C105338

If you have any concerns about our products,
you can contact us on
ProductSafety@springernature.com

In case Publisher is established outside the EU,
the EU authorized representative is:
Springer Nature Customer Service Center GmbH
Europaplatz 3, 69115 Heidelberg, Germany

Printed by Libri Plureos GmbH
in Hamburg, Germany